Fish and Fisheries of Indian Reservoirs

About the Author

Dr.V.B. Sakhare is Director, Post-Graduate Studies, Yogeshwari Mahavidyalaya, Ambajogai. He has 15 years' experience as an outstanding teacher and researcher. He is recipient of fellowship of Indian Association of Aquatic Biologists, Hyderabad. He has done pioneering work in the field of Reservoir Fisheries and Limnology. Dr. Sakhare has successfully organized *National Conference on Emerging Trends in Fisheries and Aquaculture* (*ETFA-2012*), National Workshop on Techniques of Scientific Writing (TSW-2012), *National Conference on Current Perspectives in Limnology* (*NCCPL-2009*) and *Regional Workshop* on *Water Quality Assessment (Implications in potability, productivity and pollution control).*

Dr. Sakhare has been editing an international journal *'Ecology and Fisheries'* (ISSN 0974-6323). He is member of Editorial Advisory Board of *Journal of Science Information* (ISSN 2229-5836) and *E-international Scientific Research Journal (EISRJ)* published by BCTA Inc. Philippines (ISSN 2094-1749). Dr. Sakhare has authored/edited few books such as *'Applied Fisheries'*, *'Reservoir Fisheries and Limnology'*, *'Reservoir Fisheries and Ecology: A Literary survey'*, *'Methodology for Water Analysis'*, *'Aquatic Ecology'*, *'Aquatic Biology and Aquaculture'*, *'Inland Fisheries'*, *' Applied Ecology'*, *'Perspectives in Ecology'* and *'Advances in Aquatic Ecology* (*Vols.* I, II, III, IV, V, VI, VII).

Dr. Sakhare has supervised a research project funded by University Grants Commission, New Delhi and he is a recognized post-graduate teacher and research guide of Dr. Babasaheb Ambedkar Marathwada University, Aurangabad, Solapur University, Solapur and J.J.T. University, Rajasthan. He has published 30 research articles and reviews in peer reviewed journals and about 70 Marathi articles in newspapers and magazines.

Dr. Sakhare has chaired a number of sessions of different seminars/symposia. He has been invited to different colleges/institutes to deliver lectures on different topics in aquatic ecology and reservoir fisheries.

Fish and Fisheries of Indian Reservoirs

By

Dr. V.B. Sakhare

Director

Post-Graduate Studies

Yogeshwari Mahavidyalaya,

Ambajogai-431517

2015

Daya Publishing House

A Division of

Astral International (P) Limited

New Delhi-110 002

Cataloging in Publication Data—DK

Courtesy: D.K. Agencies (P) Ltd. <docinfo@dkagencies.com>

Sakhare, V. B. (Vishwas Balasaheb), 1974-

Fish and fisheries of Indian reservoirs / by V.B. Sakhare.
p. cm.
Includes bibliographical references and index.

ISBN *9789390384204* (International edition)

1. Fishes—India. 2. Fisheries—India. 3. Reservoirs—India. I. Title.

DDC 597.0954 23

Published by : **Daya Publishing House®**
A Division of
Astral International Pvt. Ltd.
– ISO 9001:2008 Certified Company –
House No. 96, Gali No. 6,
Block-C, 30ft Road, Tomar Colony, Burari
New Delhi-110 084
E-mail: info@astralint.com
Website: www.astralint.com

Sales Office : 4760-61/23, Ansari Road, Darya Ganj
New Delhi-110 002 Ph. 011-23245578, 23244987

Laser Typesetting : Plus Computer

Printed at : **Thomson Press India Limited**

PREFACE

Reservoirs are man-made lakes created by impounding river water for the purpose of irrigation, hydel power generation, flood control and industrial water needs. In India a large number of river valley projects have been commissioned since independence as a part of our development activities, resulting in a chain of such artificial impoundments. These water sheets, by virtue of their sheer magnitude and due to their high biogenic production potential, constitute one of the most important resources for inland fisheries development in the country.

In India there are 19,370 reservoirs covering an area of 3.15 million ha. The area is expected to grow further to 6 million ha in due course of time of 20 years. The average productivity from all categories of reservoirs is estimated to about 20.13 kg/ha/yr, although potential exists for manifold increase as demonstrated in some reservoirs in the country. As compared to several foreign countries, the per hectare fish production in Indian reservoirs is very poor as against 88 kg in the USSR, 100 kg in Sri Lanka and 64.5 kg in large reservoirs of Thailand. The low fish production from Indian reservoirs is mainly due to unscientific management practices resulting from the inadequate knowledge of the ecology and production functions of this biotope.

The poor appreciation of the biological and limnological functions governing the production regimes of reservoirs and lack of stocking programmes have rendered below optimal utilization of the fisheries potential of the reservoirs. It is expected that through sustained supplementary stocking of quality fingerlings, augementing the fish stocks through auto stocking, adoption of appropriate mesh sizes, optimum fishing effort and enforcement of closed areas and closed seasons, the average productivity could be increased to a level of 500 kg/ha/yr from small reservoirs; 200 kg /ha/yr from medium reservoirs; and 100-150 kg/ha/yr from the large reservoirs.

The work of reservoir fisheries development in India was done much earlier in erstwhile Madras State in 1934 by the then state fisheries department after the formation of Mettur reservoir across river Cauvery. Thus the erstwhile Madras Fisheries Department was the front-runner of reservoir fisheries development in India now inherited by Tamil Nadu Fisheries Department. Actually, the systematic reservoir fishing investigations were initiated by Central Inland Fisheries Research Institute, Barrackpore, West Bengal in 1963. The institute took up a detailed study of fisheries of Tunghabadra reservoir in Karnataka and Damodar Valley Corporation reservoirs in Bihar. Later, an All India Coordinated Research Project on Ecology and Fisheries of Freshwater Reservoirs was launched by Indian Council of Agricultural Research (ICAR) through CIFRI. This project was eco-oriented so as to have in depth study of all determinants of reservoir productivity of five reservoirs located on each in five states (Uttar Pradesh, Tamil Nadu, Bihar, Andhra Pradesh and Himachal Pradesh) in different eco-climatic conditions. Prior to inception of this co-ordinated research project, fisheries of four reservoirs of Damodar Valley Corporation (Bihar) and three in Madhya Pradesh were also studied. Susequently, some more reservoirs of other states were also included in the investigation, for providing recommendations to manage them on scientific line.

In present book the existing literature on limnology and fisheries of Indian reservoirs has been reviewed by covering more than 165 reservoirs located in different parts of the country. An attempt has been made to gather authentic information on reservoir morphometry, seed stocking, fish yield, crafts and gears and socio-economic status of fishermen in different reservoirs scattered in different states of the country. The reservoir fisheries resources and their utilization are dealt with separately in respect of each state. With its application oriented and interdisciplinary approach, I hope that the students, teachers, researchers, scientists and policy-makers in India and abroad will find this book much more useful.

I place on record my sincere gratitude to Dr.Sureshji Khursale, President, Yogeshwari Education Society, Ambajogai for his constant encouragement.

I also thank my research guide Dr. P.K. Joshi, Parbhani. I am extremely grateful and indebted to him for his expert, sincere and valuable guidance and encouragement extended to me.

I take this opportunity to record my sincere thanks to all the faculty members of the Department of Zoology, Yogeshwari Mahavidyalaya, Ambajogai for their help and encouragement.

I also thank my father Shri Balasaheb Sakhare, Managing Director, Rayat Co-operative Sugar Factory Limited Karad and my brother Shri Avinash Sakhare, Agriculture Extension Officer, Aurangabad for unceasing encouragement and support.

I wish to thank my wife Surekha for her endurance during the compilation of work of this volume. She has helped me for constantly data feeding, processing and reprocessing on computer.

I thank my publisher Shri Anil Mittal of Astral International Private Limited, New Delhi for taking pains in bringing out the book.

I owe a great many thanks to a great many people who helped and supported me during the writing of this book.

—Dr. Vishwas Balasaheb Sakhare

Acronyms and Abbreviations

BMC	:	Bombay Municipal Corporation
BNHS	:	Bombay Natural History Society
BOD	:	Biochemical Oxygen Demand
CCA	:	Cultivable Command Area
CIFE	:	Central Institute of Fisheries Education
CIFRI	:	Central Inland Fisheries Research Institute
CIFT	:	Central Institute of Fisheries Technology
COD	:	Chemical Oxygen Demand
CPUE	:	Catch Per Unit Effort
DFDO	:	District Fishery Development Office
DO	:	Dissolved Oxygen
DVC	:	Damodar Valley Corporation
ESR	:	Erythrocyte Sedimentation Rate
EEZ	:	Exclusive Economic Zone
FFDA	:	Fish Farmer's Development Agency
FRL	:	Full Reservoir Level
GPP	:	Gross Primary Production
HDPE	:	High Density Polyethylene
HUDA	:	Hyderabad Urban Development Authority
IGRFD	:	Indo-German Reservoir Fisheries Development Project
IMC	:	Indian Major Carp
ISP	:	Indira Sagar Project
JBIC	:	Japan Bank of International Cooperation
LDA	:	Loktak Development Authority
LRnt	:	Low Risk – near threatened
MCH	:	Mean Corpuscular Haemoglobin

MCV	:	Mean Corpuscular Volume
MDDL	:	Minimum drawdown level
MEI	:	Morpho-Edaphic Index
MFC	:	Maharashtra Fisheries Corporation
MPFDC	:	Madhya Pradesh Fisheries Development Corporation
MSAA	:	Maharashtra State Angling Association
MSEB	:	Maharashtra State Electricity Board
MSL	:	Mean Sea Level
MSY	:	Maximum Sustainable Yield
NASS	:	Nushad Ali Sarovar Samardhini
NFDB	:	National Fisheries Development Board
NGO	:	Non Government Organisation
NPP	:	Net Primary Production
PCV	:	Patched Cell Volume
PFCs	:	Primary Fishermen's Co-operative Societies
PIL	:	Public Interest Litigation
PMSSM	:	Purna Matsyavyavasai Sahkari Sanstha
SHG	:	Self Helping Groups
SSD	:	Sardar Sarovar Dam
TDS	:	Total Dissolved Solids
TEC	:	Total Erythrocyte Count
TLC	:	Total Leucocyte Count
TMS	:	Tawa Matsya Sangh
TNFDC	:	Tamil Nadu Fisheries Development Corporation Limited
TSS	:	Total Suspended Solids
UNDP	:	United Nations Development Programme

Contents

1

INTRODUCTION

Fig. 1.1.

Fishing as an occupation is being practised in India since time immemorial and has been regarded as a supplementary enterprise of the fishermen community on the subsistence level with little external input. Fisheries sector, however, has a strategic role in food security, international trade and employment generation. With the changing consumption pattern, emerging market forces and technological developments, it has assumed added importance in India and is undergoing a rapid transformation.

Fisheries sector contributes to the national income, exports, food and nutritional security and employment generation. It is a principal source of livelihood for a large section of economically underprivileged population of the country, especially in the coastal areas. The share of agriculture and allied activities in the GDP is constantly declining. The agriculture sector is also diversifying towards high value enterprises, including fisheries. The contribution of fisheries sector to the GDP has gone up from 0.46 per cent in 1950-51 to 1.47 per cent in 2000-01 (at current prices). The share of fisheries in agricultural GDP has impressively increased during this period from a mere 0.84 per cent to 4.01 per cent. In fact, the fisheries sector is booming and contributing increasingly to the economic growth of the nation.

The fisheries sector plays an important role in the economy of India. It is a foreign exchange generator, besides providing nutritious food to the people. India is the fourth largest fish producer in the world and second in inland fish production. The main inland fishery resources include about 1.20 million ha and brackish water area, 2.38 million ha of fresh water pond and tanks, 3.15 million ha of reservoirs, besides about 1,91,000 kms of rivers and canals.

According to the National Sample Survey (NSS), the annual per capita fish consumption in India was 2.45 kg in 1983; it increased to 3.45 kg in 1999-2000. Only 35 per cent population in India was estimated to be fish eater and their annual per capita fish consumption was 9.8 kg in 1999-2000. However, wide regional variations do exist in fish consumption across regions, states and income classes India, with diversified agro-climatic regions, is endowed with potentially rich and varied aquatic resources. It is endowed with an Exclusive Economic Zone (EEZ) of 2.02 million square kilometres, a continental shelf of 0.5 million square kilometres and a long coast line of 8119 kilometres with some of the richest fishing grounds in the world. The main inland fishery resources include about 1.20 million hectares (Mha) of brackish water area, 2.38 Mha of fresh water ponds and tanks; about 1.24 Mha of reservoirs and 0.82 Mha of beels.

The vast, varied and bountiful nature of inland fishery resources of India are unique in their nature and potential in the world, making India the second largest producer of inland fish next to the China. The optimise fish yield from inland fisheries sector, recent research has been directed towards reservoirs, the man-made impoundments with assemblage of both lacustrine and fluviatile characters with huge, untapped potentials and are said to be called "Fish Mines".

Indian reservoirs have a total area of 3.15 million ha, out of which small reservoirs occupy 1.49 million ha followed by large (1.14 million ha) and medium (0.52 million ha) ones. Among variously sized reservoirs, maximum annual production is reported from small reservoirs (50 kg/ha/yr) followed by medium

(12.5 kg/ha/yr) and large (11.4 kg/ha/yr) with an average of 20.13 kg/ha/yr, although potential exists for manifold increase as demonstrated in some small, medium and large reservoirs in the country. Poor appreciation of the biological and limnological functions governing the production regimes of reservoirs and lack of stocking programmes have rendered below-optimal utilization of the fisheries potential of the reservoirs. It is expected that through sustained supplementary stocking of quality fingerlings, augmenting the fish stocks through auto stocking, adoption of appropriate mesh sizes, optimum fishing effort and enforcement of closed areas and closed seasons the average productivity could be increased to a level of 500 kg/ha/yr from small reservoirs; 200 kg/ ha/ yr from medium reservoirs; and 100-150 kg/ ha/ yr from the large reservoirs.

The total water spread area of Indian reservoirs forms about 50 per cent of the total reservoir area in South East Asia (Shetty, 1990). Indian Institute of Management, Ahmedabad (Srivastawa, *et al.* 1985) listed 975 reservoirs in India, in size range of 1000 ha to 10,000 ha covering an area of 1.7 million hectares. According to Sugunan (1995), the present total area under the reservoirs in India is 3.1 million hectares (Table 1.1).

Table 1.1. Distribution of Small, Medium and Large Reservoirs in India

Type	Small Reservoirs	Medium Reservoirs	Large Reservoirs	Total
Number	19134	180	56	19370
Area (ha)	1485557	527541	1140268	3153366

Data pulled from Sugunan (1995): FAO Report

According to Dixitulu (1999) there are 19370 reservoirs in the country with an extent of 3,153,366 ha (consisting of 19,134 nos. of small ones with an extent of 1,485,557 ha; 180 nos. of medium ones with 575,541 ha and 56 nos. of large ones with an extent of 1,140,268 ha). Kaur and Dhawan (1997) mentions total number of 550 reservoirs with a total water spread area of about 2 million hectares.

Table 1.2. A State-wise number of reservoirs in India

States	Small Reservoirs		Irrigation Tanks		Medium Reservoirs		Large Reservoirs	
	Number	Area (Ha)	Number	Area (Ha)	Number	Area (Ha)	Number	Area (Ha)
Tamil Nadu	58	15,663	8837	300278	9	19,577	2	23,222
Karnataka	46	15,253	4605	213404	16	29,078	12	1,79,556
Andhra Pradesh	98	24,178	2800	177749	32	66,429	7	1,90,151
Gujarat	115	40,099	561	44025	28	57,748	7	1,44,358
Uttar Pradesh	40	20,845		1,97,806	22	44,993	4	71,196
Madhya Pradesh	6	1,72,575			21	1,69,502	5	1,18,307

Contd...

Maharashtra	NA	1,19,515			NA	39,181	NA	1,15,054
Bihar	112	12,461			5	12,523	8	71,711
Orissa	1433	66,047			6	12,748	3	1,19,403
Kerala	21	7,957			8	15,500	1	6,160
Rajasthan	389	54,231			30	49,827	4	49,386
Himachal Pradesh	1	200			NA	NA	2	41,364
West Bengal	4	732			1	4,600	1	10,400
Haryana	4	282			NA	NA	NA	NA
NE States	4	1,639		600	2	5,835	NA	NA
Total	2331	5,51,677	16,803	9,33,862	180	5,27,541	56	11,40,268

Unfortunately, majority of reservoirs are not being scientifically managed for fisheries. Only a handful of than have so far has been harassed along scientific lines, while the others are either half-heariedly managed or not managed at all. As compared to several developed countries, the per hector fish production in Indian reservoirs is very poor, being only about 20 kg/yr as against 88 kg/yr in the USSR, 100 kg/yr in Sri Lanka and 64.5 kg/yr in large reservoirs of Thailand. Lot of progress has been made in several developed countries in deep water fishing in reservoirs and management of their fisheries.

Since the reservoirs are constructed primarily for hydel and irrigation purposes, the fish production from them is treated as of bye-product importance. This relegation is one of the reasons of poor fish yield from reservoirs. So much so, the reservoir fish industry could not make any remarkable progress as yet. Because of the country's growing demand for irrigation and power, new reservoirs have been coming up and thus this potential is expected to grow further giving ample opportunities of culture based capture fishing practices, provided these water bodies are managed scientifically for development of fisheries at an optimally sustainable level.

Reservoirs are generally classified as small (<1000 ha), medium (1000-5000 ha) and large (>5000 ha), especially in the records of the Government of India.

The work of reservoir fisheries development in India was done much earlier in erstwhile Madras state in 1934 by the State Fisheries Department after the formation of Mettur reservoir across river Cauvery. Thus the erstwhile Madras fisheries development was the front-runner of reservoir fisheries development in India now inherited by Tamil Nadu fisheries department.

Actually, the systematic reservoir fishery investigations were initiated by CIFRI, Barrackpore in 1963. The institute took up a detailed study of fisheries of Tungabhadra reservoir in Karnataka and Damodar Valley Corporation reservoirs (Konar, Tilaiya and Panchet) in Bihar. Later, an All India Coordinated Research Project on Ecology and Fisheries of fresh water reservoirs was launched by ICAR through CIFRI. This project was carried out on five reservoirs located one each in five states (Uttar Pradesh, Tamil Nadu, Andhra Pradesh, Bihar and Himachal

Pradesh) in different eco-climatic conditions. Prior to inception of this coordinated research project, fisheries of four reservoirs of Damodar Valley Corporation (Bihar) and three in Madhya Pradesh were also studied. Subsequently, some more reservoirs of other states were also included in the investigation, for providing recommendations to manage them on scientific lines.

CIFRI (1997) under the All India Coordinated Project on Ecology and Fisheries of Freshwater Reservoirs carried out investigations on Bhatghar reservoir of Pune district of Maharashtra. The reservoir was stocked annually with fry stage brought from Kolkata. The fry of *Tor khudree* were stocked by CIFRI. About 2.25 lakhs of fingerlings between 65 mm to 110 mm size of major carps were stocked by the period August/September 1988. Continued stocking by the fisheries department did not have an impact on the revival of the fishery probably due to the size at the time of stocking. The stocking of 2.25 lakhs of fingerlings during 1988 did show a marked improvement in the landings especially of *Catla catla* and *Mahseer* during 1989. The investigation recommended the need for regular stocking major carps fingerlings. The stocking done earlier *i.e.,* before 1984-85 was probably with small sized fry on account of which it could not reflect in commercial fishery. This trend was later reversed to some extent with stocking of big sized fingerlings only from 1984-85.

CIFRI (1998) Survyed Amaravathy, Polar-pornthalar, Uppar, Pillor, Gunderipallam and Varattupallam reservoirs of Tamil Nadu and revealed that the fish yield in Varattupallam was highest and lowest in Amaravathy. Taking into consideration the area of reservoirs and stocking density, it was concluded that the fishery management in Polar-porthalar is best followed by Amaravathy and Uppar while it was poor in Pillor. The actual fish yields obtained from the reservoirs are less compared to the production potentials estimated through primary production studies. This gives scope for enhancement of fish production from these reservoirs through judicious stocking and exploitation.

CIFRI (1998) also conducted the survey of nine reservoirs of Andhra Pradesh and found that most of the reservoirs have the necessary infrastructure for raising the seed and exploitation of the fishery and recommended that the cooperative societies need to be activated and given the responsibility of management.

The survey conducted by the CIFRI for reservoirs in Karnataka, Andhra Pradesh, Tamil Nadu and Madhya Pradesh states gave encouraging results in the sense that fish production potentiality of some of the reservoirs could be assessed to suggest measures for development of their fisheries. Accordingly, many of the reservoirs were found productive, facilitating the taking of necessary steps in this direction.

Most of the reservoirs are administratively controlled by irrigation, revenue, forestry or electricity departments. These departments do not pay required attention towards development of fisheries. Naturally, department of fisheries is not in a position to interfere in reservoir management. Therefore, the fishing ownership at all reservoirs should be transferred to respective state fisheries departments in all states.

National Fisheries Development Board (NFDB) has been set-up in year 2006 at Hyderabad. The impediments in integrated efforts towards stepping up fish production from reservoirs and ongoing inadequacies in management practices have motivated NFDB to intervene in reservoir fisheries development. NFDB has thus teamed up with CIFRI, the focal organization for national reservoir fisheries development to optimize overall national reservoir fisheries production, in association with state fisheries departments, to play their role as partners. It is expected that the proposed partnerships, supported by infrastructural facilities at individual reservoirs would certainly ensure the sustainable production from reservoirs. NFDB would also provide funds to the states concerned that may join hands with it to achieve the targeted goal. Since the new reservoir projects too are to be implemented taking into confidence the fishermen exploiting the fish from reservoirs, such community-based projects would certainly be beneficial to fishermen too.

After a latest report on the total reservoir area of India (3.15 million ha), many reservoirs have been formed in different states of the country. Accordingly, the total reservoir area of India must have gone up now. Regarding the existence of stagnancy in country's overall reservoir fish yield (20.13 kg/ha), it may be commented that this figure perhaps may be based on the catch statistics of a few reservoirs and the data of many of the reservoirs might not have been accounted for, due to sampling deficiencies. Moreover, many of the reservoirs could not have been exploited to a sustainable level due to improper and defective fishing. In this background, it may not be correct to say that there is stagnancy in reservoir fish yield of the country. In the light of constraints in reservoir fisheries development as discussed above, high yields may not be expected but, at the same time, some increase in fish production from reservoirs because of coverage of more of water area and an increase in fishing may have taken place, calling for a re-estimation of the present status of potential and annual production level.

Crafts and Gears: Fishing technology is the discipline which deals with the study, development and improvements of techniques for catching the fish. These techniques comprise of the direct means of capture such as fishing gears and fishing vessels, indirect means such as fish location techniques, attraction and concentration of fish for capture etc. and fishing operations which combine both direct and indirect means of capture.

Application of fishing technology has significant impact on commercial fishing in the lakes and reservoirs of the country. Good examples of such nature are: higher catch in the new designs of gill nets and trammel nets in Mettur dam, Krishnarajsagar dam and Damodar Valley Corporation Reservoirs. Framed gil nets in Hirakud reservoir have proved very successful (Sreekrishna, 1980).

New type of gears and fishing methods tested in other countries can be introduced in the Indian reservoirs after experiments. Dynamic gears like trawls and purse seines can be introduced in the reservoirs where the reservoir bottom is free from obstacles. Introduction of shore seines for bulk catching at selected areas after clearing the underwater obstacles may result in higher production. Catfishes and minor carps are dominant in many reservoirs resulting in uneconomic yields. Fishing technology could help to eradicate these undesirable species by operating selective fishing gears. Introduction of echo sounders may have greater impact not only on fish production but also to avoid damage to gears due to underwater obstacles.

Gear technology determines the efficiency of capturing fish with a given standing crop in the reservoir. Gear technology appears to be a fairly well developed science but many state fisheries departments do not provide necessary extension assistance. Fishermen have their own indigenous nets which are the result of their long experience. However, there can be various shortcomings in these indigenous nets such as incorrect relationship between mesh size and diameter of twine, incorrect hanging of the webbing to the float line, omission of lead line and breast lines, incorrect and insufficient rigging with floats and sinkers and insufficient height. Absence or shortage of appropriate fishing craft ofetn leads to the restriction of fishery to a marginal zone in the reservoir.

Use of improved fishing gear ad tackle is reported to have resulted in large increases in experimental yields when experts of FAO and UNDP visited India in the early sixties. These experts had recommended various improvements in fishing technologies, such as introduction of frame net, a variety of gill net, gill net with additional vertical lines, improved set gill nets improved design of beach seine, long lines with floating hooks, electrical fishing, fish attraction through above water and under water lights of different intensity and colour.

A reservoir is having its own special features as it differs from a natural lake or a river. Construction of a dam brings about many ecological changes both in upstream and downstream. In the upstream due to change from lotic to lentic conditions, riverine planktons are replaced by lacustrine forms, the benthic riverine fauna is replaced by lacustrine one and the migratory fishes disappear. Among the resident fish species, the running water forms become fewer or completely eliminated and slow water fish species predominate. These changes which have occurred in the new ecosystem give an advantage to predatory fishes. The turbidity level also falls as the reservoir acts as settling basin. The floating macro vegetation may come up particularly in the reservoirs situated in the tropical areas. The excessive macro vegetation becomes a source of many ecological problems as it creates deoxygenation of the water.

In the lower reaches of the river, periodic discharges of sediments from the reservoir may cause mud and silting up of vast stretches with serious consequences on the fauna. Plankton increase in lower reaches due to drift from the reservoir and reduction in turbidity. Thus, the reservoir acts as a fertility trap, reducing amount of dissolved plant nutrients, which would otherwise freely reaching in the lower reaches.

There is usually an initial spurt of plankton and benthic communities etc. This is followed by rapid establishment of aquatic vegetation due to availability of increased nutrients which are released from the decaying submerged organic matter. This results in ideal conditions for the establishment of planktophagus species. This trophic bursts is also an account of the vacant niches created by disappearance of riverine species. The wave action affects soils at different levels enhancing mineral enrichment as the water level rises. Nutrients thus released stimulate great photosynthetic activity. This initial phase of fertility lasts for two to three years. As the effect of trophic burst is over, the reservoir passes through a trophic depression phase which is caused by gradual decrease of nutrient release due to continuous sedimentation of the reservoir bed as well as increase in volume of impounded water and using up of available nutrients by aquatic vegetation. Further the nutrients arising from decomposition of vegetation become unavailable to water column because of thermal stratification. Once held up in the deeper water, the nutrients get lost from the reservoir through drawdown from penstock and tunnels. Regular withdrawal of water from reservoir deprives it continuously of nutrients brought in by inflowing water. This phase of trophic depression is characterized by low fish food reserves. Its duration is variable and it lasts for as long as 25 years. After trophic depression, the final fertility phase is reached which is at much low level than that is obtained at the stages of initial fertility. The fertility, however, regains itself depending upon the accumulation of organic substances in the bottom soil. Monitoring of species spectrum is during these phases is very important from effective management point of view. The monitoring of population in the changed context of community succession is the essence of reservoir management.

Due to sudden fluctuations in water level in the Indian reservoirs, it has been observed that plankton pulses coincides with the months of least level fluctuations and on the other hand all biotic communities are in their lowest ebb during the months of maximum level fluctuations (Sugunan, 1986). A comparatively stable reservoir level is more conducive to growth of organisms.

Impact of Dam: Dams alter river ecosystems and subsequently require development of new relationships between humankind and natural resources associated with these ecosystems. From a fishery prespective, dams and their resulting reservoirs can benefit human societies. Dams, however, usually alter traditional riverine fisheries, sometimes positively (*i.e.,* from tail water fisheries), but more commonly negatively. There typically are found shifts from river-adapted species to those more adapted to lentic environments. Species diversity in impoundments usually declines over time as river-adapted species fade from the system. Dams have had negative impact on riverine fisheries in various systems throughout the region. Dams represent a potential obstacle to migratory fish and special provisions have been made at dams in temperate regions to facilitate the movement of salmon (Clay, 1961). If no provisions have been made at a hydroelectric dam to facilitate upstream fish migration, fish are attracted to the base of the spillway and the outflow from the turbine channels. They are unable to proceed any further. The usual solution of this problem involves the construction of a stepped series of pools rising from the level of the river up to the water level behind the dam. This arrangement (fish way or fish ladder) allows the fish to move successfully from one pool to the next until they reach the top of the dam. Provisions for the downstream movement of juvenile fish are designed on a different basis. Dam construction on the Sefid river (Iran) resulted in reduced stream flow, increased water temperature and declines in food items for sturgeon (Vladykov, 1964), Reservoirs constructed on rivers emptying into terminal lakes of Central Asia and Kazakhstan severely reduced stocks of migratory fishes in the rivers, encouraging development of stocks more lacustraine in character, exacerbated precipitation/evaporation deficit ratios and have led to accelerated salination of groundwater as well as surface waters (Peter and Mitrofanow, 1998). Sandhu and Toor (1984) noted sharp declines of catches of *Hilsa ilisha* as a result of dams, barrages, weirs and anicuts on the Hooghly, Godavari, Krishna and Cauvery rivers (India), and that mahseer, *Tor putitora* and *Tor tor* no longer are found above Nangal and Talwara dams. Fishways constructed in conjunction with dams are used as fish traps by local fishers. In addition to impacts on hilsa and mahseer stocks and their associated fisheries, formation of reservoirs in India has had negative impact on snow trout (*Schizothorax*), and rohu (*Labeo*) in Himalayan streams, and catadromous eels and freshwater prawns in all major river systems.

Gill (1984) mentioned the effect of dams on fish fauna of Punjab region. Construction of barrages at Ropar, Harike, and Ferozpur has restricted migration of Indian major carps, in spite of fishways. During most of the year, little water is released into the river below the dams from the reservoirs, and fish are concentrated in pools where they are more easily captured by fishes. Fishways designed to promote fish passage past dams are used by fishers to capture fish.

The dams and barrages form physical obstruction to the migratory fishes causing physiological strain and breeding failure. The classic example is Indian shad (*Hilsa ilisha*) that has migratory range of 1500 km upstream of the estuary. Construction of Farakka barrage, 476 km from the river mouth has nearly eliminated the lucrative *Hilsa* fishery above the barrage. Gone are the days when up to 303.6t of Hilsa were caught at Allahabad (1956-57).The impact of Farakka on Hilsa had been very severe. The average Hilsa landing after commission of Farakka barrage at Allahabad, Buxar and Bhagalpur get reduced by 94.61, 98.12 and 83.05 per cent respectively (Chandra, 1989). Another bad effect has been found in the case of *Pangasius pangasius* in Ganga, Brahmaputra, Mahanadi and Godavari rivers. Dams located in the lower and middle reaches of these rivers obstructed the migration of this fish and adversely affected its population. Torrential fishes like *Glyptothorax, Leptognathus* etc. cannot survive in reservoirs and there is a chance of disappearance of their races in nature. The Gandak and Kossi valley projects have also adversely affected the fisheries of North Bihar to a very significant degree. They have affected not only the fisheries of rivers, mauns and chaurs, but indirectly culture fisheries as well as the spawn production, on which culture fisheries lean heavily, has been drastically reduced due to loss of breeding grounds. This has adversely affected the fishery as well as the seed resources of these water bodies, as also those of the main rivers.

In China Danjiangkou dam on the Hanjiang river prevents movements of eels (*Anguilla japonica*) (Liu and Yu, 1992). The dam also blocked migration of commercially important carps. Dams have reduced overall fishery yield from some African systems. Such yield reductions can be temporary or long term. For example, Lelek and El-zarka (1973) reported that two years after filling Lake Kainji, fish catches from the system were reduced by 30%.

Zhong and Power (1996) reported that the number of fish species decreased from 107 to 83 because the migration was interrupted by the Xinanjiang dam (China). The reduction of biodiversity occurred not only in the flooded section but also in the river below the dam. Quiros (1989) mentions that dam construction in the upper reaches of Latin American rivers appears to lead to the disappearance of potamodromous species stock in reservoirs and in the river upstream of the structure. The same occurs in reaches where a whole series of dams and reservoirs have been constructed.

In Australia, obstructed fish passage has led to many instances of declining populations or extinctions of species in the affected basin (Barry 1990; Mallen-Cooper and Harris, 1990). The concept of obstruction to migration is often associated with the height of the dam. However, even low weirs can constitute a major obstruction to upstream migration. Whether an obstacle can be passed or not depends on the hydraulic conditions over and at the foot of the obstacle in relation to the swimming and leaping capacities of the species concerned. The swimming and leaping capacities depend on the species, the size of the individuals, their physiological condition and water quality factors as water temperature and dissolved oxygen. Certain catadromous species have a special ability to clear obstacles during their upstream migration: in addition to speed of swimming, the young eels are able to climb through brush, or over grassy slopes, provided they are kept thoroughly wet; some species like *Gobies* possess a sucker and enlarged fins with which they can cling to the substrate and climb around the edge of waterfalls and rapids (Mitchell, 1995).

It may become desirable to construct fish ways in case a migrant fish is present in the riverine system on which the reservoir is constructed to enable the migratory fish to negotiate the dam height. A number of fish ways have been built in India, often at large expense. It appears that many of these fish ways have for various reasons proved to be failures (Yadav, 1983). It might be due to lack of knowledge of fish migration and leaping capacities of migratory fish combined with exorbitant cost of fish passes have prevented the construction of the right type of fish ways. Even outside India, many fish ways have been failures either because of improper design or incorrect placing, or because of adverse currents which prevented fish from finding or being guided to a fish way or because fish could not or did not wish to make use of the fish way.

The construction of a dam on a river can block or delay upstream fish migration and thus contribute to the decline and even the extinction of species that depend on longitudinal movements along the stream continuum during certain phases of their life cycle. Mortality resulting from the fish passage through hydraulic turbines or over spillways during their downstream migration can be significant. Experience gained shows that problems associated with downstream migration can also be a major factor affecting anadromous or catadromous fish stocks. Habitat loss or alteration, discharge modifications, changes in water quality and temperature, increased predation pressure as well as delays in migration caused by dams are significant issues. Passage through spillways may be a direct cause of injury or mortality, or an indirect cause (increased susceptibility of disorientated or shocked fish predation). The mortality rate varies greatly from one location to another. Mortalities have several causes: shearing effects, abrasion against spillway surfaces, turbulence in the stilling basin at the base of dam, sudden variations in velocity and pressure as the fish hits the water, physical impact against energy dissipaters. The manner in which energy is dissipated in the spillway can have a determinant effect on fish mortality rates.

Dam construction can dramatically affect migratory fish habitat. The consequence of river impoundment is the transformation of lotic environment to lentic habitats. Independently of free passage problems, species which spawn in relatively fast flowing reaches can be eliminated. From a study of the threatened fish of Oklahoma, Hubbs and Pigg (1976) suggested that 55% of the man induced species depletions had been caused by the loss of free flowing river habitat resulting from flooding by reservoirs, and a further 19% of the depletion was caused by the construction of dams, acting as barriers to fish migration.

About 40% of the spawning grounds in the Qiantang river above the Fuchunjiang dam were lost by flooding (Zhong and Power, 1996). On the Indus river, the construction of Gulam Mohammed dam has deprived the migratory *Hilsa ilisha* of 60% of their previous spawning areas (Welcome, 1985). On the Columbia river and its main tributary the Snake river, most spawning habitat were flooded, due to the construction of dams creating an uninterrupted series of impoundments (Raymond, 1979).

Dams can modify thermal and chemical characteristics of river water. This can affect fish species and populations downstream. Water temperature changes have often been identified as a cause of reduction in native species, particularly as a result of spawning success (Petters, 1988). Coldwater release from high dams of the Colorado river has resulted in a decline a native fish abundance (Holden and Stalnaker, 1975). Water-chemistry changes can also be significant for fish. Release of anoxic water from the hypolimnion can cause fish mortality below dams (Bradka and Rehackova, 1964).

It includes the clearing of the proposed area of submergence of forest trees and other obstructions like boulders ad buildings etc., which can damage or cause loss of fishing nets. Efficient gears like trawl nets cannot be operated with the help of mechanized boats. Further, the submerged dense jungles and boulders provide good hideouts of fish. All this results in the net loss of fish production. Although the trees represent a danger to fishing gear, fishermen prefer setting the nets among trees as fish tend to congregate in such areas and this makes commercial fishing viable. Deforestation of standing trees in reservoir basin has always remained neglected. Once submerged, the standing forest turns an external problem for fishing.

Performance of Exotic Fishes in Indian Reservoirs: The exotic fishes like *Cyprinus carpio,Ctenopharyngodon idella,Pangasius sutchi, carassius auratus, Lebistes reticulates, Tinca tinca, Hypophthalmichthys molitrix, Oreochromis mossambicus, Oreochromis niloticus, Salmo trutta fario, Salmo salar, Onchorynchus nerka,*

Onchorhynchus mykiss, Aristichthys nobilis, Clarias gariepinus, etc. were import and introduced in Indian waters. The transplantation of these fishes into Indian waters has not been a very good experience.

Tilapia is native of Africa and the Middle East. In India, the first consignment of tilapia was brought by the Central Marine Fisheries Research Institute (CMFRI), in August 1952 from Bangkok and the second by the Madras Fisheries Department in the same year from Ceylon.

Tilapia is hardy but display some of the most undesirable characteristics. It bears at a very small size is difficult to grow to a reasonable market size (100 gm+) with huge differences in growth between the sexes. In Indian reservoirs tilapia has adversely affected the indigenous gene pool.

In Powai Lake of Mumbai the major carps have been badly hit with the accidental introduction of *O. mossambicus* with the production sliding down from 33.3 kg/ha to 11.9 kg/ha (Das et al., 1990).

In Kabini reservoir tilapia has adversely affected the indigenous *Cirrhinus reba*.During the period from 1980-81 to 1984-85, tilapia has caused decrease of *Cirrhinus reba*'s share in the catch from 70 per cent to 20 per cent (Murthy *et al.,* 1986).

The accidental introduction of tilapia in Lake Jaismand (Rajasthan) has been reported by Fishing Chimes. The catch has registered more than two fold increase in month of fishing and the whole catch was only of tilapia. The size of tilapia was found reduced compared to that of early period. It is probably due to the inadequacy of food in reservoir.

According to Jhingran (1985) tilapia is unsuitable for culture along with Indian major carps because of adverse effect it causes on the growth and production of carps and its depredations on carp fry.

In Periyar Lake of Kerala, tilapia constitutes more than 25% of its catch whereas the region was earlier dominated by the abundant catch of local species *Etroplus suratensis.*

Tilapia has dominated and virtually eliminated all other fishes including the stocked Gangetic carps in a number of reservoirs of Tamil Nadu and Kerala.

Sreenivasan (1967) has reported adverse impact of tilapia on the growth rates of *Catla catla, Labeo fimbriatus* and *Cirrhinus mrigala.* He further observed adverse impact on the catch of *Chanos chanos.*

It is observed that introductions of tilapia and common carp has oblitered *Puntius spp* in Krishnarajasagr reservoir of Karnataka (Ramkrishna, 2008). Tilapias has entered the reservoirs of all the river systems of Karanataka state and affected the faunal composition of indigenous fishes in the reservoirs, posing a great challenge for its effective control and management.

Silver carp (*Hypophthalmichthys molitrix*), since its introduction in India during 1959, has entered a nymber of open waters in India, both accidentally and deliberately. Jhingran and Natarajan (1978) expressed the view that the species is ot suitable for larger water as it may harm the fishery of catla in long run due to short circuiting of the food-web.

Following accidental escape, silver carp formed a breeding population in Govind sagar reservoir in Himachal Pradesh and Kulgarhi reservoir in Madhya Pradesh (Fish base 2004; Singh, 2004). These reservoirs were known for the dominance of catla fishery whereas the invasion of silver carp altered the habitat rendering it unsuitable for sustaining catla. The food analysis of silver carp delineated that dinophyceae formed the main item of intestinal contents and its occurrence was high during December to May (65-83%) making the environment more conducive for thriving of silver carp growth (Kaushal *et al.*, 1980). Establishment of silver carp in Gobind sagar reservoir may be attributed to the presence of massive *Ceratium hirudinella* (dinoflagellate), which is a major plankton constituent of the reservoir. It is pertinent to mention that production of catla from these reservoirs has significantly declined (Sugunan, 1995; Singh, 2004). The fishery of *Catla catla* and *Tor putitora* in Govindsagar reservoir in Himachal Pradesh has been drastically curtailed by the entry of silver carp (Lakra and Singh, 2007).

In North Indian reservoirs establishment of silver carp has already affected the performance of catla and rohu because of overlapping in feeding niches (Devaraj *et al.*, 2008).

According to Karamchandani and Mishra (1980) silver carp and catla share a common niche and compete with each other for food in a reservoir ecosystem. Percentage composition of phytoplankton in the guts of both the fishes caught during the same time from Kulgarhi reservoir was more or less the same. Zooplankton, the favourite menu of catla, formed 21 per cent of the gut contents of silver carp. The authors concluded that silver carp hampered the growth of catla in the reservoir and advocated caution before its stocking in Indian reservoirs.

Common carp (*Cyprinus carpio*) has been transplanted into many countries. In India, it was introduced in 1939. There are three varieties of *Cyprinus carpio i.e.*, Scale carp (*Cyprinus carpio var. communis*), Mirror carp (*Cyprinus carpio var. specularis*) and Leather carp (*Cyprinus carpio var. nudus*).

Common carp is not suitable for stocking in Indian reservoirs, especially the larger ones, for diverse reasons. Being a sluggish fish, its chances of survival in a predator-dominated reservoir are very poor. Due to its slow movement and bottom dwelling habit, they are not frequently caught in a passive fishing gear like gill net. It is no wonder, despite a regular stocking for 13 years, not a single common carp was ever caught from Nagarjunasagar reservoir of Andhra Pradesh. The stocked fishes failed to survive among the predators. A more important disqualification is its propensities to compete with some economically important indigenous carps like *Cirrhinus mrigala, Cirrhinus cirrhosa* and *Cirrhinus reba* with which common carp shares food niche. The presence of common carp has resulted in the decline of *Cirrhinus* species in Girna reservoir (Maharashtra) and Krishnarajsagar reservoir (Karnataka).

Introduction of *Cyprinus carpio* var. *speculris* in to Dal Lake has been reported to affect the population of indigenous *Schizothorax* sp. The benthophagus feeding habit of the fish fully utilized the Dal Lake benthos. The induction of the common carp significantly destabilized the native community balance of the Dal Lake and altered the energy flow system (Das, 2008).

The mirror carp has jeopardized the survival of a number of native fish species, after its introduction in the upland lakes of Kumaon Himalayas and some reservoirs of the northeast. Das (1989) observed that introduction of common carp has brought about a sharp drastic decline in the native population of *Osteobrama belangiri* in Loktak Lake of Manipur.

The introduction of *Clarius gariepinus* is very recent by a private trader during 2000. The introduction of this fish has brought about significant loss to the indigenous fish biodiversity. The fish is highly carnivorous and can thrive in extreme environmental conditions. This can certainly be a potential threat to the fish community if established in natural water bodies. The government of India has banned this fish and the people need to know this because it can cause great damage to the environment and the biodiversity.

Aristichthys nobilis has been introduced in India without official sanction. Ministry of agriculture, Government of India, Department of Animal Husbandry and Dairying, Fisheries Division, through letter No. 31016/1/96 and FY (3) dated 19th December 1997 had requested all the states governments to destroy the existing stock of *Clarius gariepinus* and *Aristichthys nobilis*. The reason for the ban is that these may cross breed with the endemic species and cause genetic degradation of our fish fauna. National Committee on Exotic Species had not approved the introduction of *Aristichthys nobilis* in to the country.

Grass carp (*Ctenopharyngodon idella*) is a natural of rivers of China. It has been introduced in India in 1959 for culture purpose. Fish has been introduced in Himachal reservoirs during 1985-86 and appeared in catches of Pong reservoir during 1987-88.Grass carp still maintaining low profile in reservoirs probably due to negligible water weeds (Sharma,2008).

REFERENCES

- Barry, W.M., 1990. Fishways for Queenland coastal streams:an urgent review. In Proceedings of the International Symposium on Fishways'90, Gifu, Japan.
- Bradka,J. and Rehackova, V., 1964. Mass destruction of fish in the Slapy reservoirs in winter 1962-63, In : Vodni, 1-10 spodarstvi, 14 : 451-452.
- Chandra, Ravish, 1989. Riverine fisheries resources of the Ganga and the Brahmputra.In: Conservation and management of inland capture fisheries resources of India (Ed.Jhingran, A.G. and Sugunana, V.V.). Inland Fisheries Society of India, Barrackpore, pp. 33-39.
- Clay, C.H., 1961. Desigh of fish ways ad other fish facilities, Development of fisheries of Canada, Ottawa.
- Das, P., 1989. Exotic fish germplasm resources in India and their conservation. In : Exotic Aquatic species in India (Joseph, M.M.Ed.), Asian Fisheries Society, Indian Branch,Special Publication Number 1, Mangalore, pp. 49-50.
- Das, P., 2008. Challenging experiences of exotic fish introductions:An overview, pp 35-44. In : Fish Introductions in India : Status, Potential and Challenges (Eds. W.S. Lakra, A.K. Singh and S. Ayyapan), Narendra Publishing House, Delhi.

- Desai, V.R., 2008. Reservoir can boost inland fish production. *Fishing Chimes*. 28(1): 39-46.
- Devaraj, K.V., Basavaraju, Y. and Seenappa, D., 2008. Introduced food fishes of Karnataka : An overview, pp 235-241, In : Fish Introductions in India : Status, Potential and Challenges (Eds. W.S. Lakra, A.K. Singh and S. Ayyapan), Narendra Publishing House, Delhi.
- Fish base, 2004. CD-Rom World Fish Centre : 205 Bloomingdak Bldg. Salceds St. Legapse Village, Makati City, Metro Manilla, 1200, Philippines.
- Gill, A.S., 1984. Effect of dams and water-ways on fish fauna in rivers of Punjab, pp. 125-126. In : Status of wildlife in Punjab.Indian Ecological Society, Ludhiana, India
- Holden, P.B. and Stalnaker, C.B., 1975. Distribution and abundance of mainstream fishes of the middle and upper Colorado river basins, 1967-1973. In : Transactions of the American Fisheries Society, 104 : 217-231.
- Hubbs, C. and Pigg, J., 1976. The effects of impoundments on threatened fishes of Oklahoma Annals of the Oklahoma Academy of Science, 5 : 133-177.
- Karamchandani, S.J. and Mishra, D.N., 1980. Prelininary observations on the status of silver carp in relation to catla in the culture fishery of Kulgarhi reservoir. J. *Bombay Nat. Hist. Soc;* 77 : 261-269.
- Lakra, W.S. and Singh, A.K. 2007. Exotic fish introduction in Indian waters : Past experience and lesson for the future. *Fishing Chimes*. 27 (1) : 30-34.
- Liv, J.K. and Yu, Z.T., 1992. Water quality changes and effects on fish populations in the Hanjiang River, China, following hydroelectric dam construction, Regulated Rivers:Research and Management, 7(4) : 359-368.
- Mallen-Cooper, M. and Harris, J., 1990. Fishways in Mainland South-Eastern Australia. In : Proceedings of the International Symposium on Fishways'90, pp. 221-230.
- Mitchell, C., 1995. Fish passage problems in New Zealand. In: Proceedings of the International Symposium on Fishways'95, Gifu, Japan.
- Peter, T. and Mitrofanov., 1998. The impact on fish stocks of river regulation in Central Asia and Kazakhstan Lakes and Reservoirs : Research and Management, 3 : 143-164.
- Petters, G.E., 1988. Impounded rivers,Chichester, U.K. : John Wiley and Sons Ltd. Publishers, p. 326.
- Quiros, R.1989. Structures assisting the migrations of non-salmonids fish : Latin America, FAO-COPESCAL Technical Paper 5, UN FAO, Rome.
- Ramakrishna, N.R. 2008. Exotic fishes in Karnataka : Their history, status and prospects. In : Fish Introductions in India : Status, Potential and Challenges (Eds. W.S. Lakra, A.K. Singh and S. Ayyapan), 249-254,Narendra Publishing House, Delhi.
- Raymond, H.L., 1979. Effect on dams and impoundments on the migration rate of juvenile Chinook Salmon and Steelhead Trout from the Snake River, 1966-1975. In : Transcations of the American Fisheries Society, 108(6) : 509-29.
- Sharma, B.D., 2008. Status of exotic fishes,their introduction and role in Himachal Fisheries, pp 259-264. In : Fish Introductions in India : Status, Potential and Challenges (Eds. W.S. Lakra, A.K. Singh and S. Ayyapan), Narendra Publishing House, Delhi.
- Singh, A.K. 2004. Aquaculture diversification and species enhancement : problems and perspectives. In : Zoology and Human Welfare edited by Dr. Ashok Verma, Published by University of Allahabad (Dr. S.P. Mukherji Government College, Phaphamaos), 252-261.

- Sreenivasan, A., 1967. *Tilapia mossambicus* : its ecology and status in Madras State, India. *Madras J. Fish.*, 3 : 33-39.
- Sreenivasan, Y., 1980. Fishing technology in the development of reservoirs. *India Today & Tomorrow*, 8(4) : 171.
- Vladykov, V.D., 1964. Inland fisheries resources of Iran, especially of the Caspian Sea with special reference to Sturgean, Report No. 1818, FAO, Rome, p. 51.
- Yadav, Pradeep, K., 1983. Reservoir fishery management : major policy issues for government intervention. In : Fisheries development in India : some aspects of policy management (Eds. U.K. Srivastava and M. Dharma Reddy), Proceedings of National Seminar on Fisheries Development in India, April 9-11, 1982 at Ahmedabad, pp. 433-457.
- Zhong, Y. and Power, G., 1996. Environmental impacts of hydroelectric projects on fish resources in China. In : Regulated Rivers : Research and Management, 12 : 81-98.

2

ANDHRA PRADESH

Fig. 2.1. Map of Andhra Pradesh (Not to scale)

Andhra Pradesh ranks first in brackish water shrimp production, first in fresh water prawn production, second in fresh water fish production, second in total value of fish and prawn produced and fifth in marine fish production. Andhra Pradesh has a coastline of 974 kms. There are 102 reservoirs, Kolleru and Pulicat lake and about 77,000 perennial, long seasonal and short seasonal tanks with a water spread area of 8.47 lakh hectares. There are 1.50 lakh hectares of lands suitable for brackishwater shrimp culture.

The fisheries sector has contributed 2.31% to the gross state domestic product during 2008-09. The quantity of fish and prawn produced was 12.52 lakh tonnes during 2008-09. The state contribution is about Rs. 3000 crores by way of marine product exports, which is nearly 40% of these exports from India.

Table 2.1. List of Reservoirs in Andhra Pradesh

Sr.No.	Reservoir	Area (ha)	Average stocking density (fingerlings/ha)	Average fish yield (kg/ha)
1.	Komatur	520	442	14
2.	Sangaihpet	562	819	11
3.	Malkapur	400	788	26
4.	Sanigaram	560	464	19
5.	Satupally	520	246	32
6.	B. Gangaveram	485	473	30
7.	Ramadugu	360	225	19
8.	Pappannapet	250	480	67
9.	Yusufpet	310	684	56
10.	Thumbur	280	556	17
11.	G.B. Natham	280	679	30
12.	Velvadam	180	344	6
13.	Nagarjunasagar	28490	NA	NA
14.	Edulabad	244	4933	209
15.	Krishnagiri	1280	861	70

Surveys of 40 small reservoirs in the state of Andhra Pradesh were conducted by Sugunan and Katiha (2004) between 1994 and1999 to determine relationships between area, stocking density and fish yield. Area was negatively correlated with stocking density (r ¼)0.57) and fish yield (r ¼)0.62), but its correlation with efficiency of stocking was not significant for all the size groups of small reservoirs. Most of the reservoirs were perennial and managed by fishery co-operatives.The stocking material was raised in pens erected at the reservoir sites. The fish species stocked were primarily the Indian major carps, *Catla catla* (Hamilton), *Labeo rohita* (Hamilton) and *Cirrhinus mrigala (Hamilton).*The size of stocked fish at release was above 100 mm with average weight 13.50 g for *C.catla,* 15.93 g for *L. rohita* and 19.97 g for *C. mrigala.* The minimum weight of the fish caught was 0.5 kg.

NAGARJUNASAGAR RESERVOIR

Nagarjunasagar reservoir is formed due to the construction of dam across river Krishna at Lat. 16°34'N and Longitude 79°79'E. The tributaries Dindi and Peddavagu do not contribute significantly to the inflow as both these streams are dammed. The outlet channels are right and left canals, right and left chutes and 26 spillways. The silt brought in by the flood is deposited in the deepest regions of the lentic sector. The reservoir has a maximum water spread area of 28490 ha at full reservoir level of 179.832 m above MSL. The catchment area is 2,15,193 km^2. The reservoir has a gross capacity of 11,550 million m^3. Bottom of the reservoir is rocky with moderate to giant sized boulders. Muddy bottom is observed in bays. The annual inflow is about 34.34 m a ft. The outflow is about the same as inflow *i.e.,* 34.79 m a ft.

Active fishing in the reservoir started during the year 1968 (David *et al.* 1969), though it stabilized only in the year 1970. Ramakrishniah (1990) studied the ecology of Nagarjuna reservoir. The fishery of reservoir consisted mainly of indigenous species of the parent river Krishna. Though some stocking of major carps has been done during early years (less than one fingerling/ha/yr), its impact on fishery was negligible. The dominant species during 1971-72 was *Labeo fimbriatus* accounting for about 36% of the catch. Carps had an edge in the catch over catfishes until 1973-74, thereafter the catfishes dominated contributing to about 70-73% of the catch during later years. The carp fishery consisted of *Labeo fimbriatus, Labeo calbasu* and *Tor khudree,* while catfishes included *Pangasius pangasius, Mystus aor, Mystus seenghala, Silonia childrenii* and *Wallago attu.*

The standing crop of plankton of the reservoir showed two peaks-one in summer (April) and other in winter (December). The summer peak was more prominent. A small peak was observed in September-October also. The south-west monsoon and subsequent inflow of flood waters resulted in the sudden dilution in crop from the summer maxima. Plankton volume fluctuated with the changes in inflow and outflow of the reservoir. Both the peaks more or less coincided with the period of minimum water flow. From July-October when the inflow and outflow increased, the plankton declined. The summer peaks coincided with minimal mean depth, inflow and outflow. The winter peaks accompanied high mean depths with the inflow and outflow at their minimum.

The south-west monsoon prevailed during May to July with maximum rainflow in June. The north-east monsoon was short but active either in September or in October. Both the peaks in standing crop occurred in the period of minimum rains. The numerical abundance in phytoplankton increased immediately after the north-east monsoon. Nitrogen and total alkalinity showed the maximum values in summer (April-June) and minimum in monsoon (July-September). The morpho-edaphic index (MEI) was maximum in April-June in both the years (5.20 and 5.14 respectively). The annual average were 3.56 and 4.32 in the first and second year respectively.

The lotic sector recorded the least plankton. The Peddamunigala Bay showed highest plankton both numerically and volumetrically. The bay plankton, due to its high values, greatly dictated the trend of the plankton of the whole reservoir. Intermediate and lentic sectors yielded much lower crop than the bay. In all the sectors except the lotic, the plankton diminished from littoral to profundal zones.

Table 2.2. Plankton Diversity in Nagarjunasagar Reservoir in Andhra Pradesh

Myxophyceae: *Microcystis, Oscillatoria.*
Chlorophyceae: *Pediastrum, Spirogyra, Botryococcus.*
Bacillariophyceae: *Navicula, Melosira, Tabellaria, Synedra, Fragilaria.*
Rotifera: *Keratella, Conochilus, Lecane, Myfilina, Filina.*
Cladocera: *Chydorus, Daphnia, Ceriodaphnia, Diaphnosoma.*
Copepoda: *Cyclops, Diaptomus, Nauplii.*

Table 2.3. Some Metrological, Morphometric and Chemical Features of Nagarjunasagar Reservoir During September 1974 to August 1976

Sr.No.	Parameter	Range
1.	Air temperature (°C)	16.39 – 43.06
2.	Total rainfall (mm)	Nil – 366.30
3.	Mean depth (m)	31.08 – 39.18
4.	Total alkalinity (ppm)	74.90 – 137.19
5.	Nitrate (ppm)	0.510 – 1.200

In lotic zone, sand formed 59% of the soil followed by 26% silt and 15% clay. In intermediate zone, sand and clay were similar (37% each) and silt formed 26%. Lentic sector exhibited similar soil texture *viz.*, 40% clay, 39% sand and 21% silt. In Peddamunigala, sand formed 48% followed by clay 34% and 18% silt.

Peddamunigala bay had the most fertile soil with maximum values of organic carbon (1.25%), available nitrogen (18 mg/100 g) and available phosphorus (0.45 mg P_2O_5/100 g). Lotic sector was poor in fertility status with organic carbon 0.73%, available nitrogen 5.3 mg/100 g and phosphates 0.3 mg/100 g. The lentic and intermediate zones were similar and did not vary widely between them. Organic carbon was 0.75 and 0.8 in intermediate and lentic sectors. The corresponding values for available nitrogen were 13.5 and 12.5 mg/100 g for the two sectors respectively. Available phosphorus was 0.3 mg/100 g in intermediate and lentic sectors.

Soil pH varied from 7.5 to 8.1 in the reservoir more frequently remaining around 8.0. There was always a reserve of calcium carbonate (0.79 to 14.93%). Specific conductivity was high in Peddamunigala bay (av. 753.5 micro mhos/cm) and low in the other zones (290 to 453 micro mhos/cm). Sugunan and Das (1983) studied bottom macrofauna of Nagarjunasagar reservoir.

The highest density of benthic population is available in Peddamunigala bay followed by the lentic, intermediate and lotic sectors. There is a remarkable difference in the qualitative composition of bottom fauna among zones. Thirteen species of molluscs comprising six species of prosobranch, gastropods, three species of pulmonate gastropods and four species of bivalves were recorded. Insect larvae are distributed throughout the bottom with an increase in number of chironomids upto 6 m depth. May fly nymphs and oligochaetes are oriented towards the upper 15 m. Gastropods are found in all depths with peaks in 15 and 40 m representing the deep bottom populations of Peddamunigala bay and lentic/intermediate zones. Chemical status of the soil seems to influence the production of bottom macrofauna.The most important environmental feature affecting the distribution of bottom macrofauna is the changes in some hydrographic features. The incoming flood waters mechanically dislodge the benthic organisms and are sent down through the outflowing channels. The secondary succession takes place as soon as the disturbances due to water movements wean away. This is evident from the seasonal cycle of different groups.

Department of Atomic Energy has awarded a project to SRM University, Chennai,for a comprehensive study of uranium content in various environmental matrices including water and biota of Nagarjunasagar reservoir. The water from said reservoir is used for drinking purposes in the twin cities of Hyderabad and Secunderabad. Under the same project the measurement of uranium content in fish, samples has also been carried out. Due to its natural occurrence in soil and rocks of the earth crust, uranium is present in all environmental matrices such as air, water, soil, sediment, food materials and the biota. Its concentration in soil varies from 1 to 5 micro grammes per gram *i.e.,* 1 to 5 parts in a million parts of a gram; it varies in water from 1 to 3 nanogrammes per milliliter *i.e.,* 1 to 3 parts of a billion parts of a milliliter; while in fish it varies from 5 to 60 nanogrammes per gram (wet weight). In Nagarjunasagar reservoir, the uranium content in fish was found to vary from 20 to 30 nanogrammes per gram of fish (wet weight) which is within the normal concentration range in fish found in other parts of the country.

RAJAHMUNDRY RESERVOIR

The Rajahmundry dam is also known as "Dowleshwaram dam". It is about 10 kms from the Rajahmundry city. The dam was built by Sir Alfred Cotton and constructed between 1848 to 1852. The reservoir is formed by construction of dam cross river Godavari at longitude 81°45'E and latitude 17°01'N and elevation above sea level is 74 feet.

Rathod and Khedkar (2011) recorded 47 fish species belonging to nine orders from Rajahmundry reservoir (Table 2.4). Order cypriniformes was dominant with 16 species in the assemblage composition in which *Osteobram vigorsii* were found most abundant. *Lepidocephalus gunte, Labeo arzia, Labeo calbasu, Catla catla, Cyprinus carpio* and *Salmostoma navacula* were found abundant form. *Thynnichthys sandkhol, Cirrhinus mrigala, Cirrhinus reba, Puntius ticto, Puntius fraseri* and *Labeo rohita* were found less abundant. *Puntius filamentosus, Puntius vittatus, Osteobram dayi* were found rare abundant. Followed by order perciformes were 10 species in the assemblage composition in which *Glossogobius giuris giuris, Chanda nama, Parambassis ranga, Channa marulius, Channa striatus, Channa punctatus, Channa orientalis* was found abundant. *Etroplus suratensis, Oreochromis mossambicus* were found less abundant. *Anabus testudineus* was found to be rare abundant. Next order siluriformes were represented by 12 species in the assemblage composition.In order siluriformes, *Ompak bimaculatus* were found most abundant. *Mystus bleekeri, Mystus cavasius, Wallago attu, Bagarius bagarius* were found abundant. *Mystus seenghala, Aorichthys aor, Rita rita, Panagasius pangasius, Silonia children* were found less abundant. *Nandus nandus, Heteropnesutes fossilis* were found rare abundant. Followed by order synbranchiformes were three species in the assemblage composition in which *Macrognathus pancalus* were found most abundant. *Mastacembelus armatus, Macrognathus aculeateda* were found less abundant. Followed by order osteoglossiformes were two species in the assemblage composition in which *Notopterus notopterus* and *Notopterus chitala* found less abundant. Followed by order angulliformes were one species in the assemblage composition in which *Anguilla bengalensis* were found abundant. Followed by order clupiformes were one species in the assemblage composition in which *Hilsa ilisha* was found rare abundant. Followed by order beloniformes were one species in the assemblage composition in which *Xenentedon cancila* were found abundant. Next order salmoniformes were represented by one species in the assemblage composition in which *Rhinomugil corsula* were found less abundant (Table 2.4).

Table 2.4. Fish fauna of Rajahmundry Reservoir, Andhra Pradesh

Sr.No.	Family	Species
1.	Notopteridae	*Notopterus notopterus*
2.	Notopteridae	*Notopterus chitala*
3.	Anguillidae	*Angulla bengalensis*
4.	Clupeidae	*Hilsa ilisha*
5.	Cyprinidae	*Salmostoma navacula*
6.	Cyprinidae	*Cyprinus carpio carpio*
7.	Cyprinidae	*Osteobrama vigorsii*
8.	Cyprinidae	*Osteobrama dayi*
9.	Cyprinidae	*Thynnichthys sandkhol*

Contd...

Table 2.4: Contd...

10.	Cyprinidae	*Cirrhinus mirgala*
11.	Cyprinidae	*Cirrhinus reba*
12.	Cyprinidae	*Puntius Vittatus*
13.	Cyprinidae	*Puntius ticto*
14.	Cyprinidae	*Puntius fraseri*
15.	Cyprinidae	*Puntius filamentosus*
16.	Cyprinidae	*Catla catla*
17.	Cyprinidae	*Labeo rohita*
18.	Cyprinidae	*Labeo calbasu*
19.	Cyprinidae	*Labeo arzia*
20.	Cobitidae	*Lepidocephalus guntea*
21.	Heteropneustidae	*Heteropneustes fossilis*
22.	Bagiridae	*Mystus seenghala*
23.	Bagiridae	*Mystus bleekeri*
24.	Bagiridae	*Mystus cavasius*
25.	Bagiridae	*Aorichthys aor*
26.	Bagiridae	*Rita rita*
27.	Siluridae	*Wallago attu*
28.	Siluridae	*Ompak bimaculatus*
29.	Pangasiidae	*Pangasius pangasius*
30.	Sisoridae	*Bagarius bagarius*
31.	Schilbeidae	*Silonia children*
32.	Nandidae	*Nandus nandus*
33.	Gobiidae	*Glossogobius giuris giuris*
34.	Chanidiae	*Chanda nama*
35.	Chanidiae	*Parambassis ranga*
36.	Channidae	*Channa marulius*
37.	Channidae	*Channa striatus*
38.	Channidae	*Channa punctatus*
39.	Channidae	*Channa orentalis*
40.	Cichlidae	*Etroplus suratensis*
41.	Cichlidae	*Oreochromis mossambica*

Contd...

Table 2.4: Contd...

42.	*Anabantidae*	*Anabus testudineus*
43.	Mastacembelidae	*Mastacembelus armatus*
44.	Mastacembelidae	*Macrognathus pancalus*
45.	Mastacembelidae	*Macrognathus aculeateda*
46.	Beonidae	*Xanthodon cansula*
47.	Mugilidae	*Rhinomugil corsula*

EDULABAD RESERVOIR

The water-spread area of reservoir is 244.31 ha. The water of the reservoir received from river Musi through Narayan Rao canal, is used for irrigation and fisheries development. The water is highly polluted (Srinivas, 2007). Dipteran dominated in the reservoir during biomonitoring studies. The other benthic macroinvertebrates like choleoptarans, hemipterans, molluscs and oligocheates were also found in the reservoir.

Plankton: Phytoplankton represented by cyanophyceae, bacillariophyceae, euglenophyceae and chlorophyceae and cyanophyceae was dominant. Phytoplankton declined form 32 constituent species in 2000-01 to 24 numbers in 2001-02 and further to 19 species in 2002-03. Zooplankton represented by rotifera, cladocera, copepoda and ostracoda. Among zooplankton, rotifera was dominant zooplanktonic genera declined from 26 species in 2000-01 to 21 species in 2001-02 and further to 19 in 2002-03. Only 6 and 3 genera among phyto and zooplankton respectively were found to be most abundant.

Fish Fauna: Fish fauna represented by 13 species of cypriniformes, 9 species of siluriformes, 4 species each of channiformes and perciformes. The fish fauna declined from 32 species in 2000-01 to 13 species in 2001-02 and further down to 8 species by 2002-03. The fish fauna consisted of major carps, minor carps, catfishes, murrels, tilapia, and other food fishes. Of these, rohu, catla, grass carp and tilapia were found to be most abundant.

Fish kill were observed in Edulabad reservoir. These kills are traced to elevated physico-chemical parameters and dissolved oxygen levels. Abnormal increase of heavy metals was observed in the reservoir. The biomagnification investigations indicated accumulation of heavy metals in muscle and liver of fish.

Fish production: The average fish production from the reservoir was 208.83 kg/ha/yr.It was 340.8 kg/ha/yr in 2000-01,211.22 kg/ha/yr in 2001-02 and 65.77 kg/ha/yr in 2002-03. The fish production from Edulabad reservoir declined from the first to third year. The major carps dominated with 213.21 kg/ha/yr (63.58%),107.81 kg/ha/yr (51.04%) and 26.45 kg/ha/yr (40.21%) during 2000-01,2001-02, and 2002-03 respectively with a mean of 115.82 kg/ha/yr. Tilapia occupied second position with 100.21 kg/ha/yr (29.56%), 96.48 kg/ha/yr (45.68%) and 38.35 kg/ha/yr (58.30%) during the above periods respectively with a mean of 78.51 kg/ha/yr. Murrels and catfishes occupied third and fourth places respectively. Other food fishes found were minimal. The ratio between the major carps and tilapia was 1.31:1, between major carps and other food fishes was 16.93% and that between tilapia and other fishes was 19.05:1. The production was inversely proportional to each other between the major carps and tilapia.

Among major carps, catla landings dominated with 88.10 kg/ha/yr (41.32%), 48.43 kg/ha/yr (44.92%) and 11.97 kg/ha/yr (45.24%) respectively, during the study period with a mean of 49.50 kg.ha/yr. Rohu dominated after catla with 77.74 kg/ha.yr (36.46%), 40.97 kg/ha/yr (38.00%) and 10.35 kg/ha/yr (39.12%) respectively with a mean of 43.01 kg/ha/yr. Grass carps occupied third placc with a mean of 11.73 kg/ha/yr. Mrigal came fourth with a mean of 12.55 kg/ha/yr and the presence of the common carp was the least with a mean of 9.64 kg/ha/yr.

Advanced fry of major carps (950 nos; 55 mm) was stocked annually. There was no stocking with other fishes. Yet there was their contribution because of self-recruitment. The stocking densities respectively during 2000-01 to 2002-03 were 4933 fry/ha, 7278 fry/ha and 10513 fry/ha. This worked out to a mean of 7574 fry/ha. The number of advanced fry (yearlings) for production of one kg fish was estimated at 162 on an average. The actually stocked figures increased from 2000-01 (23 nos) to 2001-02 (68 nos) and to a highly impressive average in 2002-03 (397 nos).Fish production (115.82 kg/ha/yr) decreased with increase of stocking densities (7574 fry/ha).The fish production was 213.2 kg/ha/yr with stocking of 4933 advanced fingerlings/ha during 2000-01. Production declined in the next year (2001-02) to 107.81 kg/ha/yr with stocking of 7278 advanced fingerlings/ha. In the third year (2002-03) production declined to 26.45 kg/ha/yr with stocking of 10,513 advanced fingerlings/ha.

The growth rate of major carps from Edulabad reservoir was slow. They attained an average weight of one kg only at the end of third year.

All the above adverse changes were due to the pollution load in the reservoir caused by Musi river water that entered into the reservoir. The deterioration of water quality was due to accumulation of heavy metals. This was responsible for the fish kills that occurred frequently. The cumulative effect of water conditions of river Musi reflected in all aspects of fisheries of the reservoir. There was another stress-causing aspect in the reservoir and this is related to the heavy population of tilapia. Due to its voracious feeding, nature of prolific breeding, stunted growth and early maturity, tilapia competes with major carps for space and food. Due to its competition, the growth and production of major carps declined, resulting in the decline of earnings of fishermen.

The water conditions of the reservoir deteriorated due to the inflow of river Musi.This affected water quality, agriculture and fisheries. Very high stocking density maintained during 2002-03 (10513 advanced fingerlings/ha) to improve fish production, proved to be in vain. This indicates that the reservoir is not economically viable for fisheries development. Moreover, the biomagnification of heavy metals at higher levels in fish tissues indicates that the reservoir fish may not benefit for human consumption.

ALISAGAR RESERVOIR

The Alisagar reservoir is constructed across the river Manjara in Nizamabad district. The main scope of the reservoir is irrigation and drinking water. Tamlurkar and Ambore (2006) reported physico-chemical characteristics of Alisagar reservoir (Table 2.5).

Table 2.5. Physico-chemical Parameters of Alisagar Reservoir, Andhra Pradesh

Parameters	Range
Air temperature (°C)	25-36.8
Water temperature (°C)	22-33
pH	7.01-8.5
Carbon dioxide (mg/l)	00-22
Carbonate alkalinity (mg/l)	11-180
Dissolved oxygen (mg/l)	3.2-7.16

The higher values were observed during winter season. It may be due to high solubility at low temperature and less degradation of organic substances. The carbonate alkalinity was found high during winter season. The carbon dioxide values did not show any significant variations.

The highest positive correlation was observed between air temperature and water temperature, while highest negative correlation is observed between water temperature and dissolved oxygen. Air temperature and water temperature show negative correlation with pH, carbon dioxide, carbonate alkalinity and dissolved oxygen.

MANJIRA RESERVOIR

Manjira reservoir was constructed on the river Manjira,a tributary of the Godavar in South India has a barrage constructed across it near Sangareddy (17° 37′N, 78°06′E) to form reservoir to supply the potable water to Hyderabad city.The reservoir has a catchment area of about 16,780 km^2 and a maximum storage of 73.63 mm^3. Due to heavy siltation the maximum depth of the reservoir near the barrage is reached to about 9.5 m. The river bed is dotted with boulders of various sizes, some of them buried deep in the silty clay while a few lie exposed during the depletion of water level in summer.

The water of the Manjira reservoir is prone to sudden, short but fairly frequent spells of moderately high turbidity even though the mean values in Table 2.6 suggest that the water is optically less turbid. Import of silt from the catchment during floods, wind generated turbulence, water discharge across the barrage cause high turbidity in this shallow reservoir.

Table 2.6. Physico-Chemical Parameters of the Manjira Reservoir, Andhra Pradesh

Parameter	Range	Mean
Water temperature (ºC)	19.10 – 34	27.73
pH	7.60 – 9.35	8.40
Secchi disc depth (2 sdm)	0.030 – 1.065	0.472
Turbidity (Hydrazine units)	1.6 – 386	63
Suspended solids (mg/l)	0.0 – 456	71.388
Dissolved solids (mg/l)	96 – 417	236.756
Dissolved oxygen (mg/l)	2.45 – 18.59	8.774
Carbon dioxide (mg/l)	0.00 – 14.52	1.02
Carbonate alkalinity (mg/l)	0.00 – 57.00	14.375
Bicarbonate alkalinity (mg/l)	88.48 – 314.25	190.528
Chlorides (mg/l)	9.96 – 54.95	22.418
Sulphates (mg/l)	0.00 – 39.25	18.896
Silica (mg/l)	0.40 – 6.00	1.74
Total nitrogen (mg/l)	0.0022 – 7.1271	1.2647
Total phosphrus (mg/l)	0.000 – 0.190	0.0394
Sodium (mg/l)	6.5 – 135	50.595
Potassium (mg/l)	1.7 – 4.9	2.85
Calcium (mg/l)	3.21 – 32.06	17.212
Magnesium (mg/l)	0.2 – 54	16.232
Total iron (mg/l)	0.00 – 1.32	0.1465

THANDAVA RESERVOIR

The Thandava reservoir,situated near Thandava village in Nathavaram mandal of Visakhapatnam district has a water spread area of 1400 ha. The present productivity of the reservoir worked out to be 142.75 kg/ha/yr. Generally the Indian major carps are stocked in the reservoir. They are showing a satisfactory survival rate. Recently (year 2012-13) 3,33000 fingerlings of catla, 4,17,000 fingerlings of rohu and 83000 fingerlings of mrigal were stocked which gave an average stocking of 595 fingerlings/ha.

VELUGODU RESERVOIR

It is a major reservoir in Kurnool district at Lat.15°44′42″N and Long. 78°36′20″E. It was constructed to facilitate the storage of flood waters of Krishna and Pennar drawn through Telgu Ganga canal. The fisheries details of reservoir are given in Table 2.7 and Table 2.8. Previously the Indian major carps were stocked in the reservoir. But for the last seven years, Grass carp and jayanti rohu are also stocked along with fingerlings of catla and rohu. Jayanti rohu is genetically improved rohu with higher growth efficiency developed through selective breeding. Morphologically it is similar to rohu but grows much faster than the normal rohu. The Jayanti rohu, is recording 17% more growth per generation after seven generations of selection. It is being disseminated to farmers with close monitoring by Central Institute of FreshwaterAquaculture (CIFA), Bhubaneshwar. The improved rohu, popularly known as 'Jayanti', was named in 1997 (*i.e.*, 50th anniversary of Indian independence- Swarna Jayanti), has been released to several hatchery owners so that they can provide better quality seed to the fish farmers. The highest stocking of reservoir was done in 2007-08 and the minimum was in 2006-07 (Table 2.7).

Table 2.7. Seed stocking in Velugodu Reservoir of Kurnool district

Year	Catla	Rohu	Jayanti rohu	Grass carp	Total
2005-06	2.975	—	3.017	—	5.992
2006-07	—	2.050	—	—	2.050
2007-08	9.936	4.988	9.937	—	24.860
2008-09	—	—	8.800	—	8.800
2009-10	1.850	2.500	—	2.400	6.750
2010-11	13.200	—	—	—	13.200
2011-12	4.830	11.200	—	1.350	17.380

Table 2.8. Fish and prawn catch in Velugodu Reservoir

Year	Fish catch (MT)	Prawn production (MT)
2005-06	1182	—
2006-07	1062	8
2007-08	1631	32
2008-09	2446	15
2009-10	2712	10
2010-11	2693	10
2011-12	2741	33

YELERU RESERVOIR

The reservoir was constructed near village Yeleswaram in East Godavari district. Effective water spread area is 3600 ha. The morphometry of reservoir is presented in Table 2.9.

Seed was stocked with financial assistance from NFDB scheme. Fingerling stage of Indian major carps was stocked from 2010 to 2012. There was no stocking with other fish species. Details of fish seed stocked and fish catch in the reservoir are shown in Table 2.10.

Table 2.9. Morphometry of Yeleru Reservoir

Attribute	Value
1. Name of Reservoir	Yeleru Reservoir
2. Nearest city	Kakinada
3. River	Yeleru
4. Type of dam	Earthen and Gravity
5. Purpose of dam	Water storage and Irrigation
6. Year of completion	1991
7. Catchment area (Th ha)	223.2
8. Length of dam (m)	2544
9. Maximum height above foundation (m)	43
10. Maximum water level (m)	86.56
11. Full Reservoir Level (m)	86.56
12. Gross storage capacity (mcm)	682.7
13. Live storage capacity (mcm)	508.3
14. Length of spillway (m)	150
15. Number of spillway gates	10
16. Number of affected villages	

Table 2.10. Seed stocking and fish production from Yeleru Reservoir

Year	Seed stocked (in lakhs)	Fish production (tons)
2008	Not stocked	108
2009	Not stocked	126
2010	19.16	144
2011	9.975	162
2012	11.22	180

MADDUVALASA RESERVOIR

It was constructed across rivers Vegavathi and Suvarnamukhi in Srikakulam district and is located at 18036'0"N and 83°36′0″E. It was commissioned in the year 1977. About 15000 acres of agricultural land was brought under cultivation with the reservoir water. It has been a source of livelihood for many fishing communities in the villages surrounding it.

Table 2.11. Salient features of Madduvalasa Reservoir

Attribute	Value
1. Name of dam	Madduvalasa
2. Nearest city	Vijayanangarm
3. Type of dam	Earthen
4. Catchment area	253.5
5. Length of dam (m)	2320
6. Maximum height above foundation (m)	16
7. Full Reservoir Level (m)	65
8. Gross storage capacity (mcm)	95.51
9. Live storage capacity (mcm)	93.58
10. Number of spillway gates	11

During 2012-13 90, 60, 650 nos of seed was stocked. The fish production is depicted in Table 2.12.The total catch per year rose from 170 mt in 2008-09 to a peak of 185 mt in 2012-13. There is a need for strict enforcement and monitoring of illegal fishing practices, which exists in shallower regions of the reservoir. Mesh size regulation is not being observed for any fishing methods employed. Closed season for fishing is not practiced by the fishers.

Table 2.12. Fish production in Madduvalasa Reservoir

Year	Fish catch (MT)
2008-09	170
2009-10	172
2010-11	174
2011-12	180
2012-13	185

KADAM REERVOIR

It was constructed in year 1958 on river Kadam in Adilabad district. It is a medium reservoir with area of 2474 at FRL. The average water spread area of reservoir is 1842 ha with a mean depth of 8.9 m. Shallow reservoirs with high transparency are more productive (Ryder, 1965) which is also true in case of Kadam reservoir (*Das et al.* 2001). Due to high $CaCO_3$ content in soil, reservoir showed a pH of 7, 9 to 9.8 (*Das et al.* 2001). Alkalinity showed an increasing trend in pre-monsoon exhibiting clinograde distribution. Chemical stratification occurs during pre and post-monsoon with steeper oxycline and the reservoir could be categorized as productive reservoir. The reservoir is in higher category of production with a potential higher than 100 kg/ha (*Das et al.* 2001).

Fish food resources are fairly rich with zooplankton prevailing in plankton and dipteran larvae in bottom biota.Fishery consists of mostly indigenous fishes which include *Tor khudree*. The reservoir was stocked with fingerlings of Indian major carps. During year 2011-12 stocking was suspended because of nonavailability of seed. The total catch rose from 140 MT in 2008-09 to a peak of 620 MT I 2012-13. Stocking rate has been seeming good.

SRI RAM SAGAR RESERVOIR

The Sri Rama Sagar reservoir also known as the Pochampadu Project is constructed on the Godavari River. The Project is located in Adilabad district.The foundation was laid on 26 July 1963 by the late Jawaharlal Nehru, first Prime Minister of India. It has been described by The Hindu as a lifeline for a large part of Telangana. It is a hydroelectric project. Sriram Sagar serve for irrigational needs in Karimnagar, Warangal, Adilabad, Nalgonda, and Khammam districts. It also provides drinking water to Warangal town. There is a hydroelectric plant working at the dam site.

Sriram Sagar Reservoir's capacity is 75 billion cubic feet, and it has 42 floodgates. Construction of this dam was started in year 1957 and is not yet completed. This project is also locally known as 'Khustapuram dam'. Under first stage of this project nearly 1 million acres (4,000 km^2) irrigation facility is created to utilize 140 tmc water. Second stage of this project is under advanced stage of construction to irrigate 440,000 acres (1,800 km^2) using 25 tmc water. The flood flow canal project is also under implementation to irrigate 200,000 acres (810 km2) using 20 tmc water available at Pochampadu dam site. The live storage capacity of dam is limited to 90 tmc to reduce submergence area in Maharashtra up to FRL level 1,091 feet (333 m) above mean sea level as per the agreement between Maharashtra and Andhra Pradesh.

The maximum fish production for the last five years (from 2008-09 to 2012-13) was recorded at 2210 tonnes during 2012-13. The minimum production was during 2008-09 at 1170 tonnes. Recently in 2012-13 29 lakh seed of carps was stocked.The reservoir is not regularly stocked.

MIR-ALAM RESERVOIR

Mir-Alam reservoir was constructed in 1806 A.D. It is one of the oldest ma-made impoundments situated in North-Western direction of Hyderabad. The water body used to be a source of potable water up to recent times, however, today its water is mainly used for Nehru Zoo Park and the lake is being developed into a recreational zone by Hyderabad Urban Development Authority (HUDA).The morphometric features of the Mir-Alam reservoir are depicted in Table 2.13.

Table 2.13. Morphometric Features of Mir-Alam Lake, Hyderabad

1.	Surface area (km^2)	1.70
2.	Catchment area (km^2)	15.24
3.	Capacity at storage (Mcm)	8.12
4.	Maximum water spread (km^2)	1.69
5.	Maximum length (km)	2.54
6.	Maximum width/breadth (km)	1.63
7.	Mean width (km)	0.66
8.	Maximum depth (m)	31.41
9.	Mean depth (m)	4.80
10.	Maximum length of bundh (m)	1024.12

The phytoplankton diversity of Mir-Alam reservoir is reported by Siddiqi and Khan (2002).The phytoplankton was comprised of 58 species, which included 25 species of blue green algae, 28 of green algae and 5 of desmids. Out of these, 6 species of blue green algae (*Chroococcus minutes,Coelosphaerium kutzingianum, Spiurlina sp., Phormidium favaeolarum, Lyngbya limnetica* and *Anabaenopsis raciborski*) and 3 species of green algae (*Schroederia setigera, Ankistrodesmus falcatus* and *Monoraphidium convulutum*) were designated as dominant. Beside these, 6 species of blue green algae and one species of desmids were designated as common. Blue green algae dominated the phytoplankton composition, contributing early 93% of the total phytoplankton numbers (Siddiqi and Khan, 2002).

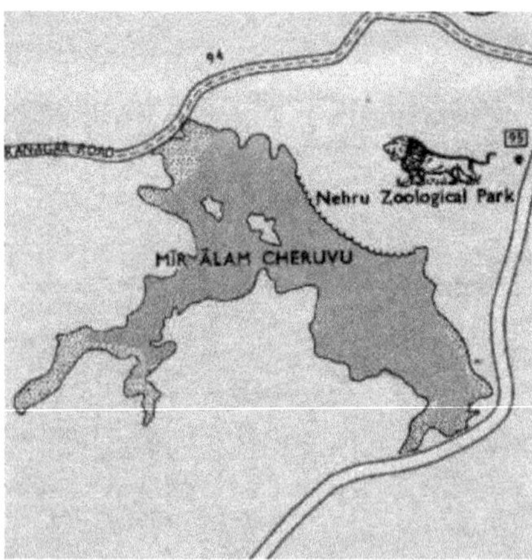

Fig. 2.2. A Map of Mir-Alam Reservoir, Hyderabad

The studies on macro-zoobenthos in Mir-Alam reservoir was carried out by Anitha *et al.* (2006). The macro-zoobenthic fauna from MirAlam lake belonged to the three major groups; annelid represented by 3 species belonging to tubificidae, naididae and lumbricidae, 4 species of arthropods belonging to chironomidae ad notonectidae ad 10 species of mollusca belonging to viviparidae, thiaridae, planorbidae, bithyniidae ad lymnaeidae (Table 2.14).

Table 2.14. Diversity of Macro-Zoobenthos in Mir-Alam reservoir, Hyderabad

ANNELIDA
Limnodrilus hoffmeisteri claparede 1862
Pristina aequiseta Bourne, 1891
Lumbricidae (Species not identified)
ARTHROPODA
Notonecta glauca
Mosquito larvae
Chironomous circumdatus kieffer
Chironomous plumosus
MOLLUSCA
Bellamya dissimilis
Bellamya bengalenis
Indoplanorbius exustus
Thiara tuberculata
Thiara scabra
Lymnaea accuminata
Gyraulus convexiuscullus
Cremoconchus coinu
Fanus ater
Gabbia stenothyroide

OSMAN SAGAR RESERVOIR

It was built to control recurrent floods ad to provide drinking water to the residents of Hyderabad city. The reservoir is located at a distance of 25 kms from Hyderabad city. It was constructed across the river Musia, a tributary of river Krishna in the year 1920. It is one of the biggest dams in the Andhra Pradesh. The reservoir has a catchment area of 740.72 km^2. The maximum length and breadth of the reservoir are 11.28 and 5.11 km respectively. Its maximum depth was recorded as 30.70 meters and the ea depth as 6.17 meters. The reservoir has a total storage capacity of 4.04 TMC at FTL. The Osman Sagar reservoir is categorized as oligotrophic in nature (Siddiqi and Khan, 2002). Shailaja and Johnson (2006) reported the values of chlorides, fluorides, calcium and magnesium in the reservoir water at 483.7, 1.30, 143.6 and 116.3 mg/l respectively. The reservoir water is primarily alkaline (Siddiqi and Khan, 2002).

Altogether, 54 species belonging to 3 groups of phytoplankton were reported (Siddiqi and Khan 2002). The phytoplankton recorded in the reservoir is depicted in Table 2.15.

Table 2.15. Phytoplakton Diversity in Osman Sagar Reservoir

CYANOPHYCEAE
Chrooccus minutes, C.turgidus, Gloeocapsa sp., *Synechocystis pevalekii, Aphaocapsa sp., Synechococus* sp., *Aphaothece* sp., *Merismopedia punctata, Coelosphaerium kutzingianum, Goephosphaeria apopina, Spirulina* sp., *Oscillatoria agusa, O.germinata, O.limnetica, O.princeps, Phormidium laminorum, P.faveolarum,Lynabya limnetica, L.perelegons, Anabaena* sp., *Anabaenapsis raciborskii, Raphidiopsis mediterranea.*
CHLOROPHYCEAE
Chlamydomonas sp. *Phacotus lenticularis, Schroederia setigera, Pediastrum boryanun, P.simplex var.duodenarium coelastrum, Chlorella vulgaris, Oocystis* sp., *Akistrodesmus falcatus, Monoraphidium convulutum, Tetraesuon limneticum, T. minimum, T. regulare, T. trigonum, Senedesmus acutiformis, S. armatus, S.bijuga, S. bijuga var.alteras, S. dimorphus, S. obliguus, S. opolicusis, S. quadricauda, cruugenia tetrapedia, Actinastriam hatzschii*
DESMIDS
Closterium sp., Cosmarium botrytis, C. laeve, C. meneghini, C. portianum, C. subexcdvatum, C. subtumidum, C. staurastrum.

Table 2.16. Ichtyofauna of Osman Sagar Reservoir

1. *Notopterus notopterus* (Pallas)
2. *Rasbora daniconius* (Hamilton)
3. *Puntius amphibious* (Valencienes)
4. *Puntius chola* (Hamilton)
5. *Puntius filamentous* (Valencinnes)
6. *Putius kolus* (Sykes)
7. *Puntius sarana* (Hamilton)
8. *Puntius ticto* (Hamilton)
9. *Puntius jerdoni* (Day)
10. *Cirrhina fulungea* (Sykes)
11. *Cirrhina mrigala* (Hamilton)
12. *Labeo rohita* (Hamilton)
13. *Cyprinus carpio var. communis* (Linnaeus)
14. *Crossocheilus latius* (Hamilton)
15. *Garra lamta* (Hamilton)
16. *Garra mullya* (Sykes)
17. *Oreonectes evezardi* (Day)
18. *Ompak bimaculatus* (Bleekeri)

SHATHAMRAJ RESERVOIR

Shathamraj is a small sized reservoir with a water spread area of 25 ha loacted in Shaikpet village near Hyderabad. The water is used for irrigation ad fisheries. Piska (2000) documented impact of stocking densities of major carp seed on fish production in Shathamraj reservoir. Seed of Indian major carps (Catla, rohu and mrigal), common carp and grass carp were stocked in the reservoir. The autostocking of other fishes was also observed. Up to 1989-90 major carp seed was not stocked. For the first time, Indian major carp seed was stocked during 1990-91, common carp seed during 1992-93 and grass carp seed during 1993-94. The grass carp seed was stocked in limited numbers. It was due to the paucity of seed availability. Seed was stocked in the form of fry stage with 2-40 mm size. The overall seed stocking in the reservoir in 10 years was 13,25,750 nos.During 1990-91 only 0.94% of this total was stocked. The seed stocking gradually increased from 1990-91 to 1998-99 (Table 2.17).

Table 2.17. Details of Seed Stocking and Fish Production in Shathamraj Reservoir During 1989-90 to 1998-99

Year	Total seed stocked (nos.)	Seed stocked nos./ha/yr	Total production (kg/ha/yr)	Fish production (kg/ha/yr)	No.of fry/yielding one kg fish production
1989-90	—	—	457.50	18.30	—
1990-91	12500	500	945	37.80	27.32
1991-92	37500	1500	2910	116.40	39.68
1992-93	87500	3500	7615	304.60	11.49
1993-94	122000	4880	13091	524	9.32
1994-95	142000	5680	14295	571.80	9.93
1995-96	186750	7470	17830	713.20	10.47
1996-97	212500	8500	22817.50	912.70	9.31
1997-98	237500	9500	20902.50	836.10	11.36
1998-99	287500	11500	13357.50	534.30	21.52
Total	**1325750**	**53030**	**114221**	**4568.84**	**11.61**

Maximum quantity of seed stocked (21.69% of the total)was during 1998-99.Among the varieties, catla seed was stocked in larger numbers when compared to the seed of others. 32.21% of seed numbers stocked was of catla which was followed by rohu (28.70%), mrigal (17.35%), common carp (16.59%) and grass carp (5.15%). Maximum amount of catla seed was stocked during 1994-95 (42.25%). Catla seed, costitutig 40% of the total was stocked during 1990-91 and a marginally higher quantity in 1993-94 (40.98%). Maximum number of rohu seed was stocked during 1992-93 (22.86%), common carp seed during 1992-93 (22.86%) ad grass carp during 1998-99 (9.57%). The seed stocked per hectare ranged from 500 to 11500 fry/ha/yr from 1990-91 to 1998-99 with an average of 5892 fry/ha/yr (Table 2.18).

High stocking densities were observed from 1994-95 onwards, when the stocking density exceeded 5000 fry/ha/yr.

The production of major carps from the reservoir ranged from 457.50 (90.40%) to 22817.50 (19.98%) kg during the period 1989-90 to 1996-97.This yield gradually increased up to 1996-97. Therefore, a fall was observed in the following two years. The overall major carp production in the reservoir in the 10 years period

Table 2.18. Details of fish seed stocked in Shathamraj Reservoir during 1989-90 to 1998-99

Major Carps	1989-90	1990-91	1991-92	1992-93	1993-94	1994-95	1995-96	1996-97	1997-98	1998-99	Total
Catla No. %	—	500040	1500040	2700030.86	500040.98	6000042.25	6000028.24	6000028.24	7000029.47	8000027.83	42700032.21
Rohu No. %	—	500040	1500040	2050023.36	3000024.59	4000028.17	6000028.24	6000028.24	7000029.47	8000027.83	38050028.70
Marigal No. %	—	250020	750020	2000022.86	2000016.39	2000014.08	4000018.82	4000018.82	4000016.84	5000017.39	23000017.35
C.carp No. %	—	—	—	2000022.86	2000016.39	2000014.08	4000018.82	4000018.82	4000016.84	5000017.39	22000016.59
G.carp No. %	—	—	—	—	20001.64	20001.41	125005.88	125005.88	175007.37	275009.57	682505.15
Total No. %	—	125000.94	75002.83	375002.83	1220009.20	14200010.71	21250016.03	21250016.03	23750017.91	28750021.69	1325750 —

Table 2.19. Composition of fish landings in Shathamraj Reservoir during 1989-90 to 1998-99

Major carps	1989-90	1990-91	1991-92	1992-93	1993-94	1994-95	1995-96	1996-97	1997-98	1998-99	Total
Catla kg %	211.346.19	381.940.40	1222.242	2977.539.10	544841.92	572640.06	5438.130.50	7050.630.9	6124.429.30	4020.630.1	38600.6033.79
Rohu kg %	151.1333.03	255.927.08	84329	235330.9	271020.70	376026.30	3993.922.40	4860.121.3	4305.920.60	2484.518.6	25718.3322.52
Mrigal kg %	95.0720.58	122.6012.97	203.77	837.711	216016.50	274419.20	3262.918.3	4038.717.7	407619.50	2511.218.8	20051.8717.56
C.carp kg %	— —	184.6019.53	640.222	1446.819	256819.62	192220.44	4778.526.8	5202.422.8	4431.421.2	2831.821.20	24005.7021.02
G.carp kg %	——	——	——	——	2051.56	1431.00	356.62.00	1665.27.3	1964.89.4	1509.411.30	5844.505.12
Total kg %	457.513.43	945.0028.01	2910	761579.5	1309194.94	1429592.92	1783095.33	22817.597.26	20902.598.02	13357.596.89	11422189.04

was 114221 kg with an average of 456.88 kg/ha/yr. The fish yield ranged from 18.30 to 912.70 kg/ha/yr during 1989-90 to 1996-97, which increased gradually up to 1996-97.Thereafter decline was observed in 1997-98 and 1998-99. The major carp production increased significantly from 13.43% to 98.02% during 1997-98 and 1989-90 respectively.

Maximum amount of fish production was observed in the case of catla when compared to other major carps (Table 2.19). Catla yielded 38600.60 kg during the 10 year period,which formed 33.79% of the total with an average of 544 kg/ha/ yr.Catla was followed by rohu (25718.33 kg (22.52%) and an average of 102.87 kg/ha/yr, common carp 24005.70 kg (21.02%) with an average of 96.02 kg/ha/ yr,mrigal (20051.87 kg (17.56) with an average of 80.21 kg/ha/yr and grass carp 5844.50 kg (5.12%) with an average of 23.38 kg/ha/yr.

Maximum overall weight of catla production was 7050.60 kg during 1996-97.Production of over 5000 kg per annum was observed from 1993-94 to 1997-98 (Table 2.3). The maximum yields observed in 1996-97 were :Rohu-4860.10 kg, Mrigal -4038.70 kg; and common arp 5202.40 kg. In 1997-98 the yield of grass carp was 1964.80 kg.

The ratio of number of fry for the production of one kilogram fish is depicted in Table 2.17. The average number of fry ranged from 9.31 to 39.68 nos. for the production of one kilogram of fish in 1990-91 and 1996-97 respectively. It took a higher number of fry for production of one kg of fish in 1990-91, 1991-92 and 1998-99 but in other years the average number of fry that gave production of one kg was less (9.31 to 11.49). The overall figure of number of fry to produce one kilogram fish was 11.61.

Studies on physcio-chemical parameters of water and soil, gross primary productivity, plankton and fish production indicated that Shathamraj reservoir is a highly productive reservoir (Devi, 1997).

The fish production in Shathamraj reservoir increased gradually with the increase of stocking density. However, after the saturation point, *i.e.,* very high or over-stocking, the fish production decreased.

IBRAHIMBAGH RESERVOIR

Ibrahimbagh reservoir is located at 17°23′ Lat. 28°22′ Long. It is 50 ha in extent ad about 12 km from Hyderabad city. The primary fishermen's co-operative society, Shaikpet is engaged in culture fishery development of the reservoir.

Plankton diversity: The communities of cyanophyceae, chlorophyceae, bacillariophyceae ad euglenophyceae constitute the phytoplankton bulk in Ibrahimbagh reservoir. Among the phytoplankton community, the cyanophyceae was found to e rich and dominated. Cyanophyceae was found 70.94 in 1993-94 and 70.64% in 1994-95 among the phytoplankton. The common cyanophyceae community observed in the reservoir were *Microcystis, Anabaena, Oscillatoria* and *Merismopedia*. High quantum cyanophyceae plankton was observed in pre monsoon period. The values were found to be 16943 and 17137 in the years 1993-94 and 1994-95 respectively. The maximum value was observed in February (*i.e.,* 18890 org./lit.) during 1993-94 and April (19048 org./lit.) during 1994-95.

The minimum value was found in October (6814 org./lit.) during 1993-94 and in the same month (7121 org./lit.) during the second year.

The least contribution of plankton observed was chlorophyceae (2.01% ad 1.95%) for year 1993-94 and 1994-95 respectively. Amog the chlorophyceae *Chlamydomoas* was the predominant genus. *Spirogyra, Chlamydomoas, Volvox* and *Eudorina* were found to be common in the chlorophyceae. The total genera found in Ibrahimbagh reservoir was eight.

High chlorophyceae plankton was recorded in premonsoon period (430845 org./lit.) and low in post monsoon period (176 and 278 org./lit.).

The maximum amount of chlorophyceae was observed in May during 193-94 (455 org/lit) and 1994-95 (432 org./lit.).

The second dominating plankton was the bacillariophyceae, which contributes 14.33% (1993-94) and 14.65% (1994-95) in Ibrahimbagh reservoir (Fig. 2.3 and Fig. 2.4).

The premonsoon season was found to be favourable (*i.e.,* 3164.25 org./lit. and 3240 org/lit for year 1993-94 and 1994-95 respectively). The rich bacillariophyceae planktonic blooms were observed during both the years. The planktonic blooms of this group were observed to be decreased considerably with the onset of monsoon, and further reduction in the blooming was also observed in post-monsoon.

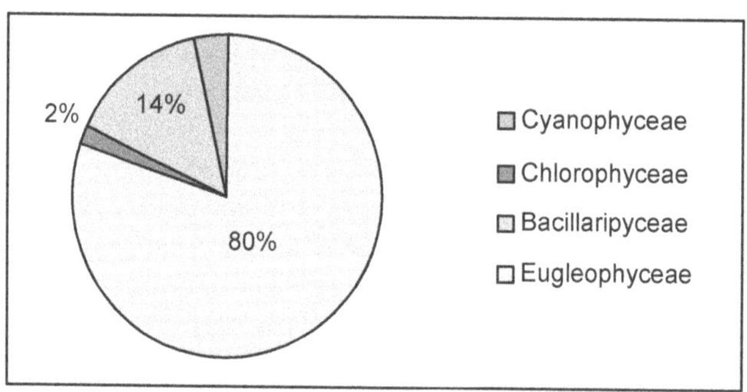

Fig. 2.3. The percentage composition of phytoplankton in Ibrahimbagh reservoir during 1993-94

The bacillariophyceae was foud to e represented by nine genera. Among these genera thecommon plankton blooms were *Navicula, Synedra, Cymbella, Nitzschia, Gamphonema, Cyclostella* and *Diatoma.*

Euglenophyceae occupied third position among phytoplankton. Eugenophyceae constitute 3.72% and 3.75% in Ibrahimbagh reservoir. Large amounts of these organisms were found during pre monsoon period *i.e.,* 785 org/lit in 1993-94 and 773 org./lit. in 1994-95. The *Euglena maxima* was found in August and minima was in October month.

Chlorophyceae domination was least among the phytoplankton population. The most dominant group was cyanophyceae followed by bacillariophyceae. The euglenophyceae was also less dominant group of the phytoplankton, but slightly more than chlorophyceae.

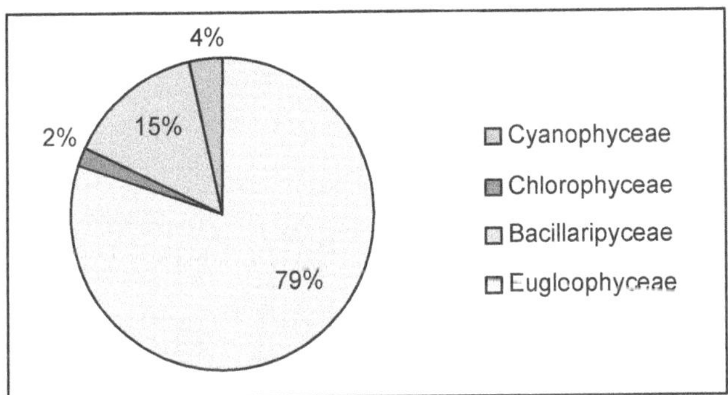

Fig. 2.4. The percentage composition of phytoplankton in Ibrahimbagh reservoir during 1994-94

Oscillatoria was the most dominant form among all plankton. It showed a continuous upward trend from December to September. The second most dominant phytoplankton was *Nitzchia.*

The zooplankton was represented by protozoa, rotifera, cladocera, copepoda and ostracoda.Protozoa was the prominent group among zooplanktonic organisms followed by those of rotifer, copepod, cladocera and ostracoda. Protozoa accounted for about 93.66% and 93.18% in Ibrahimbagh during 1993-94 and 1994-95 respectively. In protozoa eight genera were found. *Amoeba, Actinophrys, Cryptomonas, Difflugia, Colpidium, Paramecium, Vorticella, Coleps* and *Citonotus* were present during the two years investigation. High amounts of protozoa were observed in pre-monsoon period *i.e.,* 7960 organisms/liter and 7904 organisms/ liter during 1993-94 and 1994-95 respectively. The protozoan were less in monsoon in Ibrahimbagh during 1993-94 (4000 organisms/liter) and 1994-95 (3566 organisms/liter). Maximum protozoans were observed in November during 1993-94 (9841organisms/liter) and 1994-95 (95978 organisms/liter). Rotifera was the second dominant group among the zooplankton. Five rotiferan genera were found in the Ibrahimbagh reservoir. The rotifers accounted for about 4.25% and 4.41% during 1993-94 ad 1994-95 respectively. The maxima of rotifer genera was observed in premonsoon period *i.e.,* 554 organisms/liter ad 534 organisms/litre during 1993-94 ad 1994-95 respectively.

Cladocera contributed 1.14% and 0.56% in total zooplankton. The cladocera group was represented by *Daphnia, Cerodaphnia, Daphniosoma, Leptodera, Macrothrix, Alova, Sida* and *Loctona.* The common genera observed were *Daphnia, Cerodaphia, Daphniosoma, Macrothrix* and *Leptodera.* Maximum clodocerans were found during post monsoon period during 1993-94 (185 org./lit.)and premonsoon period during 1994-95(78 org/lit) in Ibrahimbagh. Highest number of cladocerans were observed in March.

Cyclops, Dioptomus and *Helidiaptomus* were the common genera of copepod. Copepods were found in third place among zooplankton and constitute 1.64% and 1.59% for year 1993-94 and 1994-95 respectively (Fig. 2.5 and Fig. 2.6). The highest concentration of copepod was observed in the monsoon period. A considerable increase was noticed from pre-monsoon to monsoon while a decrease was noticed from monsoon to post monsoon during 1993-95.

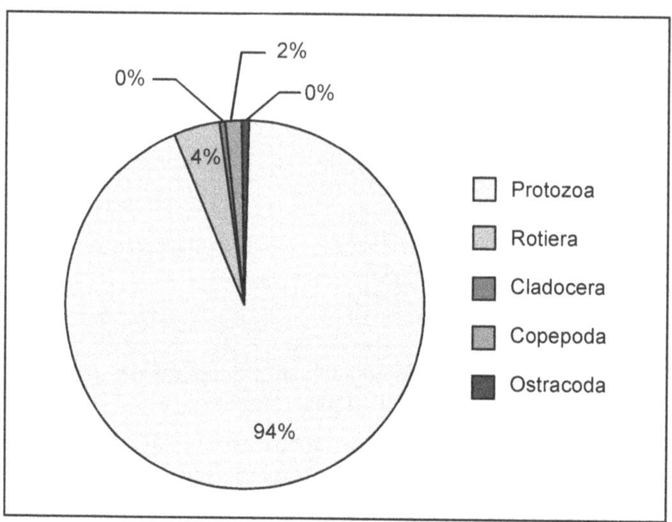

Fig. 2.5. The percentage composition of zooplakton in Ibrahimbagh reserviour during 1993-94

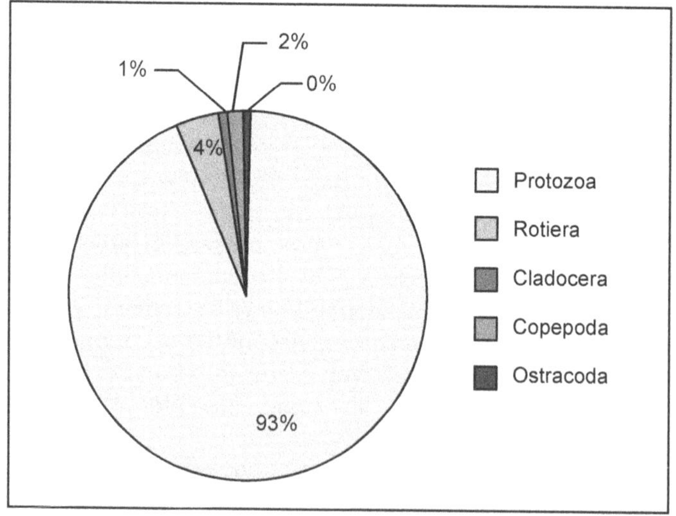

Fig. 2.6. The percentage composition of zooplakton in Ibrahimbagh reservoir during 1994-95

Among crustacean, ostracoda was the least group. Ostracoda constitutes only 0.31% and 0.26% in Ibrahimbagh during 1993-94 and 1994-95 respectively.

The analysis of gross primary productivity (113.64 $mgcm^{-3}$ hr^{-1}) indicated that this reservoir was highly productive. The highest productivity was observed during pre-monsoon and the lowest during post-monsoon period. Primary productivity was noted to be directly proportional to carbonates, bicarbonates, total alkalinity, phosphates and fluctuations of phytoplankton.

Aquatic weeds: The common aquatic weeds recorded in the reservoir water were *Eichhornia, Hydrilla, Hypomea* and *Vallisneria.*

Fish Fauna: Fishes belonging to 4 orders and 28 species were found in Ibrahimbagh reservoir. These were classified into major carps, minor carps, murrels, catfishes, other fishes and trash fishes. Major carps dominated the catches from the reservoir. The list of fishes recorded in the reservoir is given in Table 2.20.

Table 2.20. List of fishes found in Ibrahimbagh Reservoir of Hyderabad

Common Name	Order	Scientific Names
Major Carps	Cypriniformes Suborder: Cyprini	*Catla catla* *Labeo rohita* *Cirrhina mrigala* *Ctenopharygodon idella*
Common Carp Murrels	Cypriiformes Channiformes	*Cyprinus carpio* *Channa striatus* *Channa punctatus* *Channa gachua*
Catfishes	Cypriniformes Suborder : Siluri	*Heteropneustes fossilis* *Clarias batrachus* *Wallago attu* *Ompak bimaculatus* *Mystus seenghala* *Mystus aor* *Mystus vittatus*
Minor Carp	Cypriniformes Suborder : Cyprini	*Puntius ticto* *Puntius sarana* *Puntius amphibious*
Other Fishes	Clupiformes Perciformes	*Notopterus* spp. *Tilapia mossambica* *Gobius giuris*
Trash Fishes	Cypriniformes Suborder : Cyprini Cyprinidontiformes	*Salmostoma bacilia* *Salmostoma clupeodes* *Amblypharyngodon mola* *Esomus* spp. *Gambusia* spp.

Fish Catches and Production: Fish catch was absent during July and August.This may be due to the stocking of major carp seed during these months. Major carps dominated the reservoir. Major carp percentage was found to be 90.73% in 1993-94 and 93.50% in 1994-95.

The catches recorded in Ibrahimagh during 1993-95 included fish and prawn. The maximum catch of major carps was recorded in May 1994-95 *i.e.,* 4700 kgs.

The minimum major carp catches were recorded in October 1993-94 *i.e.,* 174 kg. The catches of major carps were more from January to May.

In Ibrahimbagh 15905 kg major carps were caught during January to May 1993-94 and 17880 kg during 1994-95. Out of the total catches 17322 kg, in Ibrahimbagh in 1993-94 91.82% of major carps were caught from January to May. On the similar lines during the year 1994-95 out of the total catch (19435 kgs) 17880 kgs were harvested *i.e.,* 92% of the catch was in the moths of January to May.

In Ibrahimbagh, the percentage of major carps varied from 73.65 (November) to 93.23% (April) during 1993-94 and from 73.92% (November) to 97.25% (April) during 1994-95.

Murrels occupied the third place *i.e.,* 3.70% and 2.59% during 1993-94 and 1994-95 in Ibrahimbagh reservoir. Among the murrels *Channa striatus* was the dominated species. In Ibrahimbagh the maximum murrels catch were recorded in March and May during 1993-94 (150 kgs), whereas during 1994-95 in January (100 kg).

Catfishes accounted for about 4.39% during 1993-94 and 31.08% during 1994-95 of the total fish production. Among catfishes clarias and ompak were dominating. The maximum catfishes were found in May (175 kg) during the year 1993-94, whereas during 1994-95 the maximum was recorded in February (113 kg).

Maximum catches were recorded in the November (53 kg) in 1993-94 and in May (70 kg) in 1994-95 and minimum was recorded in June during both the years. The percentage ranged from 2.68 (February) to 15.50 (November) during the year 1993-94 and 1.44 (April) to 15.50 during the year 1994-95.

The total catch of minor carps was 34 kg in 1993-94, 44.5 kg in 1994-95.

The total contribution of the minor carps catch to the total production was very much low as only 0.18% in the year 1993-94 and 0.21% in 1994-95. Maximum minor carps were caught in January (15 kg) during 1993-94 and January (10 kg) and May (10 kg) during 1994-95.

The maximum trash fish were caught in January 1994 (8 kg) and October 1994 (7 kg). Minimum trash fish was caught in November 1993 (2 kg) and March and April 1995 (2.5 kg). No trash fish catches were found in June 1994 and 1995.Trash fish accounted for only 0.20% in 1993-94 and 0.21% in 1994-95.

Prawns contributed a negligible amount to the total fish production (0.1% and 0.08%) in both the years. In 1993-94, the total catch of prawns were recorded 18.75 kg. The maximum prawn catch was found in May (4 kg) while minimum was found in November (1 kg) in the year 1993-94. In the year 1994-95, the total catch was recorded 13.75 kg and maximum in the month of May (2.50 kg) and minimum in the month of September (0.5 kg).

Fishing Efforts: In Ibrahimbagh the total number of fishing days during 1993-94 were 105 and 104 days during 1994-95. The average number of fishing days/month were 8.75 and 8.67 during 1993-94 and 1994-95 respectively. Highest number of days engaged during May in both the years. Maximum number of days engaged in fishing during March to May in both the years, least number of days engaged fishing in November during both the years. Mostly the fishing was done during holidays, because most of fishermen were working in government and private organisations or in business.

Table 2.21. Total fish catch in Ibrahimbagh Reservoir during 1993-95

Total	1993-94		1994-95	
	Total	Percentage	Total	Percentage
1. Major carps	17322	90.73	19435	93.5
2. Murrels	705	3.70	540	2.59
3. Catfishes	836	4.39	640.5	3.08
4. Minor carps	34	0.18	44.5	0.21
5. Other fishes	137	0.7	69	0.33
6. Trash fishes	39	0.20	44.5	0.21
7. Prawns	18.75	0.1	13.75	0.08
Total	**19091.75**		**20787.25**	
Catch/ha/yr	**381.80**		**415.74**	

In Ibrahimbagh total number of fishermen engaged in fishing were 364 and 373 during 1993-94 and 1994-95 respectively. The average number of fishermen per month were 30.33 and 31.08 during 1993-94 and 1994-95 respectively. Maximum number of fishermen engaged in May during both the years. More number of fishermen were engaged during February to May in both the years.

The average fish catch per day was calculated as 1405.50 ad 1497.13 during 1993-94 and 1994-95 respectively. The average catch per day per month was 117.13 kg in 1993-94 and 124.76 in 1994-95. The maximum average catch per day were foud in May in both the years. The more catch/day was found during January to June in both the years.

The average catch of fish/fishermen/month was 38.10 kg in first year and 41.07 kg in second year.

Stocking Densities: The stocking of major carps was over 2 lakhs in both the years in Ibrahimbagh. The average stocking rate was 4450 fry/ha. The catla stocking was high, the percentage varied from 40.98 to 52.08. The catla was followed by rohu, mrigal, common carp and grass carp. The stocking rate was least in case of grass carp.

Fishing Craft: The fish is being exploited in Ibrahimbagh reservoir by the members of fishermen co-operative society, Shaikpet Ltd. The traditional catomarian which is improved with thermocole is being used. Coracle is also used for fishing. It is a wide basket made of a skeleton of M.S. Iron spring/Bar welded into required shape the outer surface covered with a tarpolene to enable floating water. This is also operated with oar certain quantity nets and fish caught will be carried of this.

Table 2.22. The stocking Rate of carps in Ibrahimbagh Reservoir during 1993-95

Carps	1993-94	Percentage	1994-95	Percentage
Catla	1,00,000	48.78	1,25,000	52.08
Rohu	50,000	24.39	50,000	20.83
Mrigal	30,000	14.63	30,000	12.50
Common carp	20,000	9.76	30,000	12.50
Grass carp	5000	2.44	5000	2.08
Total	**2,05,000**		**2,40,000**	
No./ha	4100		4800	

Tackle used for Fishing: Fishing is done by gill nets, cast nets, Pondi Jal ad traps. The nets used by the members is made of nylon/high density polythene. Normally a cast net is used for single operation at a walk able depth. Another net is a gill net, which is fixed at particular point in water which is stretched to its length and kept floating with buoyant material (floats).

Pondi Jal is a entangled net operated by a group of members. This will name a typical structure having floats ad sinker, head and foot ropes. The hand net is being used for transportation of fish for short distance.

HUSSAINSAGAR

Fig. 2.7. Hussainsagar Lake of Hyderabad

Hussainsagar, although popularly termed as a lake, is a man made reservoir constructed in year 1562 mainly to store drinking water take from the River Musi, a tributary of Krishna. The lake is situated between twin cities of Hyderabad and Secunderabad in Andhra Pradesh. The lake area receives 800 mm annual rainfall.

Table 2.23. Physico-graphic features of Lake Hussainsagar

1.	Year of construction	1562
2.	Average depth	5.2 meters
3.	Basin area	240 km^2
4.	Depth variable	1to 12 meters
5.	Catchment Area	67 km^2
6.	Shore length	14 km
7.	Storage volume (spill)	$28.6x10^6$ m^3
8.	Maximum operating level	514.93 meters
9.	Water spread area	5.7 km^2
10.	Normal Operating Level	513.43 meters
11.	Capacity	27.1 million m^3

Located 15°N78°E, the lake is 510 meters above the mean sea level. The lake basin bounded on west by Banjara hills. The watershed of Husainsagar is divided into four sub basins viz. Kukatpally, Dullapally, Bowanpally and Yusufguda. The lake hydrology is sustained by four feeding channels (nallahs); Kukatpally, Picket, Banjara and Balkapur. Of the four inlets, Kukatpally with total catchment of 168 km^2 passes through two major industrial areas *viz.* Kukatpally and Balanagar and is the main feeding channel of the lake. Asia's biggest monolithic Buddha statue standing majestically on the rock of Gibraltar in the centre of the lake, has emerged as the major attraction like the statue of liberty in USA. The extensive development of lake environment with the help of national and international funding has bought Hyderabad on the tourism map of the world and the lake has emerged as the landmark representing socio-cultural ethos of the historical city and the region.

One of the major impacts of urbanization ad industrialization on the lake was in the form of poor water quality due to pollution. Since beginning of the last century when lake basin was undergoing rapid change very little attention was paid to this vital issue. Thus failure of lake basin management in terms of proper sewerage system and industrial waste disposal infrastructure was responsible for rapid degradation of lake environment.

Soil erosion due to construction activities in the catchment generates a lot of silt which along with the surface runoff ultimately ends up into the lake. Similarly, in festivals like Durga Puja and Ganesh by immersion of massive idols of the deities in the lake. These events apart from addition of tons of silt also pollute the lake by floral offerings, paints. Pigments, wooden and iron frames that go into making of idols, year after year.

Hussainsagar has been a subject of extensive research for the last 55 years. Earlier reports on its water quality clearly indicate that up to 1950 it was virtually pollution free. Subsequent degradation of the ecosystem is directly linked to nutrient loading from domestic sewage, the magnitude of which could be assessed from a study that has estimated daily flux of 1041 kg of phosphates and 1204 kg nitrates into the lake (Kodarkar *et al.*1991; Reddy *et al.* 2002). Some of the precitible manifestations of eutrophication are (1) Algal blooms and wild growth of aquatic macrophytes like water hyacinth (2) Breeding of vectors (3) Foul smells and (4) Fish kills (Kodarkar *et al.*, 1991). The pollution has also an adverse effect on the biodiversity of the lake and consequent disruptions in bio-geochemical cycling of organic load on the lake.

Fishermen and small dairy farmers are traditional lake dependent socio-economically backward communities and degradation of the lake has direct effect on their livelihood. Since studies in 1980, that indicated contaminated nature of fish, fishing activities in the Hussainsagar are banned and once flourishing fishermen community has almost disappeared.

Since 1990 after pollution of Hussainsagar reached an alarming proportion, culminatiig into massive fish kills of 1993, otherwise complacement state government was forced into action. Credit for this turn around also should go to civil society movements and judicial interventions in response to Public interest Litigation (PIL). Apart from earlier actions for conservation of the lake, Hyderabad Urban Development Authority (HUDA) with support from Japan Bank of International Cooperation (JBIC) has initiated an ambitious project on Hussainsagar Lake and the catchment area improvement in 2006. The project cost was about 370 crore rupees.

Fish Fauna: Husainsagar lake once use to harbor 27 species of fishes (Babu Rao and Siva Reddy, 1984). However recent survey has reported loss of sensitive species and existence of only few species of hardy catfishes belonging to the genera *Channa,Clarias,Mystus* and *Heteropneustes*. There is a need to reassess the situation ad introduction of large scale composite fish culture. To begin with some controlled cage culture can be introduced to begin with and based on the results fisheries should be introduced on large scale.

Rahimullah (1943) had earlier recorded 23 species of fishes including major carps from Hussainsagar. Amog these, 12 species do not encounter among the fish fauna of the lake at present. These include fishes like *Mastacembelus* spp., *Chela clupeiodes, Rasbora buchanani, Puntius kolus, Brachydaio reria, Labeo calbasu, Labeo fimbriatus, Danio aequippinatus* and the major carps.

In recent years, after the eutrophication had set in, 27 species of fishes were recorded from the same water body (Babu Rao and Siva Reddy, 1984a). On the other had hardy species like *Heteropneustes fossilis, Clarias batrachus, Mystus bleekeri, Garra mullya, Gambusia affinis, Poecilia reticulate, Notopterus notopterus* etc. formed the additional component. From these records, it can be well concluded that the sensitive species were replaced by hardy species with the eutrophication of the reservoir.

Biology of Fishes

***Channa punctatus*:** Biological studies of *Channa punctatus* were made from Hussainsagar (Siva Reddy, 1985). The length weight data, when fitted to the conventional curvilinear regression equation of $W = CL^n$, the exponent 'n' expressing the increase of weight of fish with length, was found to be 2.85996.

The fecundity in this species ranged from 1004 to 9869 at the eutrophic Hussainsagar (Siva Reddy, 1985).

Among the gut contents, insects and insect larvae formed the main food in Hussainsagar specimens (Siva Reddy, 1985). Hussainsagar waters being polluted, the insects and insect larvae had a thick population. Moreover, since *Channa punctatus* being a carnivorous fish, insects and insect larvae naturally dominated among the gut contents. The oligochaete annelids formed the other important component of the food of Hussainsagar murrels.

***Heteropneusles fossilis*:** Insect larvae, insects and insect eggs formed the major food items of this species too collected from the lake. Crustaceans, plant materials and algae, gastropods, bryzoans etc. were the other groups of food items in order of abundance.

***Mystus vittatus*:** In the length-weight relationship formula $W = CL^n$, the growth exponent 'n' was 0.7875 for Hussainssgar specimens (Siva Reddy, 1985). This indicates the subnormal increase in weight of specimens with the increase of length of fish. Further, there is no significant correlation between the length and weight data in the case of specimens.

The fecundity ranged between 11388 and 30553. The correlation coefficient between total length of fish and total weight of fish on one hand and fecundity o the other was found to be not significant.

Water quality studies made in the feeding channel, Kukatpally nallah by Reddy *et al.* (1986) showed very high values of conductivity, total solids, total dissolved solids, suspended solids, biochemical oxygen demand, chemical oxygen demand .chlorides and low or nil oxygen values. Heavy metals like zinc, lead, chromium, cobalt, cadmium, copper and nickel were also recorded. Thus, the water is completely deteriorated making it unfit for irrigation, consumption, fisheries and recreational activities.

Hydrobiological study of Hussainsagar by Babu Rao *et al.* (1981) indicated depletion of oxygen from 10.6 mg/l to almost nil in some parts of the lake. Total alkalinity varied from trace to 750 mg/l indicating completely disturbed buffer systems of the lake. Chlorides also showed a steady increase compared to its earlier values.

Considerable number of abnormal fishes have also been reported from the Hussainsagar (Babu Rao and Siva Reddy, 1984). Pollution (toxicity) has direct or indirect responsibility in inducing abnormalities in the early formative stages of fish.

Fish Kills: Kodarkar *et al.* (1992) observed fish kills during year 1984-86 which revealed mass mortality of *Notopterus notopterus* and few specimens of *Puntius* spp and *Channa* spp. It is presumed that these fish kills may be the cumulative effect of human activity. Several theories ad hypotheses revolving round the 'fish kills' in Hussainsagar have been propounded, either I the form of unpublished reports, discussions with eminent scientists and academicians both at national and international level.

One of the common observation expressed in connection with storm is, these fish kills might have occurred due to sudden dropout in the atmospheric and water temperature of the lake brought about by the severe storm, which might have created a thermal shock leading to mass mortality of fishes. The second one is sudden depletion of dissolved oxygen in the surface water of the lake followed by the storm, might have created upward migration of the fish species which need more oxygen for survival and in the course, their gills might have been clogged due to algal blooms on the surface leading to mass mortality. However, post-mortem studies on selected fishes did not support these findings, as none of the fish, examined showed 'clogging of gills' due to algae.

One more aspect, propounded by several experts is upwelling of the water in the lake might have lead to highly turbid conditions. This turbid water might be containing particles of finely divided particles of less than 200 M in size and are colloidal in nature. These particles due to their low density, do not settle and clogging the gills of the fishes, as fishes need more oxygen for breathing followed by depletion of oxygen in the surface waters. With regard to depletion of oxygen in the surface waters of the lake, several explanations have been available. According to few influx of organic matter and sewage brought by the storm water as well as sudden change in the photosynthetic activity followed by high community respiration, might have lead to depletion of oxygen level in the surface waters. These fluctuations associated with critical values of certain toxic elements like of ammonia and/or hydrogen sulphide might have leaded to 'fish kills' in the lake.

PALAKURTHY TANK

Palakurthy tank is a historical freshwater body, which was constructed at the grand period of Kakatiya dynasty, adjacent to Sri Someshwara Laxminarayana Temple at Palakurthy village of Warangal district. The present investigation was an attempt to Ramu Gundu (2011) documented the fish fauna of the Palakurthy reservoir.

Palakurthy tank is a historical freshwater body which was located Palakurthy village (Old name Palkuriki) adjacent to Sri Someshwara Laxminarasimha Temple of Warangal district. The tank constructed at historical period of Kakatiya dynasty. The main scope of this tank is irrigation and fishing purpose.

The 16 species of fishes recorded in this study belonging to 4 orders and are given in Table 2.24. Among them the order cypriniformes was dominant with 6 species to be followed by order siluriniformes with 4 species and the order perciformes and channiformes each with 3 species recorded total 16 species from Palakurthy tank.

Table 2.24. Fish fauna of Palakurthy Reservoir

Sr.No.	Name of fish
1.	*Catla cattla* (Hamilton-Buchanan)
2.	*Cirrhinus mrigala* (Hamilton-Buchanan)
3.	*Labeo rohita* (Hamilton-Buchanan)
4.	*Labeo potail* (Sykes)

5.	*Cyprinus carpio carpio* (Linnaeus)
6.	*Amblypharyngodon microlepis* (Bleeker)
7.	*Mystus vittatus* (Bloch)
8.	*Mystus bleekeri* (Day)
9.	*Wallago attu* (Schneider)
10.	*Clarius batrachus* (Linnaeus)
11.	*Channa orientalis* (Bloch)
12.	*Channa punctatus* (Bloch)
13.	*Channa striatus* (Bloch)
14.	*Chanda nama* (Hamilton)
15.	*Glossogobius giuris giuris* (Hamilton)
16.	*Mastacembelus armatus* (Lacepede)

POCHARAM LAKE

The largest and the most important reservoir of the Medak District is the Pocharam reservoir (water spread area 16.835 km^2, with a depth of about 6-7 m) formed by damming the Aleru River. It was constructed between 1916 and 1922 (18008′N & 77057′E) about 100 km north-west of Hyderabad in Medak and Nizamabad districts. Owing to its unique location and the presence of forested tracts, the vertebrate faunal diversity is rich in comparison to other lakes. From Pocharam Lake 25 fish species were recorded (Rao *et al.* 2011). One specimen of *Ompak bimaculatus* was reported in Pocharam Lake. This is generally not cultured in lakes but might have made its entry through the Godavari river system.

Table 2.25. Fish diversity of Pocharam Lake of Medak district

1. *Notopterus notopterus* (Pallas)
2. *Salmostoma bacaila* (Hamilton)
3. *Chela laubuca* (Hamilton)
4. *Parluciosma daniconius* (Hamilton)
5. *Osteobrama vigorsii* (Hamilton)
6. *Puntius sophore* (Hamilton)
7. *Puntius ticto* (Hamilton)
8. *Catla catla* (Hamilton)
9. *Labeo rohita* (Hamilton)
10. *Schistura d.denisoni* (Day)
11. *Mystus vittatus* (Hamilton)
12. *Mystus cavaisus* (Hamilton)
13. *Sperata seenghala* (Sykes)
14. *Ompak bimaculatus* (Bloch)
15. *Wallago attu* (Schneider)
16. *Xenentedon cancila* (Hamilton)
17. *Macrognathus pancalus* (Bloch)

18. *Masacembelus armatus* (Hamilton)
19. *Chanda nama* (Hamilton)
20. *Paraambassis ranga* (Hamilton)
21. *Etrolpus maculates* (Bloch)
22. *Glossogobius giuris* (Hamilton)
23. *Polycanthus fasciatus* (Schneider)
24. *Channa punctauts* (Bloch)
25. *Channa striatus* (Bloch)

WYRA LAKE

Wyra Lake is located to the north of Wyra Town, about 25 km south of Khammam between Khammam and Kothagudem towns in Khammam District. It was constructed in 1930. The lake is located 2 km off the Hyderabad-Visakhapatnam highway, and is surrounded by greenery. Its water is unpolluted and potable and serves nearly 20,000 acres for cultivation and provides drinking water for 70 villages. The water spread area of the lake is about 19.166 km^2. Rao *et al* (2011) recorded 21 species of fishes from Wyra Lake. *Etroplus suratensis,* a shoaling fish, is very common in the lake. Dense vegetation, many water plants, hiding places and open swimming areas are suitable for this species. Another fish which is reported only in Wyra Lake, *Rhinomugil corsula* which thrives well in estuarine waters, is also reported from the Krishna River basin. The mixing of waters of the Krishna-Godavari river basins explains the occurrence of *Rhinomugil corsula* in Wyra.

Table 2.26. Fish diversity in Wyra Lake in Khammam district

1. *Notopterus notopterus* (Pallas)
2. *Parluciosma daniconius* (Hamilton)
3. *Puntius sophore* (Hamilton)
4. *Puntius ticto* (Hamilton)
5. *Catla catla* (Hamilton)
6. *Labeo rohita* (Hamilton)
7. *Mystus vittatus* (Hamilton)
8. *Mystus cavaisus* (Hamilton)
9. *Sperata seenghala* (Sykes)
10. *Wallago attu* (Schneider)
11. *Xenentedon cancila* (Hamilton)
12. *Macrognathus pancalus* (Bloch)
13. *Masacembelus armatus* (Hamilton)
14. *Chanda nama* (Hamilton)
15. *Paraambassis ranga* (Hamilton)
16. *Etrolpus maculates* (Bloch)
17. *Etroplus suratensis* (Bloch)
18. *Rhinomugil corsula*
19. *Glossogobius giuris* (Hamilton)
20. *Channa punctauts* (Bloch)
21. *Channa striatus* (Bloch)

MYLARAM RESERVOIR

The Mylaram reservoir is one of the minor reservoirs in state. It is situated in between Latitude 79°36′00″ North to 79°38″0″ North and Longitude 18° 6′15″ East to 18°8′35″ East. The reservoir was constructed for irrigation.

Ram *et al.* (paper available on www.researchgate.net) studied fish fauna of Mylaram reservoir in Warangal district and reported 31 fish species belonging to 6 orders. The order cypriniformes was dominant with 13 species followed by order siluriformes with 7 species. While the order perciformes was represented with 6 species and channiformes with 3 species and remaining orders oseteoglossiformes and antheriniformes were represented with one species each.

KAMALPUR LAKE

The Kamalpur lake is with 700 million cubic meters of water storage capacity. The lake water is used for irrigation. The lake receives most of water from small canals and the agricultural wastages. The lake site is bounded at L 14°36′N) (78°42′E) at Kamlpur village in Krimnagar district.

Thirupathaih *et al.* (2011) observed 18 species of zooplankton belonging to four different orders (Table 2.27). Rotifera was the dominant group composing 34.02% of total zooplanktons, cladocera 28%, copepoda 31.95% and protozoa 5.12%. The maximum abundance of total zooplankton was recorded in the month of January and February. The community size of zooplankton was the highest in winter while lowest density of zooplankton was in summer due to the higher temperature. Among the all zooplankton rotifera formed the dominant group followed by copepoda, cladocera and protozoa.

Table 2.27. Zooplankton diversity in Kamalpur lake of Karimnagar district

Rotifera: *Keratella cochlearis, Keratella tropica, Euchlanis dilatata, Synchaeta oblonga, Philodina resoeola, Brachionus quadridentatus, Brachionus angularis, Polyarthra.*
Copepoda: *Diaptomus* sps, *Cyclops vicinus.*
Cladocera: *Leptodora kindtti, Moina bachiata, Dhapnia carinata, Dhapnina longispina, Bosmia longirostris, Alona rectangular.*
Protozoa: *Paramecium cadatum, Verticella companula.*

SOMASILA RESERVOIR

This is an operational major project built on river Pennar at Somasila village in Atmakur taluka of SPSR Nellore district. The dam is located at Latitude 14°29′15″N and Longitude 79°1825″E. The maximum height above deepest foundation level is 38 m.The gross storage capacity at full reservoir level 100.58 m and live storage capacity of the reservoir are 2210 mcm and 1994 mcm respectively. The reservoir water spread area is 212.28 km^2. The details of reservoir mophometry are given in Table 2.28.

Table 2.28. Morphometry of Somasila Reservoir

1. Type of project	Major
2. Location	Somasila village
3. Year of approval by planning commission	1973
4. Name of the river	Pennar
5. Gross command Area (Th ha)	107.31
6. Culturable Command Area (Th ha)	39.3
7. Ultimate Irrigation Potential (Th ha)	38.48
8. Water spread area (km^2)	212.28
9. Gross storage capacity (m)	100.58
10. Live storage capacity (mcm)	2210

Fig. 2.8. Somasila Reservoir of Nellore district

As per the Somasila project report-1984, prepared by the Irrigation Department, Govt. of Andhra Pradesh, the rate of silting considered for the fixation of DSL of Somasila reservoir is 0.019 ha.m/sq.km/year. The silt volume that gets deposited below MDDL during its life period of 100 years is 171.32 mm^3 out of the total likely silt deposit of 398 mm^3 and the remaining silt will get settled in the live storage capacity of the reservoir *i.e.,* between MDDL and FRL. The live storage of the reservoir between MDDL and FRL being 1994 mm^3, the utilizable live storage of the reservoir at the end of 100 years would be 1767 mm^3. With the introduction of the link canal, no effect on the sedimentation of Somasila reservoir is anticipated since the water delivered to Somasila reservoir through Srisailam – Pennar and Nagarjunasagar – Somasila links will be silt free. The average annual evaporation losses from Somasila reservoir are assessed as 266 mm^3 (9.4 TMC).

Table 2.29. Fish production in Somasila reservoir

Year	Fish production (tonns)
2008	1837
2009	970
2010	1560
2011	1371
2012	2108

The details of the fish production from reservoir for the period of 2008 to 2012 is shown in Table 2.29. The study of the data reveals that the maximum fish production for the last five years (from 2008 to 2012) was recorded at 2108 tonnes during year 2012. The minimum production was in year 2009, at 970 tonnes. The fish catches are mainly constituted by *Catla catla, Labeo rohita, Cirrhinus mrigala,Rasbora daniconius, Channa marulius, Channa striatus, Channa gachua, Mastacembelus armatus, Ambassis ranga, Mystus gulio, Mystus seenghala, Mystus vittatus* and *Wallago attu.*

Table 2.30. Seed stocking in Somasila reservoir

Year	Quantity of seed stocked (in lakhs)
2007-08	40
2008-09	12.24
2009-10	0.00
2010-11	8.10
2011-12	53.12

During 2007-08 reservoir was stocked with 40,00,000 fish seed of Indian major carps,while during 2008-09 12,24000 seed was stocked. During 2010-11 and 2011-12 reservoir was stocked with 8,10,000 and 53,12,000 fish seed respectively (Table 2.30). The stocking of Indian major carp seed is done in the ratio of 60 : 20 : 20 *i.e.,* catla, rohu and mrigal comprise 60%, 20% and 20% of the seed stocked.

KANDALERU RESERVOIR

It is the world's biggest earthen dam with 68 TMC of storage spread over 11 square kms. This huge reservoir is part of Telgu Ganga Project meant for supplying Krishna river water to Chennai citizens in the south Indian state of Tamil Nadu. It was constructed across the river Kandaleru. This river orginates in Velugonda hills and flows across Gudur.

Table 2.31. Fish yield in Kandaleru reservoir

Year	Fish production (in tonnes)
2008	1950
2009	1951
2010	1932
2011	1730
2012	2231

The fish production from reservoir ranged from 1730 tonns (in year 2011) to 2231 tonns (in year 2012) with an annual average of 1958.80 tonnes.

KANGIRI RESERVOIR

This reservoir located at north west area of Buchireddypalem, about 20 km from SPSR Nellore. Water is being fed from Somasila dam, the water from this reservoir is supplied to Buchireddypalem, Kovur, Vidavalur, Kodavalure, allure mandals for irrigation. The project was designed by Sir Arthur Cotton. The reservoir is located at Latitude 14°35′21″N and Longitude 79°51′28″E. The average water spread area of reservoir is 4610 ha and water spread area of reservoir at full reservoir level is 6230 ha. Recently in year 2011-12 reservoir was stocked with 58,52,000 seed of Indian major carps.

Table 2.32. Fish catch in Kangiri reservoir

Year	Fish production (in tonns)
2008	1672
2009	1942
2010	1853
2011	1871
2012	1865

The fish production for five years i.e., from year 2008 to 2012 fluctuated from 1672 to 1942 tonnes with an average of 1840.6 tonnes/year.

REFERENCES

- Anitha, G., Kodarkar, M.S. and Chandrasekhar, S.V.A., 2006. Studies on macrozoobenthos in Mir-Alam lake, Hyderabad, Andhra Pardesh. In : Ecology of Lakes and Reservoirs (Ed. V.B. Sakhare), Daya Publishing House, Delhi, pp. 1-15.
- Babu Rao, M., 1990. Effect of eutrophication on the biology of fishes in reservoirs-A case study of Hussainsagar.p. 48-52. In : Jhingran, Arun G. and Unnithan, V.K. (eds.). Reservoir Fisheries in India. Proceedings of the National Workshop on Reservoir Fisheries, 3-4 January, 1990. Special Publication 3, Asian Fisheries Society, Indian Branch, Mangalore, India.
- Babu Rao M. and Siva Reddy, Y., 1984. Fish fauna of Hussainsagar. Jatu. 2 : 1-6.
- Das, A.K., Ch. Gopalakrishnayya and M. Ramakrishiah, 2001. Guidelines for Management of Andhra Pradesh Reservoirs. *Fishing Chimes*. 21(5) : 25–28.
- David, A., Roy, P., Govind, B.V., Rajagopal, K.V. and Bannerjee, R.K., 1969. Limnology and Fisheries of Tungabhadra reservoir. Bull. Cent. Int. Fish. Res. Inst., Barrackpore, 13 : 188p.
- Devi, Sarla, B., 1997. Present status, potentialities, management and economics of fisheries of two minor reservoirs of Hyderabad, Ph.D. Thesis, Osmania University, Hyderabad.
- Kodarkar, M.S., Muley, E.V., and Vasant Rao, 1992. Kukatpally–Hussainsagar (Ecological studies on industrially polluted stream and its impact on freshwater lake in Hyderabad, Publication No. 1, Indian Association of Aquatic Biologists, Hyderabad.

- Kodarkar, M.S., 2008. conservation of lakes (Applied Limnology for Global Perception and Local Actions), Indian Association of Aquatic Biologists, Hyderabad, p.101.
- Rahimullah, M.1943. Fish survey of Hyberabad State. Part II. Fishes of Hyderabad city and its suburbs. *J.Bombay Nat.Hist.Soc.,* XLIV (1) : 88–91.
- Rao, C.A.N., J. Deepa and M. Hakeel., 2011. Comparative account on ichthyofauna of Pocharam and Wyra lakes of Andhra Pradesh, India. *Journal of Threatened Taxa* 3(2): 1564-1566.
- Piska, Ravishankar. 2000. Impact of stocking densities of fish production in a minor reservoir. *Fishing Chimes.*20 (9) : 38-41.
- Ramakrishniah, M., 1990. Observations on different trophic phases in Nagarjunasagar reservoir, A tropical impoundment in Andhra Pradesh. In : Contributions to the Fisheries of Inland Openwater Systems of India, Part I (Edited by A. G. Jhingran, V. K. Unnithan and Amitabha Ghosh), Inland Fisheries Society of India, pp. 80-84.
- Ramu, G., Ravinder, B., Narsima Ramulu K. and Benarjee, G., 2009. The fish fauna of Mylaram reservoir in Warangal district, Andhra Pradesh, paper available on www.researchgate.net
- Ramu Gundu, 2011. Ichthyofaunal diversity of Palakurthy tank, a historical freshwater body of Warangal district. IndiaStudyChannel.com
- Shailaja, K. and Johnson, Mary Esther Cynthia, 2006. Elemental analysis of certain plants growing around Osman Sagar lake, Hyderabad, Andhra Pradesh. *J. Aqua. Biol.* 21 (1) : 231-234.
- Sheshagiri Rao and Khan, M.A., 1982. Ecobiology of *Corvospongilla alpidosa* (Annandale 1908) (Porifera: Spongillidae) in the Manjira reservoir, Sangareddy, Andhra Pradesh.Proc.Indian Acad.Aci.(Anim. Sci.), 91(6) : 553-562.
- Siddiqi, S.Z. and Kha, R.A., 2002.Comparative limnology of few man-made lakes in and around Hyderabad, India. Rec. Zool. Surv. India. Occ. Paper No. 203 : 1-64 (Published by the Director, Zool . Surv . India, Kolkata).
- Siva Reddy, Y.1985. The biology of a few common fishes from Hussainsagar Lake, Hyderabad, India, Ph.D. Dissertation, Andhra University, Waltair.
- Sreenivasan, A. 2000. Fish production from a medium-sized hardwater reservoir (Krishagiri) in Tamil Nadu. *Fishing Chimes.* 20(5) : 32-34.
- Srinivas, Ch., 2006. Fisheires of Edulabad reservoir in Andhra Pradesh. Ph.D. Thesis, Osmania University, Hyderabad.
- Srinivas, Ch., 2007. Fisheries of Edulabad reservoir in Andhra Pradesh (Annual production at 66 kg/ha, despite adverse impact of pollution). *Fishing Chimes.* 26 (10): 105 and 107.
- Sugunan, V.V. and Das. R.K., 1983. Studies on the bottom macrofauna of Nagarjunasagar reservoir, Andhra Pradesh, India. *J. Inland Fish, Soc. India, 15*(1 and 2) : 1-12.
- Sugunan, V.V., 1980. Seasonal fluctuations in plankton of Nagarjunasagar reservoir, A.P., India. *J. Inland Fish. Soc. India,* 12(1) : 79-91.
- Sugunan, V.V. and Katiha, P.K., 2004. Impact of stocking on yield in small reservoirs in Andhra Pradesh, India. *Fisheries Management and Ecology.* 11 : 65-69.
- Tamlurkar, H.L. and Ambore, N.E., 2006. Correlation co-efficients of some physico-chemical characteristics of Alisagar dam water, district Nizamabad (A.P.), India. *J.Aqua.Biol.* 21(2) : 115-118.
- Thirupathaiah, M., Samatha, Ch. and Sammiah, Ch., 2011. Diversity of zooplankton in freshwater lake of Kamalpur, Karimnagar district, (A.P.) India. The *EcoScan,* 5(3 and 4) : 195-197.

3

CHATTISGARH

Fig. 3.1. Map of Chattisgarh (Not to scale)

Chhattisgarh has an area of 1.35, 194 sq.km which is 4.14% of the geographical area of India. Its average annual rainfall is 1470 mm. The state embraces four river systems containing 31 big and small rivers which debouch into the sea from both east and west coasts of the country. The length of the rivers is about 3573 km. Mahanadi river system covers 12 out of 16 districts of the state. Godavari covers 4 districts. Ganga river system covers part of Sarguja and Jashpur districts. Similarly Narmada system covers part of Rajnandgaon and Kawaradha districts.

It is estimated that 1.547 lakh ha of water area comprising ponds and reservoir is available for development of fisheries in the state. Among reservoirs, the small sized category of them comprises 99.13% by numbers. Since its birth in November 2000, the state has witnessed an upsurge in production of fish from 83,833 mt in 2000-01 to 1,20,072 in 2004-05. About 92% is contributed by the private sector.

Among reservoirs, 0.1% fish production is contributed by large reservoirs,9.73% by medium ad 89.15% by small reservoirs. The small category comprises about 50% of the reservoir area. Average annual fish yield from small reservoirs is fairly good but that from medium and large reservoirs is not satisfactory as shown hereunder:

Small Reservoirs: 139 kg/ha

Medium Reservoirs: 13.56 kg/ha

Large Reservoirs: 0.76 kg/ha

Under stocking, use of smaller size of fish seed and inadequate fishing effort have been identified as some of the reasons for poor yield from reservoirs (Singh and Tuli, 2006).

During year 2011-12, National Fisheries Development Board, Hyderabad sanctioned Rs. 18.46 lakh to Department of Fisheries, Chhattisgarh for seed stocking of in two reservoirs of the state (Table 3.1).

Table 3.1. Financial assistance from NFDB for stocking reservoirs in Chhattisgarh

Sr.No.	Name of the reservoir	Sanction for second year stocking fingerlings					
		EWSA (ha)	Category	No. of fingerlings sanctioned for 1st year stocking (year 2009 - 10)	No. of fingerlings eligible for 2nd year stocking	Amount released as 1st installment	Amount released of 2nd installment
1.	Sikasar	1174	Medium	11.74	5.87	2.935	2.935
2.	Manohar Sagar	2518	Medium	25.18	12.59	6.295	6.295
	Total	3692		36.92	18.46	9.23	9.23

RAVISHANKAR SAGAR RESERVOIR

The reservoir is also known as 'Gangrel reservoir'. It was constructed in 1978.The reservoir was formed by damming the river Mahanadi in Raipur district of Chhattisgarh state. The water-spread area of reservoir is 9540 ha. The reservoir is named after late Pandit Ravishankar Shukla, the first Chief Minister of Madhya Pradesh.Work for development of its fishery was taken up in 1983.

Water Drainage of Reservoir: The river Mahanadi, from its origin to the location of Ravishankar Sagar reservoir, covers a stretch of 115 km. Halfway of this stretch, the river was also impounded earlier in 1962 to form a reservoir at Dadhawa in Baster district of Chhattisgarh. Thus, apart from receiving the water from its catchment area, Ravishankar Sagar reservoir also gets water from Dudhawa and down below from another reservoir *i.e.,* Murumsilli, located on southern side of Ravishankar Sagar reservoir. Thus, the water stored in Ravishankar Sagar reservoir from the above mentioned resources, is regularly taken out through irrigation and Bhilai feeder canals. Regarding drainage system of the reservoir, it is active throughout, rarely attaining stagnancy in its water level. Thus, Ravishankar Sagar reservoir is a balancing intermediary water body with the receipt of monsoon inflow and discharge from Dudhwa and draw down below for irrigation and for supply to Bhilai Steel Plant. This process continues round the year making the reservoir a 'fluviatile' lake with a lesser period of water retention.

Water Quality: Based on the total alkalinity (66.8 mg/l) and total hardness (49.5 mg/l), the reservoir is classified as medium productive reservoir (Desai *et al.* 2007). The mean values of phosphates and nitrates also support this classification. The water temperature and chemical parameters of Ravishankar Sagar reservoir did not show well marked variation from surface to bottom. The condition of water was seen holomixed during most of the year, not showing district thermal or chemical stratification even in summer.

Biotic Communities: There was an overall dominance of zooplankton (80 to 88%) in the reservoir. The occurrence of phytoplankton was low due to factors like constantly flowing water of the reservoir and grazing effect of zooplankton and feeding activity of minnows. The benthic population was mainly constituted by Dipteran larvae (45 to 57%), followed by Gastropods (20 to 32%), Cadisworms, Oligochaetes, Bivalves and Mayflies.The macrophytes were represented by *Hydrilla, Vallisneria, Najas* and *Potamogeton.*

Fish Fauna: Desai *et al* (2007) reported 48 species of fishes. The fish fauna mainly comprises *Catla catla, Labeo rohita, Cirrhinus mrigala, Gudusia chapra, Mystus aor,* and *Mystus seenghala*. Weedfishes were abundant in the total fishery of the reservoir, contributing 42% in 1989-90. Later its population was reduced to 31% (1990-91) and 17% (1991-92). This may be due to intensive drag net fishing. Among weed fishes, *Gudusia chapra* (56%) was the most dominant.

Fish Production: The fishing in reservoir was started from 1983-84. The percentage composition of fish catch, total fish catch and fish production/ha is depicted in Table 3.2. The fish production/ha varied from 0.05 kg/ha (1984-85) to 8.30 kg/ha (1991-92).

The fishery of the reservoir is mainly constituted by catfish followed by minnows and major carps. Among major carps, *Cirrhinus mrigala* was the most dominant followed by *Catla catla* and *Labeo rohita*. The catfishes were represented by *Mystus aor, Mystus seenghala* and *Wallago attu*. The catfishes are well established in the reservoir.

Table 3.2. Fish production of Ravishankar Sagar reservoir

Percentage composition					
Year	Fish catch (t)	Yield/ha (kg)	Major carps	Catfish	Miscellaneous fishes
1983-84	1.4	0.22	7.9	65.7	26.4
1984-85	0.3	0.05	38.5	42.5	19.0
1985-86	4.2	0.67	5.3	34.9	59.8
1986-87	9.3	1.47	15.8	68.4	15.8
1987-88	13.0	2.00	23.1	44.5	32.4
1988-89	17.7	2.80	32.5	46.9	20.6
1989-90	41.2	6.50	32.7	23.7	43.6
1990-91	50.0	7.90	40.6	28.5	30.9
1991-92	53.0	8.30	24.5	52.0	23.5
1992-93	16.7	2.60	11.9	55.9	32.2

Stocking: During the period 1986-87 to 1992-93 the reservoir was stocked annually at an average of 195 nos/ha (30-100 mm) in the ratio of catla 3; rohu 5; mrigal 2.The stocking rate from 1986-87 to 1988-89 varied from 195 to 315 nos/ha, which certainly helped to boost the fishery to the tune of 53 t in 1991-92 after a duration of 3-4 years. This becomes clear after noting the status of age groups of catla, rohu and mrigal encountered in commercial catch. While *Catla catla* (length 712 mm) was of 3 years of age. *Labeo rohita* and *Cirrhinus mrigala* (lengths 587 and 572 mm respectively) were of 4 years of age. In spite of giving more emphasis on its stocking *Labeo rohita* could not contribute to the fishery so well perhaps because the reservoir conditions were not congenial for it. On the contrary, *Cirrhinus mrigala* with less stocking support formed a consistent fishery of importance probably on account of its good recruitment in the reservoir, adjusting itself with its low water level during monsoons. Fishery of *Catla catla* can be further improved with adequate stocking in view of dominance of zooplankton in the reservoir, the ideal food of *Catla catla*. Desai *et al* (2007) recommended stocking of 130 fingerlings/ha/annum.

Breeding: The monsoon fishing was done in July 1989. The examination of gonads of major carps at the time showed that the breeding which commenced in the first week of July (first flood) was completed by 25th July (second flood). The reservoir water level was 338 m, which does not inundate the Mahod site. Among the fry of major carps collected from intermediate sector of the reservoir,

Cirrhinus mrigala was the most dominant followed by *Catla catla* and *Labeo rohita*. Thus, *Cirrhinus mrigala* not requiring higher upward migration, could breed well in middle sector of the reservoir as reflected from its established fishery but breeding of *Catla catla* and *Labeo rohita* is restricted for want of congenial breeding conditions. The breeding of major carps, which was early in July 1989 (6-7-1989), was delayed in the following year (it took place on 20-7-1990) for want to adequate water inflow. The breeding in July 1990 was recorded at reservoir water level of 341m, still low to inundate the site at Mahod. Though the breeding was better in July 1990 than in July 1989,after exploiting good many brooders during monsoon fishing (also done in July 1990), good natural recruitment of the fish could not be expected. In July 1991, at higher reservoir water level of 347 m and with availability of higher monsoon inflow, the site at Mahod was flooded and inundated with water to permit migration of fish for breeding. Accordingly, the eggs (ova diameter 5.3 mm) collected form Mahod were found to be mostly of Catla (76%). It was not possible earlier in 1989 and 1990 when successful breeding of *Catla catla* could not take place. In July 1991, again the breeding of fish was hampered due to inadequate monsoon inflow and low reservoir water level. It is thus concluded from these observations that breeding of catla and rohu in the reservoir largely depends on monsoon water inflow of July and submergence of lotic sector of the reservoir under water. These favourable breeding conditions may not be available regularly every year due to erratic rain and draw down of water for irrigation when, with the exception of mrigal, the breeding success of catla and rohu may be unpredictable.

Potential Fish Yield: Desai *et al* (2007) estimated a potential fish production of 25 kg/ha/yr. As against this, the actual maximum fish production from this reservoir was 8 kg.ha/yr. Thus, only 33% of the potential is actually being harvested from the reservoir and hence there is still scope to develop a sustainable fishery.

REFERENCES

- Desai, V.R., Shrivastava, N.P. and Kumar, D., 2007. Constraints in reservoir fishery development: A case study of Ravishankar Sagar reservoir. *Fishing Chimes*. 27 (4) : 19-22.
- Singh, P.P. and Tuli, R.P., 2006. Fisheries Development in Chhattisgarh. *Fishing Chimes*. 26 (1) : 214-217.

4

GUJARAT

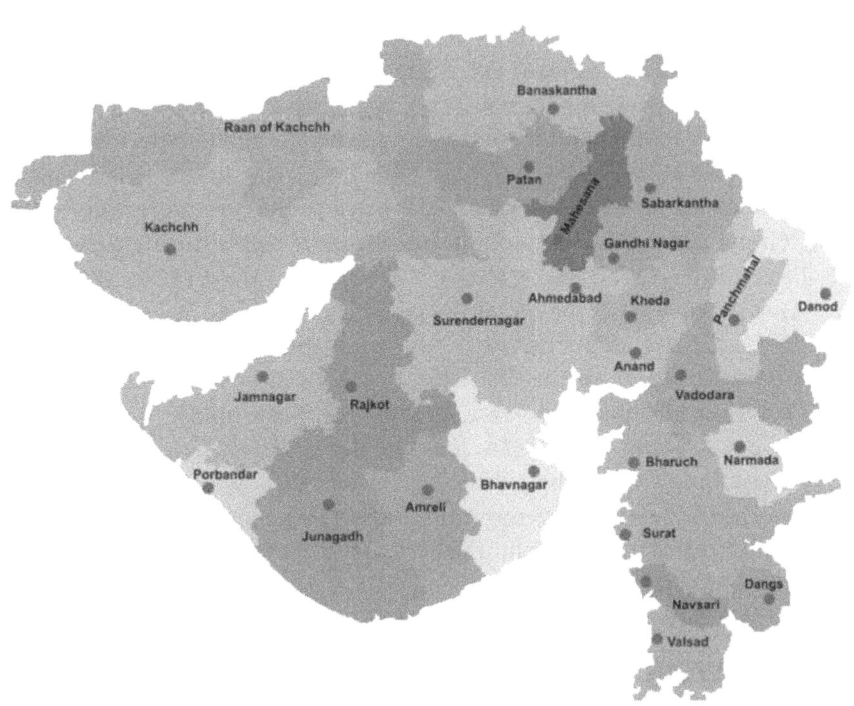

Fig. 4.1. Map of Gujarat (Not to scale)

Inland fisheries resources of Gujarat include rivers, lakes, ponds, tanks, estuaries and brackish waters. Total length of rivers and canals in the state is 3,865 km.The area of major and medium reservoirs available is about 2.55 lakh ha, ponds and tanks 0.22 lakh ha, estuarine waters of 0.21 lakh ha, brackish water inland area of 3.76 lakh ha and waterlogged area of 6,000 ha. In village ponds and reservoirs, mainly three species of major carps *viz.*, catla, mrigal and rohu are cultured. Exotic carps like common carp and grass carp have also been introduced in these village ponds. Composite fish culture is done in a scientific manner in small perennial village ponds, yielding good results.

There are 50 large, 38 medium and 1547 small sized reservoirs I the state. Fish production from reservoirs during year 2007-08 was 18,261 mt.

The state government has declared a new leasing policy for reservoirs of the state vide GR No. FDX-112003-1618-T dated 25th February 2004. Under the new policy, the reservoirs are to be leased out for a period of five years first, which will be reviewed and further extended for another five years. Thus a long term policy has been evolved to facilitate the lease to undertake culture-cum-capture fisheries production in a systematic and scientific manner. Some of the salient features of new policy are as follows:

1. Water sheets situated in tribal area have been reserved for the tribal co-ops/groups formed under SHG and to be given to them at pre-determined price.
2. Reservoirs up to 20 ha have been earmarked for FFDA beneficiaries and to e given to them based on predetermined price.
3. Average water spread areas of six years are taken to assess the effective water spread area and lease price is calculated against this effective water spread area.
4. Local co-op societies are given 20% price preference o the highest tender.
5. Compensation on lease amount price is permissible during draught years.
6. Price preference is given to the SGSY groups and SHG of BPL families.
7. Apex co-operative society of the state is to be given one reservoir in each of the districts with a view to strengthening the co-op marketing infrastructure in all the districts.
8. Provision for apeeal on lease amount has been made.
9. If lease is given in the later party of the year concession is given in the lease amount of the first year.
10. Affected tribals of Ukai reservoir and other major dam have been settled around reservoirs, accepting fishery activity as a part time occupation for their livelihood.
11. 716 fish landing centers have emerged.

UKAI RESERVOIR

Ukai Dam, constructed across the Tapti River, is the largest reservoir in Gujarat. It is also known as Vallabh Sagar. Constructed in 1971, the dam is meant for irrigation, power generation and flood control. Having a water spread of about 52,000 hectare, its capacity is almost same as that of the Bhakra Nangal Dam. The site is located 94 km from Surat.

The storage capacity of Ukai dam is almost 46% of the total capacity of all the other existing dams in Gujarat if put together. Ukai Dam is thought to be the mega project in Gujarat. During last 40 years, the actual irrigation potential is attained through all the major and medium water resources projects in the State, which comprises only 14 million hectares. The water is used by Left bank canal from Ukai and Left and Right bank canals from Kakrapar weir 20 km downstream of dam. The weir lifts and reuses the flow from 4 units of 75 MW capacity hydropower (6000 cusecs/unit of hydropower). The Ukai Left bank canal inlet is at RL 270′ and administration is keen to keep Ukai reservoir at 342′. The 4,052 m main dam is earthen with 425 m long spillway with capacity of 21 lakh cusecs (59,000 cumecs). The spill way has 22 gates x 48′6″ high. The FRL for dam is 345′ (105.16 m) and MWL as per design is 351′. The top of dam is 361′ (110 m). The main objective of Ukai project is to obtain the optimum irrigation and hydropower through simultaneously it also helps to achieve partial control over effects of heavy floods.

Fig. 4.2. Sluice gates of Ukai Reservoir

Table 4.1. Salient features of Ukai Reservoir

1. Nearest city	Songadh
2. River	Tapi
3. Basin	Tapi
4. Type of dam	Earthen and Gravity
5. Purpose	Irrigation, hydroelectric and flood control
6. Year of construction	1972
7. Catchment Area (Th ha)	6222.5
8. Length of dam (m)	4926.79

Contd...

Table 4.1: Contd...

9. Maximum heigh above foundation (m)	68.58
10. Maximum water level (m)	106.985
11. Full Reservoir level (m)	105.1
12. Gross storage capacity (mcm)	7497000
13. Live storage capacity (mcm)	6615000
14. Type of spillway	Ogee
15. Length of spillway (m)	425
16. Type of spillway gates	Radial
17. No.of spillway gates	22
18. Number of villages affected	98

SARDAR SAROVAR PROJECT

The Sardar Sarovar Project is a multipurpose reservoir with scope for massive economic development of the state of Gujarat. It is the largest dam and part of the Narmada Valley Project, a large hydraulic engineering project involving the construction of a series of large irrigation and hydroelectric multi-purpose dams on the Narmada River. The project took form in 1979 as part of a development scheme to increase irrigation and produce hydroelectricity.

Fig. 4.3. Sardar Sarovar Project in Gujarat

It is the 30th largest dams planned on river Narmada, Sardar Sarovar Dam (SSD) is the largest structure to be built. It has a proposed final height of 163 m

(535 ft) from foundation. The project will irrigate more than 18,000 km^2 (6,900 sq mi), most of it in drought prone areas of Kutch and Saurashtra. The dam's main power plant houses six 200 MW Francis pump-turbines to generate electricity and afford a pumped-storage capability. Additionally, a power plant on the intake for the main canal contains five 50 MW Kaplan turbine-generators. The total installed capacity of the power facilities is 1,450 MW. Critics maintain that its negative environmental impacts outweigh its benefits. It has created discord between its government planners and the citizens group "Narmada Bachao Andolan".

The dam is one of India's most controversial dam projects and its environmental impact and net costs and benefits are widely debated. The World Bank was initially a founder of the SSD, but withdrew in 1994. The Narmada Dam has been the centre of controversy and protest since the late 1980s. Singh (1993) reported 84 fish species from Sardar Sarovar dam.

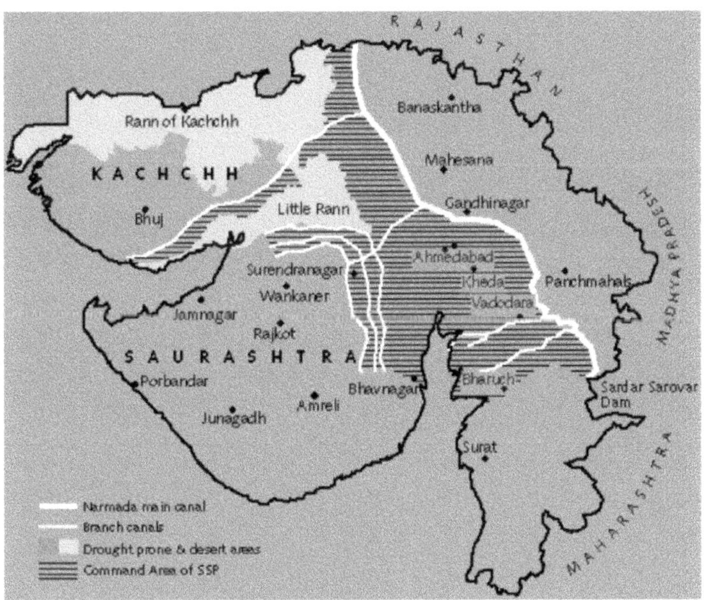

Fig. 4.4. Location of Sardar Sarovar Reservoir

Table 4.2. Morphometry of Sardar Sarovar dam

Type of dam	Gravity and concrete
Length of dam (m)	1210
River	Narmada
Spillway capacity (m^3/s)	84949
Capacity (m^3)	9,500,000,000
Active capacity (m^3)	5,800,000,000

Contd...

Table 4.2: Contd...

Catchment Area (km^2)	88000
Surface Area(km^2)	375.33
Normal elevation (m)	138
Length of reservoir (km)	214
Maximum reservoir width (km)	16.10

REFERENCES

- Singh, A.K., 1993. Pre impoundment studies on Sardar Sarovar Area of Narmada River (West Zone) with special reference to fisheries, Ph.D. Thesis, Vikram University, Ujjain, India.

5

Himachal Pradesh

Fig. 5.1. Map of Himachal Pradesh (Not to scale)

Himachal Pradesh is a state in Northern India. It is spread over 55,670 km^2, and is bordered by Jammu and Kashmir on the north, Punjab on the west and south-west, Haryana and Uttar Pradesh on the south, Uttarakhand on the southeast and by the Tibet Autonomous Region on the east. Himachal Pradesh is known to be abundant in natural beauty. Due to the abundance of perennial rivers, Himachal also sells hydro-electricity to other states such as Delhi, Punjab and Rajasthan. The economy of the state is highly dependent on three sources: hydroelectric power, tourism and agriculture.

Himachal Pradesh is a mountainous state with an area of 55,673 km^2. The state lies touching inner Himalayas in the north with altitude varying from 350 m to 6675 m MSL. The average temperature in different zones or regions manifests a wide variation ranging from 0°C to 42°C. The state has a vast network of fisheries resources in the form of rivers, streams, man-made reservoirs, natural lakes, ponds and raceways. In order to develop ad harness the fisheries potential of these resources, the department of fisheries was set up during 1950 as a wing of forest department. The important man-made reservoirs in the state are Gobind Sagar, Maharana Pratap Sagar, Pandoh and Chamera.

Table 5.1. Fisheries Resources of Himachal Pradesh

1. Riverine length Trout oriented Generally oriented	3600 kms-600 km 2400 km
2. Reservoirs	42200 ha
3. Ponds	1000 ha
4. High altitude lakes	725 ha

Himachal Pradesh has two large reservoirs, namely Gobind Sagar and Maharana Pratap Sagar. There are also three small ones *i.e.,* Chamera I and II and Pandoh. The total water area of these reservoirs is 42,000 ha at FRL.Two more reservoirs Mallana and Koldam are likely to add the reservoir tally of the state, increasing the total area to approximately 3000 ha.

A specific fishery policy for the development and exploitation of fish from the reservoirs in the state is in operation since 1976. Erstwhile fishermen of the riverine portions, forming part of the reservoir now and the dam out see families are organized into co-operative societies. After the development of fisheries they are allowed to undertake fishing operations and marketing of their fish as per provisions of Himachal Pradesh Act 1976. A closed fishing season of two moths from 1st June to 31st July is observed every year. The fish landed by the members of co-operative societies is brought on the fixed landing centers set-up by the department. Variety wise, the fish is graded in terms of weight, graded and then handed over to the concerned contractor on the rates fixed in the beginning of the year, on cash payment. Further marketing of this fish is done by the contractors on a valid waypass issued by the Department of Fisheries. There are 725 ha of water spread area covered by natural lakes in the high hills of the state. These have been stocked with either common carp or brown trout seed.

PONG RESERVOIR

Pong reservoir is a artificial reservoir spread over 15662 ha across the river Beas in Kangra district of Himachal Pradesh. The reservoir is with a catchment area of 12,561 sq.km. Completed during 1974, the Pong reservoir was constructed mainly for power generation, irrigation and flood control. Fish production aspects were of given ay reckoning during the planning or pre-impoundment stages, though the river Beashad a wide array of resident and migratory fish fauna. The salient features of the reservoir are depicted in Table 5.2.

Table 5.2. Salient Features of Pong Reservoir, H.P.

1.	River	Beas
2.	Water source	melted snow and monsoon run-off
3.	Type of dam	earthen
4.	Height of dam (meters)	132.59
5.	Altitude height (m above MSL)	435.86
6.	Year of commissioning	1974
7.	Catchment Area	12562
8.	Water Spread Area at FRL (ha)	24529
9.	Water Spread Area at DSL (ha)	14312
10.	Average Waterspread Area (ha)	14600
11.	Mean Depth(m)	75
12.	Total Length (km)	41.8
13.	Widest stretch (km)	19
14.	Shoreline Development Index	560
15.	Annual Water Level fluctuation (m)	384-433
16.	Maximum water level fluctuation (m)	117
17.	Gross Storage Capacity (Mill.Cub.m.)	8570
18.	Live Storage Capacity (Mill.Cub.m.)	77771
19.	Inflow (Mill.Cub.m.)	8215-13134
20.	Outflow (Mill.Cub.m.)	6555-13641

In view of the biogenic capacity of the ecosystem and systematic stocking over a number of years with mirror carp and Indian major carps, the catch structure of the reservoir completely altered ad the carps accounted for as high as 61.6% of the total ladings during 1987-88. The per hectare yield of the reservoir also increased significantly from a meager 6.5 kg/ha during 1976-77 to 53.1 kg/ha during 1987-88. Catfishes and carps accounted for 70.1% and 29.9% of the total production of the reservoir during 1982-83 against 27.8% and 67.6% during 1989-90 respectively.

Catch Trends: A total of 23 fish species belonging to six families have been encountered in Pong reservoir. The major fishes currently encountered in order of abundance are *Mystus seenghala, Tor putitora, Labeo rohita, Cyprinus carpio, Catla catla, Cirrhinus mrigala, Labeo calbasu, Chana* spp., *Labeo dero* etc.

A perusal of indigenous carp composition in the total catches from 1976 to 1999-2000 indicated that they dominated overcatfishes till 1991-92. From this year carp catches declined. During 1991-92, the arps *viz., Labeo rohita, Cirrhinus mrigala,Labeo calbasu* and *Catla catla* accounted 46.3% (225 tonnes) of the total ladings while catfishes *viz., Mystus seenghala, Wallago attu* accounted for 37.9% (60 tonnes) and others 2.4% (12 tones). Against this, during 1999-2000 the percentage composition of indigenous carps, catfishes, mirror carp, *Tor putitora* and others was 18.83% (84 tonnes), 56.98% (258 tones), 2.89% (13 tonnes), 19.88% (90 tonnes), 0.9 % (4 tonnes) respectively.

Among the Indian major carps, *Labeo rohita* is the dominant fish. The highest catch (339 tonnes) of this fish was recorded during 1987-88. For the last five years (1995-96 to 1999-2000), the catches have fluctuated between 36 to 65 tonnes. During 1999-2000 the landings of *Labeo rohita* was 65 tonnes. *Catla catla* has always kept a low profile I the reservoir *i.e.,* the maximum (22 tonnes) was recorded during 1996-97, in the recent years. The catches during 1999-2000 were 11 tonnes. *Cirrhinus mrigala* ranged from 2 to 8 tonnes only. *Labeo calbasu* too used to have a significant presence in the reservoir during eighties, the highest being 85 tonnes during 1982-83. The catches of this fish plummeted to a level of 3 tonnes during 1997-98.

Now Pong reservoir may be categorized as 'Mystus reservoir'. *Mystus seeghala* has been showing constant increase during the last decade. The highest catch (257 tones) was recorded during 1999-2000 accounting for 57% of the catch. *Wallago attu,* however, suffered a decline and an all time low (104 tonnes) was recorded during 1999-2000.

Mirror carp composition is quite erratic. It is presumably due to the fact that ideal breeding grounds are on-existent in the water body. However, regular stocking has helped in the revival of mirror carp fishery ad for last eight years there has been a progressive increase in its presence followed by an all time low of two tones during 1992-93, mainly to show an increase to 17 tonnes during 1997-98.The catches during 1999-2000 were 13 tonnes. Mahseer is a highly precious fish. It is probably the only reservoir in the country, which provides the opportunity of mahseer angling. The catches of mahseer in the reservoir have shown remarkable consistency during the past decade and ladings have fluctuated between 50 to 90 tonnes a year. The highest catch of mahseer was recorded during 1999-2000 *i.e.,* 90 tonnes.

Employment Generation: Under the scheme for generating employment to unemployed, every year some interested unemployed youth are being enrolled in the fishermen's cooperative societies and they are the issued fishing licenses to earn money by conducting fishing in the reservoir. Employment opportunities provided under this scheme during last four years are as under:

Table 5.3. Year wise employment opportunities in Pong Reservoir

Sr.No.	Year	Number
1.	1996-97	167
2.	1997-98	163
3.	1998-99	102
4.	1999-2000	174

Seed Stocking: The seed stocking was started during 1974-75, when the first consignment of 1.30 lakh fry of minor carps was released. Since then regular stocking is being done. Stocking has been mainly confined to the seed of mirror carp ad Indian major carps, *Labeo rohita, Catla catla* and *Cirrhinus mrigala*. The seed of other species *viz., Tor putitora, Schizothorax spp* and *Labeo dero* could not e stocked. It is due to the absence of large scale seed production technologies. The fisheries department has set up improvised four fish seed farms located near the reservoir site, namely Kangra, Deoli, Gagret and Nalagarh. Fish seed stocking done in the reservoir during the last six years is given in the Table 5.4.

Table 5.4. Seed stocking in Pong Reservoir

Sr.No.	Year	C.carp	L.rohita	C.catla	C.mirgala	IMC mix	Total
1.	1995-96	3.30	—	—	—	—	3.30
2.	1996-97	3.00	—	—	—	—	3.00
3.	1997-98	1.44	9.35	1.00	—	—	11.79
4.	1998-99	2.25	8.10	2.00	—	—	12.35
5.	1999-00	2.50	6.0	1.00	—	2.50	12.00
6	2000-01	1.00	8.23	2.25	—	2.45	13.93

Fish Production: An analysis of production figures for 24 years (1976 to 1999-2000) indicates that there is consistency in yield rate which works out to 30 kg/ha. The yield ranged between 6 to 53 kg/ha. The average yield during 1999-2000 was 29 kg/ha.

Revenue: The value of fish caught increased from Rs. 5.18 to 218.3 lakh. During 1995-96 the value of fish was Rs. 112.7 lakhs, whereas during 1999-2000 fish worth Rs. 218.3 lakhs was harvested from the reservoir.

Remarks: Among the positive decisions taken by the department, the most important was to stock the reservoir with seed of Indian major carps. This helped in the establishment of *Labeo rohita* catch, contributing as high as 42.5% of the catches during 1989-90. The amendment to state fisheries act 1976 (Act No. 16), enforced of mesh regulation, organization of fishermen under the cooperative fold, promulgation of closed season, setting of fishermen during the initial stages from outside the state, initiation of fishermen's welfare scheme etc., were other well conceived measures which helped in boosting the reservoir fisheries activities and providing vocation to the displaced inhabitants of the reservoir.

Thus, the Pong reservoir is a good example of using a reservoir for food production and generating employment avenues. Fisheries development in Pong reservoir has helped in settlement and providing livelihood to families uprooted due to the impoundment.

Table 5.6. Information on fishermen, Gear and Crafts from Pong Reservoir during 1995-96 to 1999-2000

Particulars	Unit	1995-96	1996-97	1997-98	1998-99	1999-2000
No.of fishermen's societies	Nos.	12	13	13	13	14
Membership	Nos.	1439	1631	1782	1825	1943
No.of gill net licenses issued	Nos.	2142	2424	2704	2756	2934
No. of fishermen registered	Nos.	1071	1212	1352	1378	1467
No. of Anglers registered	Nos.	157	199	126	221	191
Fish Production	Mt.	329.7	3973	414.8	359.8	653.1
No. of fishes caught	Nos.	231736	275660	286533	245597	335031
Value of fish caught	Lakhs	112.7	169.1	178.9	160.1	218.3
Royalty fee	Lakhs	17.10	25.16	26.81	24.04	32.73
No. of illegal fishing cases registered	Nos.	384	408	413	419	443
Compensation realized	Rs.	57820	67635	65360	64715	72045
Value confiscated fish auctioned	Rs.	10509	10569	5967	7732	12688
Total Reveue	Lakhs	19.11	27.45	29.14	26.39	35.42

GOBINDSAGAR RESERVOIR

Gobindsagar reservoir came into existence during 1963, with the construction of a dam across river Sutlaj at village Bhakra in Bilaspur district of Himachal Pradesh. It is located within the geographical coordinates of 31°25′ Latitude North and 36°25′ Longitude East. The reservoir catchment area is 56980 km^2. At full storage level, the reservoir has a surface area of 16867 ha, while at the minimum dead storage level it is of 5063 ha. The average surface area is about 10,000 ha The annual water fluctuation varies from 41 to 61 m, the maximum amplitude being 70 m. The total length of reservoir is 168 km with the widest stretch of 6 km near the dam. It is ine of the deepest (165 m) reservoirs in India and is fed by warm monsoon run-off and with cold snow-melt.

Fish Fauna at Pre-Impoundment Stage: No preliminary information of fishery of its riverine waters was available. It took considerable time for the fisheries department to extend the application of Indian Fisheries Act and Frame rules under the act for enforcement in the state. The available records indicate about the presence of *Barilius* sp. *Cirrhina reba, Labeo bata, Labeo dero, Labeo dyochilus,*

Labeo calbasu, Puntius spp., *Schizothorax* sp., *Tor putitora, Mystus seenghala, Mastacembelus* sp., *Noemacheilus sp.* in parts of river Sutlaj, falling in Bilaspur and Una districts which now cover the reservoir area.

Fish Fauna at Post Impoundment Stage: Altogether 51 species of fishes belonging to 9 families were accounted by Kumar (1990). The list of fishes in reservoir is depicted in Table 5.7.

Table 5.7: Fish fauna of Gobindsagar Reservoir

Family : Cyprinidae: *Labeo bata, Labeo calbasu, Labeo dero, Labeo dyocheilus, Labeo rohita, Catla catla, Cirrhinus reba, Cirrhinus mrigala, Cyprinus carpio ver.nudus, Cyprinus carpio var.communis, Cyprinus carpio var, specularis, Ctenopharyngodon idella, Hypopthalmichthys molitrix, Barilius barila, Barilius bedelisis, Barilius vagra, Barilius modestus, Oxygaster bacaila, Rasbora daicoius, Carassius auratus, Crossocheilus latius latius, Shizothorax richardsonii, Schizothorax plagiostomus, Tor putitora, Garra gotyla, Garra lamta, Putius sarana, Puntius ticto, Puntius chola, Puntius sophore, Puntius punjabensis.*
Family : Cobitidae: *Botia dario, Botia dayi, Botia bindi, Botia lohachata, Noemacheilus botia, Noemacheilus rupicola, Noemacheilus montanus, Noemacheilus horal, Noemacheilus kangrae.*
Family : Bagridae: *Mystus seenghala.*
Family : Schilbeidae: *Clupisoma garua.*
Family : Sisoridae: *Glyptothorax pectinopterus, Glyptothorax cavia.*
Family : Belonidae: *Xenentedon cancila.*
Family : Ophiocephalidae: *Channa punctatusChanna gachua.*
Family : Mastacembelidae: *Mastacembelus armatus armatus.*
Family : Salmonidae: *Salmo trutta fario.*

Stocking: During 1967-71, the state fisheries department stocked 3500 gravid spawners and 0.6 million fingerlings of Indian major carps. By 1969, Indian major carps established themselves in the reservoir and started breeding. The stocking continued mainly with the seed of mirror carp. In order to produce the seed, the state fisheries department set up three fish seed farms, near the reservoir site at Deoli, Nalagarh and Alsu. The year wise details of fish seed stocking are give in Table 5.8.

Table 5.8: Fish seed ranching in Gobindsagar reservoir (quantity in lakhs)

Year	Mirror carp	Indian major carp
1976-77	6.26	—
1977-78	6.43	—
1978-79	8.68	1.69
1979-80	7.44	—

Contd...

Table 4.2: Contd...

1980-81	9.18	8.4
1981-82	12.99	1.25
1982-83	14.44	—
1983-84	15.33	3.2
1984-85	20.2	2
1985-86	21.95	2
1986-87	22.75	—
1987-88	17.22	10.3
1988-89	36.55	13.49
1989-90	25.97	3.72
1990-91	35.81	2.19
1991-92	59.53	0.55
1992-93	54.56	2.5
1993-94	33.48	—
1994-95	15.01	1.103
1995-96	59.69	10.64
1996-97	37.09	8.5
1997-98	14.83	9.88
1998-99	13.9	11.3
1999-2000	15.32	8.5
2000-01	15.1	8.4
2001-02	6.0	9.0
2002-03	6.2	5.7
2003-04	6.35	10
2004-05	7.0	50

Fisheries: The fish production for the period of 1976-77 to 2003-04 varied from 377 tonnes (1986-87) to 1202 tonnes (2002-03). The average fish production from the reservoir was about 1000 tonnes per annum or 100 kg/ha/yr, which is the highest in the country for a large reservoir. The reservoir has a predominant fauna of the exotic silver carp (about 60-65% of the total catch) followed by Indian major carps (20-25%), mahseer species (8-10%) and minor carps (8-10%). The Indian major carps are dominated by *Catla catla*. In recent years exotic carps have monopolized the reservoir. Since their introduction to the reservoir, the exotic carps have been increasing at the rate of 8.2% per year, on an average basis.

The Department of Fisheries, Government of Himachal Pradesh implements several welfare and production oriented schemes for the benefit of the reservoir fishermen. These centrally sponsored schemes include savings cum relief scheme, group insurance scheme for active fishermen and the housing scheme.

Besides closed season, which is implemented in the reservoir for two months (01 June to 31st July) every year, areas suitable for natural breeding in the reservoir are also protected to allow the fishes to breed and thereby help in auto stocking of the reservoir. To maintain species balance, supplementary stocking is done from time to time.

About 3000 registered fishermen inhabit the periphery of the reservoir of which about 1900 are active. These fishermen are grouped into 16 co-operative societies, which are then grouped into an apex body called the Bilaspur Fisheries Marketing and Supply Federation. The federation assists the fishermen in disposal and marketing of fish (including retail marketing, if necessary) and its assets include a cold storage, an ice plant and refrigerated vans. The Gobindsagar fish is marketed in major towns/cities in Punjab, Delhi and Jammu and Kashmir and also the bordering areas in Uttar Pradesh and Haryana.

A reservoir development committee which was established in 1976 took appraisal of the whole range of reservoir fisheries problems and made some drastic policy decisions in order to re-organize the fishing and marketing procedures. Grading the fish was started as the first step, followed by fixation of rates each year. It became mandatory to fix the rates at the beginning of the year for each category of fish. A decision was also taken to appoint contractors for sale of fish outside the state. The fish caught from the reservoir by the fishermen of primary societies are now brought to the fixed lading centers where representatives of the fisheries federation receive the fish and record the category and species in the presence of fisheries officer. The catch is the handed over to the contractors. These contractors are jointly called by tenders at the beginning of the year and those who bid the highest price are given the contract for the whole year. The contractors make weekly payment to the federation, besides keeping a fixed deposit of Rs. 5,00,000 to 6,00,000 as 'advances'. The federation makes the payment once in 15 days to the societies after deducting 15% royalty of the department ad 5% commission. The societies make payment to the fishermen after deducting a marginal commission which varies from society to society and is fixed by the general house. The general house body is selected by the members of the society. The federation acts as intermediary between the societies and contractors. The federation is fully responsible for making payment to the fishermen and even if the contractors do not pay on time, the federation has its own funds and is bound to make the payment to the fishermen from this fund. To avoid conflict between societies regarding the area of operation, the department has divided the reservoir into eight beats.

Fishermen who are members of the cooperative societies are issued licenses through the societies by the respective fisheries officers of the landing centres.There are 11 fishermen cooperative societies functioning in the Gobindsagar reservoir with a total stretch of 1000 active fishermen. They have full time fishermen.

Trials being conducted by Central Institute of Fisheries Education, Mumbai at Bhakra landing station on seed rearing in cages have given quite encouraging results. The programme needs enlargement by installation of more such cages and raising of advanced fingerlings in bulk quantities matching with the need of the reservoir. By raising the size of stocked seed the survival rate of growing juveniles could enhance thereby contributing significantly to the total biomass of fish in the reservoir.

Kaushal and Tyagi (1990) studied food and feeding habits of *Cyprinus carpio*.Food habits of *Cyprinus carpio* exhibited a wide food spectrum with macrovegetation as the basic food and benthic macroinvertebrates as subsidiary food. The fish showed relatively higher utilization of benthic forms from January to May concurring with receding water level. Food habits of *Labeo calbasu* suggested that the fish feeds on variety of food available in the reservoir with greater emphasis on dynophyceae. The occurrence of large quantity of detritus, sand and silt revealed the bottom feeding of the fish.

Crafts and Gears used: Fishermen have their own boats, generally of the size 15x3x2 ft costing approximately Rs. 2000/-. The fishermen use mainly nylon gill ets. Each fishermen has on an average 3-4 gill nets of 108 to 140 mm mesh size which last for 1-2 years. On an average, 40% of the fishermen have some education. Their monthly income ranges from Rs. 800 to 8000/-. The department has initiated a series of schemes for the benefit of the fishermen fishing in the reservoir. 50% subsidy is provided by the department to a maximum of Rs. 5000/- for the purchase of fishing equipments, nets, boats and tents. The scheme is highly popular and has helped in raising the total fish catch.

Preliminary observations on thermal stratification were reported by Sarkar *et al.* (1977) and Kaushal *et al.* (1984) in Gobindsagar reservoir. Singh (1990) carried out detailed depth-wise study on thermal and chemical stratifications in Gobindsagar reservoir during the years 1986 and 1987. As the winter months were over, the surface water temperature was elevated, but the temperature remained low at subsurface or bottom layers. Barring winter months *i.e.,* from November to February, thermal and chemical stratifications were observed in the lake during both the years of observation.Thermal stratification was observed from May to October in 1986 (except in the month of September) and March to October in 1987 (except in the month of May) at various depths of observation. The differences between surface and bottom temperature ranged from 3 (April 1987) to 12.5 (August 1986).

The chemical stratification in respect of pH and dissolved oxygen was always associated with thermal stratification. Though the klinograde distribution of dissolved oxygen (oxycline) was observed, it reduced abruptly wherever thermocline was observed. The differences in pH and dissolved oxygen from surface to bottom fluctuated from 0.2 to 0.9 ppm to 7.04 ppm respectively. Chemical stratification in respect of pH and dissolved oxygen will have a bearing on the movement of the fishes, restricted by a sudden fall in the pH and dissolved oxygen in the stratified region of the reservoir. A critical perusal of the data indicates a step reduction in dissolved oxygen from epilimnion to hypolimnion.

This shows a klinograde distribution suggestive of the productive nature of the reservoir (Singh, 1990).

Jha (1990) described the abundance and fluctuation of periphyton in Gobindsagar reservoir. The abundance of periphytic population revealed a bimodal pattern of fluctuation. The primary peak was recorded in March (4456 U cm^{-2}), while the secondary peak in September (3792 U m^{-2}). July and August were the months of least abundance. During the summer months (April to June) the values remained almost static. The community structure of periphyton revealed the dominance of bacillriophycean flora. The diversity of the population was found to be regulated by the abundance of diatoms.

The primary peak recorded greater abundance of tax like-*Melosira varias, Cyclotella comensis, C.menishiniana, Stephanodiscus nigaro, Cylindrotheca gracilia, Synedra ulna var.danica, Achnantheca exigu, Caceniea dieculus, Strauneia smithii, Cymbella ventricesa, C.microcephala, Navicula hasta, Gomphonema aguatum.* The other forms of importance were *Diatoma elongatum, D.anceps, Fragillaria pinnata, Synedra rumpens, S.pulchella, S.offinis, Navicula capitata, N.anglica, Gomphonema clavaceum, G.accumunatum,Pinnularia braunii,* and *Gammtophora serpentine.*

The secondary pulse during post monsoon was mainly consisted of the greater proliferation of species like *Amphora ovalis, Synerra pulchella, Fragillria japonica, F.vaucheriae, Achnanthes microcephala, Anomoeoneiss sphaerophora, Cymbella lanceolata, C.prostrata, Navicula popula, N.radiosa, N.grevelli, Gomphonema gracilus, Hantzschia amphioxys* etc.

However, *Diatoma anceps, Tabellaria fenestrate, Cyclotella comota, Navicula pygmea, Gomphenema clavaceum* etc. were the other forms contributed sizably to the formation of this peak.

Table 5.9. Avergae number (unit cm^{-2}) of periphyton in Gobindsagar reservoir

January	2998 (2525)
February	3188 (2731)
March	4456 (3560)
April	2765 (2015)
May	2930 (2178)
June	2740 (2040)
July	1750 (1235)
August	1837 (1400)
September	3792 (2618)
October	3466 (3038)
November	2734 (1825)
December	2927(2125)

CHAMERA RESERVOIR

Due to various reasons the fishery of this reservoir could not be developed on expected lines. Taking the potential of this reservoir, it is important that requisite seed of suitable species and desired size be stocked in this water body. It is important that prior to stocking seed be reared at the Sultanpur Fish Farm or in cages. Mahseer and Silver carp are the suggested species.

REFERENCES

- Jha,B.C.,1990. The periphyton of Gobindsagar reservoir and allied waters ,Himachal Pradesh.1. Abundance and periodicity of diatoms. In: Contributions to the Fisheries of Inland open water systems in India, Part I (Ed. A.G. Jhingran, V.K. Unnithan and Amitabha Ghosh), Inland Fisheries Society of India, pp. 122-129.
- Kaushal, D.K., Joshi, H.C. and Rama Rao, Y., 1984. Observation on the vertical distribution of plankton in Gobindsagar, Himachal Pradesh. *Nat. Acad. Sci. Letters,* 7(2) : 65-69.
- Kaushal, D.K. and Tyagi, A.P., 1990.A study of the macroinvertebrates of Gobindsagar reservoir in relation to fish and fisheries. In : Contributions to the Fisheries of Inland open water systems in India, Part I (Ed. A.G. Jhingran, V.K. Unnithan and Amitabha Ghosh), Inland Fisheries Society of India, pp. 72-79.
- Kuldip Kumar., 1990.Management and Development of Gobindsagar Reservoir—A Case Study, pp.13-20. In : Jhingran, Arun G. and V.K. Unithan (eds). Reservoir Fisheries in India. Proceedings of the National Workshop on Reservoir Fisheries, 3-4 January, 1990, special publication 3, Asian fisheries Society, Indian Branch, Mangalore, India.
- Sarkar, S.K., Govind, B.V. and Natarajan, A.V., 1977. A note on some distinctive linological features of Gobindsagar reservoir, Himachal Pradesh. *Indian J. Anim. Sci.* 47(7) : 435-439.
- Singh, Gurucharan, 2001.Status of Development of fisheries of Pong Reservoir (Himachal Pradesh). *Fishing Chimes*. 21(1) : 88-90.
- Singh, H.P., 1990.Studies on thermal and chemical stratifications in Gobindsagar reservoir, Himachal Pradesh. In : Contributions to the fisheries of inland open water systems in India, Part I (Ed. A.G. Jhingran, V.K. Unnithan and Amitabha Ghosh), Inland Fisheries Society of India, pp. 171-176.

6

JAMMU AND KASHMIR

Fig. 6.1. Map of Jammu and Kashmir (Not to scale)

The state of Jammu and Kashmir has three diverse agrioclimatic zone *viz.* Jammu, Kashmir and Ladakh with a tremendous potential for development of warm water and coldwater fisheries. All these three regions of the state are bestowed with a network of natural water bodies like rivers, streams, lakes, reservoirs, ponds, swamps etc., spread over an area of about 57000 ha; out of which about 34000 ha are in the shape of lakes, marshy areas, ponds and reservoirs. Further another 23,000 ha extent in the shape of river systems.

The fish production from the reservoirs is very low varying from 5 kg/ha/yr to 250 kg/ha/yr. By promoting cage culture, production can be stepped up from reservoirs to 25-50 tonnes/ha/year (Charak and Fayaz, 2006).

DAL LAKE

The Dal lake of Kashmir situated in the north-east of Srinagar at mean latitude of 34°7′ N and longitude of 74°52′E at an average altitude of 1583 m. The top crust of the lake has been observed to freeze when the mercury falls to – 11°C. Early spring and summers are the wet periods when maximum rainfall occurs and average annual rainfall recorded is 655 mm. It is in this season that the snow melts in the high catchments results in maximum discharge in Dachigam and Dora nallah which flows into the lake. Dal lake comprises of five basins *viz.* Hazrat bal, Bod-dal, Gabribal, Nageen and Brari-Nambal. A perennial inflow channel known as Telbal Nallah enters the lake from the north and supplies 80 percent of the water from high altitude lake called Marsar lake (Quadri and Yusuf, 1980). With the lake basin itself there are number of springs (Kundangar *et al.*, 1995) which act as permanent water source to the lake. Dal lake has been a great tourist attraction in the past. However, once beautiful waterbody has been the worst victim of the anthropogenic pressures during the second half of the 20th century and has undergone tremendous ecological changes (Quadri and Yusuf, 2008).

Water Quality: Murtaza *et al.* (2010) studied impact of pollutants on physico-chemical characteristic of Dal lake. Remarkable changes have taken place in the water quality over the last few years which are revealed by comparing the present water quality with that of the past period.

Table 6.1. Water quality of Dal Lake

Parameters	Hazaratbal basin		Nishant basin		Nehrupark basin		Nigeen basin	
	2007	2008	2007	2008	207	2008	2007	2008
Air temp. (°C)	18.2	18.8	18.5	23.8	19.4	22.8	17.2	16.8
Water temp. (°C)	16.4	15.4	16.6	16.1	17.4	18.1	15.3	15.7
pH	7.8	7.5	8.1	7.8	7.9	7.2	8.0	7.8
Dissolved oxygen (mg/l)	7.0	6.5	7.3	7.8	6.4	6.1	6.5	6.2

Contd...

Table 6.1: Contd...

Total alkalinity (mg/l)	102	105.2	56	45	87	84.8	125	126.2
Silicate (mg/l)	4.0	3.6	3.1	3.5	3.8	4.0	4.2	4.5
Total phosphorus (mg/l)	504	583	507	608	535	521	570	532
Ammoniacal nitrogen (mg/l)	366	345	263	223	662	698	461	348

Source: Lakes and waterways development Authority and Murtaza *et al.* (2010)

Hazratbal Basin: The data presented in Table 6.1 indicates that air temperature, total alkalinity, and total phosphorus increased significantly. However water temperature, pH dissolved oxygen, Silicate and ammoniacal nitrogen showed a progressive decline. The significant increase in the air temperature, total alkalinity, and ammonical nitrogen could be attributed to continued drought conditions while the decrease in pH, water temperature, dissolved oxygen, silicate and ammonical nitrogen may be due to efficient utilization of these elements by the aquatic weeds (Pandit *et al.*, 2001).

Nishant Basin: The data shown in Table 6.1 depicts that air temperature, dissolved oxygen, silicates, and total phosphorus have increased while water temperature, pH, total alkalinity and ammonical nitrogen decreased. The progressive increase in water quality parameters could be attributed to hyper-eutrophication and continued drought conditions while a decrease in water temperature, pH, total alkalinity and ammonical nitrogen could be due to heavy growth of aquatic weeds which efficiently utilize the ammoniacal nitrogen in the lake (Kaloo *et al.* 1995).

Nehru Park Basin: The data presented in Table 6.1 shows that air temperature, water temperature, silicates and ammonical nitrogen have increased while as pH, dissolved oxygen, total alkalinity and total phosphorous showed a significant decrease. The faster rate of increase in air temperature, silicate and ammoniacal nitrogen could be ascribed to eutrophication and efficient utilization and trapping of sunlight by the lake (Jenson and Anderson, 1992) while as the decrease in pH, dissolved oxygen, total alkalinity and total phosphorus may be due to efficient utilization of phosphorus by macrophytic vegetation in the lake (Lijiklema, 1994).

Nigeen Basin: The data shown in Table 6.1 depicts that water temperature, total alkalinity and silicate have increased while air temperature, pH, dissolved oxygen, silicates, total phosphorus and ammonical nitrogen have progressively decreased. The increase in the above mentioned parameters could be attributed to continued drought conditions and eutrophication in the lake (Mohan, 1991) while the decrease in air temperature, pH, dissolved oxygen, silicates and total phosphorus may be due to thermal stratification in the lake ecosystem and efficient utilization of the elements by macrophytic vegetation in the lake (Pandit and Yousuf, 2002).

Plant species: Quadri and Yousuf (2008) reported 31 species of plant species from Dal lake. Emergent were *Alisma Plantago-aquatica, Corex* sp., *Cyperus defformes, Lycopus europus, Myriophyllum verticillatum, Nasturtium officinale, Phragmites australis, Polygonum hydropiper, Polygonum amphibium, Saggitaria saggitifolia, Scirpus triquenter, Sium latijugum, Sparganium ramosum* and *Typha angustata,* while the rooted floating leaf type were *Nelumbo nucifera, Eichhornia crassipes, Hydrocharis dubia, Nymphaea alba, Nymphaea mexicana, Nymphoides peltatum, Potamogeton natans* and *Trapa natans.* Submerged species *Potamogeton crispus, Potamogeton lucens, Potamegoton natans, Myriophllum spicatum, Hydrilla verticillata,* and *Ceratophyllum demersum.* Free floating plants observed in Dal lake were *Lemna* spp., *Azolla pinnata* and *Salvinia natans.*

Plankton Diversity: Dal lake has a mixed microflora dominated by diatoms followed by blue green and green algae. The phytoplankton are represented by chlorophyceae, bacillariophyceae and myxophyceae. About 150 phytoplankton genera have been identified (Ahmad, 1980). The maximum population of phytoplankton has been recorded in May to June and minimum in December to January. The secondary peak occurs in the month of September.

Copepods, rotifers and cladocerans are the principal components of the zooplankton. Copepods forms the largest group, next being rotifers and third cladocerans. Zooplankton shows maximum population in spring and autumn and poor population in summer and winter. Ostracods having 2 genera has been found absent in late spring, late summer and early autumn.

Fish Fauna: Ahmad (1980) accounted thirty-seven species of fish from Dal lake. The important species which supports good fishery are *Schizothorax esocinus, Schizothorax niger, Schizothorax curvifrons, Labeo dero, Crossocheilus* sp, *Botia birdi, Nemacheilus kashmirensis, Gambussia affinis, Glyptothorax* sp. *and Glyptosternum.*

The common carp (*Cyprinus carpio*) constitute major catch of commercial fishery of Dal lake. Common carp introduced in the year 1957, constitutes about 70% of the total catches whereas *Schizothorax* contributes for about 10% and other fishes including catfishes form 20% of the total landings. The catch composition of *Crossocheilus latius* in lake has increased from 18.8% during 1968-69 to 27.1% during 1970-71 in the total fish catch. The total fish production in 1979 was 702 tons.

Jan *et al.* (2012) studied haematological and biochemical indices of *Cyprinus carpio* collected from Dal Lake. The objective of the study was to investigate and establish the reference values of haematological and biochemical ranges for the common carp, *Cyprinus carpio.* The findings for erythrocytes number was 2.17 + 0.3 × 10 mm^3, haemoglobin 6.29 + 0.46 g/dl, packed cell volume (PCV) 18.9 + 1.4%, mean corpuscular volume (MCV) 88.2 + 11.7 ft and mean corpuscular haemoglobin (MCH) 29.4 + 3.9 pg. The total plasma protein, glucose, uric acid and urea levels were 4.1 + 0.7 g/dl, 76.7 + 4.9 mg/l dl, 1.6 + 0.4 mg/dl and 22.5 + 3.5 mg/dl respectively. A correlation matrix was established to compare the degree of association among the biochemical and haematological indices. A positive correlation was observed among weight, length, MCV, and MCH, as well as between length and packed cell volume. The blood glucose level was positively correlated with weight and length, whereas the total plasma was negatively correlated with haemoglobin.The RBCs count was positively correlated with haemoglobin and negatively correlated with MCV and MCH.

Crafts and Gears used in Fishing: The gears used in Dal lake fishing are cast net, scoop net, hook and line, pool and line and narsoo. The cast net is the main fishing gear used in the lake. It is of 6 pieces with 4.8 meters in diameter having mesh size 1 to 2 cm, bar to bar with iron or lead sinkers along the peripheral cord weighing about 5 kg per net. The net is operated from boat by one fisherman while another fisherman rows the boat. Besides this, Scoop net and hook and line are most commonly used by the fishermen. Narsoo, a multiple head spear having 5 to 7 heads, is used in deep pockets of the lake. The nets are fabricated locally by the fisherman, mostly of nylon twine, but cotton twine is also used.

The fishing crafts consists of open wooden boat of about 20 feet long and 4 feet centre width, having capacity of carrying 2 to 3 persons, besides fish catch of about 50 kg. The boats are mostly constructed from Devdar wood.

Socio-Economic Conditions of Fisherman Population: The fishermen of Dal lake are extremely poor and generally illiterate and not aware of scientific exploitation of fishery resources. Only 5.4% of the total fisher children go to the schools (Ahmad, 1980). The main reason is poverty. About 120 fishermen families are totally dependent on catches they get from the lake. The estimated average fish catch per boat per day is about 6 kg. The catches mostly comprises of common carp.

ANCHAR LAKE

The Anchar Lake has an area of 6.6 km^2 and is located about 5 km to the west of Kashmir University campus. It is a shallow water body with a maximum depth of 3 meters and the most part of the lake is infested with macrophytes. Yousuf and Gagroo (1990) studied protozoan community and physico-chemical features of the Anchar Lake for the period of April 1987 to March 1988.They reported 21 species of protozoa, out of which 12 were flagellates and remaining nine ciliates.

Table 6.2. Protozoan community in Anchar Lake

Flagellates: *Synura* sp., *Chilomonas paramecium, Euglena acus, Euglena* sp., *Phacus pleurnectes, P.longicauda, P. torta, Peranema* sp., *Gonium socials, G. pectorales, Pleodorina* sp., *Ploytoma* sp.
Ciliates: *Euplotes patella, Chilodonella cucullus, Vorticella campanula, Paramecium* sp., *Lionotus fasciola, Stylonychia* sp., *aspidisca* sp., *Coleps hirtus, Colpidium compylum.*

The protozoans recorded from the lake were mostly stenothermal. Not a single species was present throughout the year. *Chilomonas paramecium, Peranema* sp, *Colpidium campylum* and *Stylonychila* sp. belonged to warm stenothermal group, being present during summer. *Synura* sp., *Euglena* sp., *Phacus* spp., *Gonium* spp., *Vorticella* sp., *Euplotes* and a few unidentified ciliates and flagellates were cold sternothermals occurring from November to March. *Aspidisca* sp., *Polytoma sp and Pleodorina sp* avoided both bery high and very low temperatures and were recorded only during autumn. *Chilodonella* sp., *Coleps* sp. and *Lionotus* sp. were, however found to resist both high and low temperatures. When relationship between the protozoan abundance and the ambient temperatures was statistically analysed the two did not show any significant correlation.

In Anchar Lake water was alkaline and pH fluctuated within a arrow range of 7.70 to 8.84.

In Anchar Lake a significant positive relationship was found between hydrogen ion concentration and the protozoan population.

During the warmer season the concentration of oxygen in water was low. During the colder period from September to March, oxygen was always in a super saturated condition.

Phosphorus and nitrogen were in high concentration during winter. From autumn onwards volume of water in the lake shows a gradual decrease and the water level reaches the lowest ebb during winter when certain areas of the lake become transformed into marsh land and are inaccessible. However, the inflow of sewage from the catchment area continues throughout the year and decomposition of organic matter also continues round the year. As a consequence, phosphorus and nitrogen ion go on accumulating in water. Their removal by the aquatic plants during winter is very low because most of the plant community is dead ad even those which survive the cold winter show very low metabolic activity. The flagellates were more abundant during winter when the nutrient concentration in the lake was much higher than during summer. However, ciliates did not record an increase in their population in winter which is attribute to the fact that most of the protozoans belonging to ciliophora in the lake were warm stenothermal and therefore, in spite of high nutrient concentration during winter, their population density was limited.

Yousuf ad Gagroo (1990) concluded that the protozoa community in Anchar Lake, which is in a advanced stage of eutrophication, is well adapted to its environmet and the community structure and population abundance of this group of organisms is a good indication of the trophic nature of this water body.

SALAL RESERVIOR

The Salal reservoir, first man made medium reservoir in state is constructed by damming river Chenab-an important tributary of Indus drainage system, at Thanpal (Reasi) at an elevation of 496 mtrs. above MSL. A reservoir is an artificial water body formed by damming a river for the purpose of irrigation, power generation, drinking water supply and flood control. According to the area the reservoirs have been classified into 3 categories small (< 1000 ha), Medium (1000 – 5000 ha) and Big (> 5000 ha). The Salal reservoir has an area of 1000 ha and falls under medium reservoir with total catchment area 21479 sq. kms. The present paper deals with the physico-chemical conditions and indigenous fish fauna of Salal reservoir to facilitate formulation of various measures for judicious exploitation and further enhancement of yield from the reservoir. It has been observed that all the physico-chemical parameters are in the productive range. However, among the indigenous fish fauna it has been observed that the Mahseer- an important sport and food fish is dwindling due to construction of dam, which restrict the upward migration of the fish for breeding. There are two beautiful tributaries namely "Rud & Ans", which were the main breeding grounds for Mahseer; but construction of dam has hampered the migration of Mahseer to these tributaries. Hence effecting the breeding of fish. This fish can certainly be protected by adopting conservation and rehabilitation measures and if need felt Mahseer seed being produced at Mahseer hatchery Anji (Reasi), can be stocked in the different water bodies so as to conserve the migratory Mahseer "*Tor putitora*".

Table 6.3. Morphometry of Salal Reservoir

1.	Location:	
	(*a*) State	Jammu & Kashmir
	(*b*) Division	Jammu
	(*c*) District	Udhampur
	(*d*) Latitude	30°.08′N
	(*e*) Longitude	74°.50′E
2.	Dam Elevation	496 mtrs. Above MSL.
3.	River	Chenab
4.	Catchment area	21479 Sq. kms.
5.	Max. discharge	11895 Cums.
6.	Min. discharge	91 Cums
7.	Mean annual run off (at Dam site)	17315 million cu. Mtrs.
8.	Shore line	487.68 mtrs.
9.	Reservoir capacity at NRL	284 million Cu. Mtrs.
10.	Dead storage level	487.40 Cu. Mtrs.
11.	Full reservoir level	494 Cu. Mtrs.
12.	Sub merged area	935 sq. kms.
13.	Max. length of reservoir	40 kms.
14.	Water spread area	1000 ha
15.	Average Depth	36 mtrs.

The Salal reservoir is constructed by damming river Chenab at Thanpal (Reasi) in District Udhampur within the territorial jurisdiction of Jammu and Kashmir State. It is situated at an elevation of 496 mtrs. above MSL at latitude 30°.08′ N and longitude 74°.50′ E. The details of morphometric features of the Salal reservoir are furnished in Table 6.3.

Charak and Fayaz (2007) studied fisheries of Salal reservoir.The result of various physico-chemical parameters of reservoir are presented in Table 6.4. The air temperature ranged from 14.0° to 30.5° C. January is the coldest month of the year. From January onward, the day temperature increase progressively and reached to maximum in July. The water temperature ranged from 8.30 to 18.40°C; being maximum in June and minimum in January.

Table 6.4. Physico-chemical profile of Salal Reservoir

S.No.	Parameters	Range
1	Air temperature (°C)	14.00 – 30.50
2	Water temperature (°C)	8.30 – 18.40
3	pH	7.40 – 8.50
4	Dissolved Oxygen (mg/l)	7.20 – 10.30
5	Free Carbon Dioxide (mg/l)	0.80 – 1.45
6	Carbonate (mg/l)	0.00 – 12.0
7	Bicarbonate (mg/l)	99.4 – 216.50
8	Chloride (mg/l)	0.98 – 5.93
9	Calcium (mg/l)	14.30 – 22.80
10	Magnesium (mg/l)	2.40 – 9.96
11	Sulphate (mg/l)	11.62 – 34.56

The pH was in the range of 7.4 to 8.5. The minimum and maximum values of pH were recorded in the months of July and April respectively. This pH ranges clearly indicates the alkaline nature of the water.

The dissolved oxygen, which is an important chemical factor that reflects the health and productive status of a water body, ranged from 7.20 to 10.30 mg/l, being maximum in December and minimum in July. The slight increased in dissolved oxygen contents during winter months may be due to overall cumulative effects of low temperature and increase in transparency.

The availability of free carbon dioxide ranged from 0.80 to 1.45 mg/l *i.e.,* negligible, which is another indicator of suitability of water for fish production. The presence of free carbon dioxide in negligible quantity that too during summer shows poor rate of decomposition at the bottom.

The carbonate values were observed to be poor ranged from zero in summer to maximum of 12.0 mg/l in December, whereas, the bicarbonate values ranged from 99.4 mg/l to 216.50 mg/l being minimum in January and maximum in September.

The value of chloride contents were observed between 0.98 to 5.98 mg/l. As the chloride contents were very low in the reservoir reflects that there is very less amount of organic waste of animal origin. The calcium, magnesium and sulphate were observed within the range from 14.30 to 22.80 mg/l ; 2.10 to 9.96

mg/l and 11.62 to 34.56 mg/l respectively. On the basis of above observation, it may be assumed that the reservoir water is free from any kind of pollution.

Charak and Fayaz (2007) reported the occurrence of 13 species of phytoplankton and 9 species of zooplankton. The phytoplankton recorded from the reservoir belongs to chlorophyceae, bacillariophyceae and cynophyceae. Chlorophyceae was emerged as largest group and was represented by 7 genera *viz; Volvox, microspora, Uronema, Spyrogyra, Pediastrum, Closteridium and Treubaria.* Bacillarophyceae is represented by *Pinnularia, Navicula, Gomphonema, Synedra and tabellaria.* Where as the cynophyceae was represented by one genera *i.e., Oscillatoria* species. The phytoplankton population was found to increase during post monsoon and reached to maximum during winter. The decrease in phytoplankton population during the summer and monsoon as observed in Salal reservoir may be ascribed to drifting impact of rain during monsoon along with high turbidity and dilution of water.

Zooplankton of the reservoir was represented by 9 genera, belonging to two groups viz; rotifera and cladocera. The rotifers were represented by *Brachionus, Keratella, Anuraeopsis, Lepedella, Monostyla, Testidunella* and *Lecane* species; whereas the cladocerans were represented by *Macrothrix* and *Allona.* The member of cladocera recorded their peak during winter and minima during monsoon. The fall in numerical count of zooplankton during monsoon may be due to the drifting of these organisms under the impact of accelerated water currents resulted as a consequence of water influx from the catchment areas. Turbidity caused by surface run off and soil erosion due to rains too severely affect the zooplankton population.

In total 10 species of fishes *viz., Tor putitora, Tor tor, Schizothorax plagiostomus, S. progastus, Glyptothorax pectinopterus, Nemacheilus* spp., *Crossocheilus* spp., *Garra gotyla, Garra lamta and Barilius bendelisis* have been reported from the reservoir. The paucity of fish fauna in the reservoir may be due to mountain barrier separating the area from rest of the state. The fish yield from the reservoir is also low, which can be ascribed due to irrational use of fishing gears and large scale exploitation of brooders and juveniles in the past.

It has been observed that among the existing fish fauna, the Mahseer fishery in the reservoir is vanishing due to construction of dam, which affects the migratory behavior of the *Tor* Species, since there is no fish pass. The construction of dam has created a complete barrier for fish to migrate up stream. Earlier this fish use to ascend nullah Rud and Ans for breeding purpose and therefore suffered to a severe setback due to construction of dam. Earlier, the mahseer was found in abundance in the area, but now, the catches reveals that the mahseer catch is declining. However the previous authentic data on fish catch is not available, but the personal observation and local inquiries revealed that the fish population of the mahseer is dwindling gradually and is at the verge of extinction, if certain conservative steps are not taken. This fish can certainly be protected by adopting suitable conservation and rehabilitation measures and if need felt, mahseer seed is produced and raised up to the size of fingerlings can be released in the reservoir under river matching programme, so as to conserve the *Tor putitora.* For this purpose, the State Fisheries Department has well established mahseer hatchery at Anji (Reasi), where the artificial promotion of *Tor putitora* is undertaken by stripping method. In addition to propagation of mahseer, the strict conservation measures involve the check of illegal fishing of brooders and juveniles. Observance of close season if declared would facilitate further propagation of the fish. The propagation of mahseer seed on large scale and stocking of the same in the other selected areas will go a long way to increase the mahseer population.

Considering the urgent requirement to boost fish production from this reservoir, major thrust should be on stocking on a well designed approach. Stocking of quality fish seed on scientific line is another important tool in reservoir fisheries management. When the reservoir is formed, new trophic niches are created and indigenous fish faunas will not be able to utilize these niches and hence suitable species needs to be stocked. Such species should be introduced after thorough investigation of behaviour, adaptability, growth and reproduction, which can acclimatize in prevailing biotic and abiotic factors. The species should be non-competitive to indigenous species but should have fast growth rate. The stocking of reservoir with proper number and size of yearlings of cultivable indigenous fishes like rohu,catla,mrigal is the essential feature for the development of the reservoir fisheries. The stocking of yearlings of proper size and also with the desirable fishes to some extent would solve the problems of survival, so that the stocking stock became a part of natural fishery for ensuring the continuity of their kind. The yield from the reservoir can also be enhanced by adopting pen and cage culture. The fish species suitable for stocking are *Catla, Labeo rohita; L. calbasu; Cirrhina* sps. However, the exotic carps, common as well as Chinese *i.e., Cyprinus carpio* sps, (Common carp) *Ctenopharyngodon idella* (grass carp) and *Hypophthalmichthys molitrix* (Silver carp) have not found suitable for introduction.

To conclude, important measures such as strict vigilance to check capture of brooders and juveniles, closed season fishes and Judicious exploitation through use of desirable net are some of the important measures to ensure effective conservation and management of the reservoir for enhancing the fish yield. The management should be based on fish stock manipulation, though it is advisable to restore the selective stocking for correcting the imbalance, if any, in species composition. Regular stocking of the reservoir with fast growing, high yielding and adaptable species of yearlings for improving the density will definitely step up fish production. However, the precautions should be taken to select and stock such species which could cover all the niches of the reservoir and also to reorient the quality and quantity of the fish stock for better fish yield.

The reservoir being virgin since no exotic species has been added so far. It is recommended that the reservoir be exclusive developed for sport fisheries, where mahseer could be the main thrust area. There are hardly few water bodies left, where sport loving Angler's could get "thrilling mahseer" particularly "golden mahseer" *Tor putitora* to play with. This reservoir which is still a virgin reservoir with beautiful golden mahseer; if managed properly by adopting suitable conservation measures and management methods could become a beautiful spot for Angler's by way of putting sincere efforts. The mahseer which has got trapped in this man made water body and if able to acclimatize/establish in this new environment, and also in two beautiful streams; which were earlier also providing very good breeding grounds, could not only preserve the endangered mahseer but also develop the recreational fishery and enhance the fish yield. Besides, *Schizothorax sps,* another important fish for angling as well as the commercial value could also be established.

Table 6.5. Fish fauna of Salal reservoir

No.	Species
1.	*Tor putitora*
2.	*Tor tor*
3.	*Schizothorax plagiostomus*
4.	*Schizothorax progastus*
5.	*Glyptothorax pectinopterus*
6.	*Namaichelus* sps.
7.	*Crossocheilus* sps.
8.	*Garra gotyla gotyla*
9.	*Garra lamta*
10.	*Barilius benedilius*

Fig. 6.2. A panoramic view of Salal reservoir

Fig. 6.3. Another view of Salal reservoir

Fig. 6.4. View near dam site

RANJIT SAGAR RESERVOIR

Ranjit Sagar Reservoir with an installed 600 MW and 160 meter high earth core cum gravel shell dam is agigantic multipurpose river valley project. The project is named after Maharaja Ranjit Singh, the renowed ruler of Punjab. The reservoir is located in a gorge section near village Thein of Jammu and Kashmir and as such it is also known as Thein Reservoir. It is 24 km upstream of Madhopur Head Works. Ranjit Sagar Reservoir spreading in area of 8700 sq.km. falls in three states of Jammu and Kashmir, Punjab and Himachal Pradesh. This project is an embodiments of interstate relationship. The project was commissioned on

12th August, 2000 and was dedicated to the nation by the then Hon'ble Prime Minister of India on 4th March, 2001. The Ranjit Sagar Dam is the highest earth core-cum-gravel shell dam in India. The power plant has the second biggest Hydro-Turbine in India.

Table 6.6. Salient Features Ranjit Sagar Reservoir

Catchment area	6086 sq. km
Reservoir area	87.00 sq. km
Gross Storage capacity	3280 million cum
Live storage capacity	2344 million cum
Dam Type	Earth core-cum-gravel shell dam
Top level of the Dam	EL 540.00 m
Maximum height of dam	160.00 m
Length at Top of the dam	617.00 m
Width at top of the dam	14.00 m
Maximum width at base of the dam	669.2 m
Normal reservoir level	527.91 m
Clear water-way of spillway	109 m.
Crest level of spillway	EL 511 .7 m
Maximum outflow	24637 cumecs
Spillway design flood	20678 cumecs
No. of Penstock Headers	2
No. of Penstock Branches	4
Dia of each Penstock Header	8.5 m
Dia of each Penstock Branches	5.17 m
Type of turbines	Vertical Shaft Francis
Maximum net head	121.9 m
Minimum net head	76.0 m

National Fisheries Development Board, Hyderabad extended financial assistance to Fisheries Department, Government of Punjab for seed stocking of Indian Major Carps of 80-110 mm size in Ranjitsagar reservoir. The office has sanctioned Rs. 15 lakhs and released 50% as first installment *i.e.,* Rs. 7.50 lakh against Ranjitsagar reservoir for second consecutive year of stocking for the year 2011-12.

REFERENCES

- Ahmad Manzoor, M., 1980. Dal lake : A Rich Fishery Resources of Kashmir. India *Today and Tomorrow*. 8(4) : 184-186.
- Charak and Fayaz, F.A., 2006. Strategies for the development of fisheries in Jammu & Kashmir. *Fishing Chimes*. 26 (1) : 204-206.
- Charak and Fayaz, F.A., 2007. Preliminary study on status of fisheries in Salal reservoir and suggestive measures for development. In : Advances in Aquatic Ecology (Ed. V.B. Sakhare), pp. 124-131, Daya Publishing House, Delhi.
- Jan Ulfat, G.Mustafa Shah and Aijaz Ahmad Bhat, 2012. Haematological and biochemical indices of *Cyprinus carpio*, collected from Dal Lake, Kashmir. *Fishing Chimes*. 31(2) : 38-40.
- Kaloo, Z.A., Pandit, A.K. and Zutshi, D.P., 1995. Nutrient status and phytoplankton dynamics of Dal Lake under *Salvinia hatans* and obnoxious weed growth. *Oriental Science*. 1 : 74.85.
- Lijklema, L., 1994. Nutrient dynamics in shallow lakes : effects of changes in loading and role of sediment water interactions. *Hydrobiol*. 275 : 335-348.
- Mohan, K.S., 1991. Limnology of a mid lake station of lake Mir Alam Hyderabad. In: Current Trends in Limnology (Ed.K. Shastree), pp. 121-132, vol.1. Narendra Publication House, New Delhi.
- Murtaza, S, Malik Ashaf Aziz, Syed M. Fazil Ali and S.A. Hussain, 2010. Impact of pollutants on physico-chemical characteristics of Dal lake under temperature conditions of Kashmir www.forestynepal.org, pp 1-3.
- Pandit, A.K., Rather, S.A. and Bhat, S.A., 2001. Limnological features of Freshwater of Uri, Kashmir. *J.Res.* and *Dev.* pp. 22-29.
- Pandit, A.K. and Yousuf, A.R.,2002.Trophic status of Kashmir Himalayan lakes as depicted by water chemistry. *J.Res* and *Dev.* 2 : 1-12.
- Quadri Humaira and Yousuf, A.R., 2008. Dal Lake Ecosystem: Conservation Strategies and Problems. Proceeding of Taal 2007 : The 12th World Lake Conference (Editors: Sengupta, M. and Dalwani, R.) pp 1453-1457.
- Quadri, M.Y. and Yousuf, A.R.,1980.Limnological studies on Lake Malpur Sar. 1. The *Biotope Geobios* pp117-119.
- Yousuf, A.R. and Gagroo, Sunita, 1990. Ecology of protozoa in AcharLake, Kashmir. In : Contributions to the Fisheries of Inland Open Water sytams in India, Part I (Edited by A.G. Jhingran, V.K. Unnithan and Amitabha Ghosh), Inland Fisheries Society of India, pp52-57.

7

JHARKHAND

Fig. 7.1. Map of Jharkhand (Not to scale)

The 28th state of the Indian Union was brought into existence by the Bihar reorganization Act on 15th November 2000. Jharkhand is famous for its rich mineral resources like Uranium, Mica, Bauxite, Granite, Gold, Silver, Graphite, Magnetite, Dolomite, Fireclay, Quartz, Fieldspar, Coal (32% of India), Iron, Copper

(25% of India) etc. Forests and woodlands occupy more than 29% of the state which is amongst the highest in India.

Jharkhand is a landlocked state endowed with resources such as ponds, lakes, reservoirs, streams and rivers. Jharkhand state has an extensive wealth of lacustrine fishery resources, consisting of 99 reservoirs of various categories covering a water spread area of about 73,727 hecatres. Unfortunately, majority of these water bodies are not being scientifically managed. These reservoirs could be utilized for intensive and semi-intensive fish and giant freshwater prawn, and ornamental fish production, maintaining a capture-culture balance. The reservoir fish production in the state is only 5 kg/ha/yr. This unimpressed and low fish production potential may be due to low retrieval of the fish stocks and also may be due to uncontrolled poaching that takes place in most of the reservoirs (Singh and Ahmad, 2006).

The reservoirs in state are somewhat medium-productive and can be developed to produce substantial quantity of fish by resorting to proper stocking and manipulation of the various species of fish population and judicious exploitation through effective crafts and gears. The reservoirs in state, if managed on scientific lines, would surely boost fish production from the state.

Breeding of Indian major carps has not so far been noticed in the reservoirs of state. This seems to be for the reason that upstream of these reservoirs are having steeper slopes, rocky beds and are prone to flash floods. They do not provide favourable habitat for spawning of the desirable fishes that are economically important. Trash and predatory fishes are most self-generating and prolific breeders.

Gill nets fabricated using nylon monofilament/twine are the most common fishing gear used in the reservoirs. The most effective gear for catching majority of the fish species is, in fact a passive gear *i.e.,* surface gill nets of different mesh sizes. The gill nets are fixed in the evening and allowed to remain in the same position overnight and the fish catches are hauled in the morning.

The reservoirs of Jharkhand state can be classified into four distinct categories. There is only one large reservoir, namely Maithan Dam with water spread area of 11,491 ha.It is located in Dhanbad. The other large reservoirs, five in number, namely Canal Dam (10,000 ha) located in Santhal Paragana ,Panchet Dam (7,511 ha) in Dhanbad, Tilaiya Dam (6,475 ha) and Mayurakshi Dam (6,434 ha) in Hazaribagh and Tenughat Dam (6,000 ha) in Giridih District have a total water spread area of 36,720 ha. There are two medium sized reservoirs, namely Getalsud (3,500 ha) in Ranchi District ad Konar Dam (2,792 ha) in Hazaribagh District. So far as small reservoirs are concerned, together they constitute the bulk in terms of numbers (total area of small reservoirs: 15,000 ha).

The average fish production of Jharkhand reservoirs is in the order of less than 5 kg per hectare. This low fish production may be due to low retrieval of the fish stocks ad also may be due to uncontrolled poaching that takes place in most of the reservoirs. For the present an average production of 15 kg/ha/yr can be expected from the reservoirs of state (Singh and Ahmad, 2006). On this basis, production for a water spread area of 73,727 ha of reservoirs, comes to only 1100t/year.

Table 7.1. Important Reservoirs in Jharkhand

Sr.No.	Name	District	Area (ha)
1	Getalsud	Ranchi	3500
2	Hatia dam	Ranchi	176
3	Zatratu, karrah	Ranchi	300
4	Kalkari	Ranchi	992
5	Konar	Hazaribagh	2792
6	Mayurakshi	Hazaribagh	6734
7	Tilaiya	Hazaribagh	6457
8	Gonda dam	Hazaribagh	175
9	Tenughat	Giridih	6000
10	Canal	Duma and Jamtara	10,000
11	Masanjore	Duma and Jamtara	3846
12	Sitarampur	East Singhbhum	516
13	Nakti Jalasay	West Singhbhum & Saraikela Kharsawan	534
14	Maithan	Dhanbad	20360
15	Panchet	Dhanbad	7640
16	Nandini Irrigation	Lohardagga	305
17	Anraj	Garhwa	200
18	Balha	Garhwa	120
19	Batane	Palamau & Latehar	300
20	Maloya	Palamau & Latehar	250

The reservoirs in the state are somewhat medium productive and can be developed to produce substantial quantity of fish by resorting to proper stocking and manipulation of the various species of fish population and judicious exploitation through effective crafts and gears.

Breeding of the Indian major carps, rohu (*Labeo rohita*), Catla (*Catla catla*) and Mrigal (*Cirrhinus mrigala*) has not so far been noticed in the reservoirs of the state (Singh and Ahad, 2006). This seems to be for the reason that upstreams of these reservoirs are having steeper slopes, rocky beds ad are prone to flash floods. They do not provide favourable habitat for spawning of the desirable fishes that are economically important (Singh and Ahmad, 2006).

Gill nets fabricated from nylon monofilament are commonly used for fishing. The surface gill nets are usually fixed in the evening ad allowed to remain in the same position overnight and the fish catches are hauled in the morning.

MASANJORE RESERVOIR

The Masajore reservoir is one of the important man-made impoundment in Santhal Paragana in the state of Jharkhand. The dam was constructed across the Mayurakshi river at Masanjore. The reservoir is positioned at 23040′ and 25018′ North Latitude and 86°28′ and 87°57′ East Longitude. The Mayurakshi is a slow running river. It was constructed in year 1956 by the munificence and technical cooperation of Canada, hence also known as 'Canada dam'. The water spread area of the reservoir is about 71 sq.km. The catchment area of the reservoir is spread over 204 square kilometer.

Table 7.2. Meteorological conditions and water quality of Masanjore Reservoir

Sr.No.	Parameters	Range
1	Air temperature (°C)	8 – 35
2	Water temperature (°C)	19.2 – 30.6
3	Humidity (%)	32 – 97
4	Rainfall (mm)	Nil – 496
5	Wind Velocity (km/h)	0.9 – 9.1
6	pH	7.3 – 8.2
7	Dissolved oxygen (mg/*l*)	4.1 – 8.4
8	Free carbondioxide (mg/*l*)	Nil – 2.8
9	Carbonate alkalinity (mg/*l*)	3.0 – 3.5
10	Bicarbonate alkalinity (mg/*l*)	54 – 82
11	Chloride (mg/*l*)	32 – 64
12	Silicate (mg/*l*)	18 – 48
13	Phosphate (mg/*l*)	0.009 – 0.029
14	Nitrate (mg/*l*)	0.34 – 0.95

Phytoplakton diversity of the reservoir included *Cymbella turgid, Melosira graulata, Synedra* sp., *Nitzschia* sp., *Spirogyra* sp., *Scenedesmus* sp., *Cosmarium* sp., *Mougiotia* sp., *Phacus* sp., and *Ceratium* sp.

Zooplankton were represented by *Cyclops* sp., *diaptomus* sp., *Daphnia* sp., *Keratella* sp., *Brachionus* sp., *Asplachna* sp., *Polyarthra sp* and *Filinia* sp.

Indian major carps *(Catla catla, Labeo rohita, Labeo calbasu, Cirrhinus mrigala)* and common carp (*Cyprinus carpio*) were stocked since 1976. The average annual yield was recorded at 12.8 kg/ha. The details of fish landings are furnished in Table 7.3. The population of *Labeo calbasu, Cirrhinus mrigala* and *Cyprinus carpio* constituted more than 90 per cent of the commercial catches.

The common fishes recorded in the reservoir are *Catla catla, Labeo rohita, Labeo calbasu, Cirrhinus mrigala, Cyprinus carpio, Wallag attu, Channa* sp., and *Mystus seenghala.*

Table 7.3. Fish catches of Masanjore Reservoir during 1988-96

Year	Total catch (kg)	Yield (kg/ha)
1988-89	7215	23.5
1989-90	707	2.3
1990-91	2361	7.2
1991-92	132	0.4
1992-93	2957	9.6
1993-94	9634	31.4
1994-95	3918	12.8
1995-96	4504	14.7

REFERENCES

- Kumar Arvind and Verma, P.K., 2002. Ecological status of Masanjore reservoir in relation to fisheries management, Santhal Paragana (Jharkhand), India. In : Ecology and Conservation of Lakes, Reservoirs and River. (Editor : Arvind Kumar) Vol. 2, Ashish Publishing House, New Delhi, pp. 1-12.
- Singh, Arun Kumar and Ahmad Syed Hasib, 2006. Potential and prospects of fishery development in Jharkhand state. *Fishing Chimes*. 26 (1) : 192-203.

8

KARNATAKA

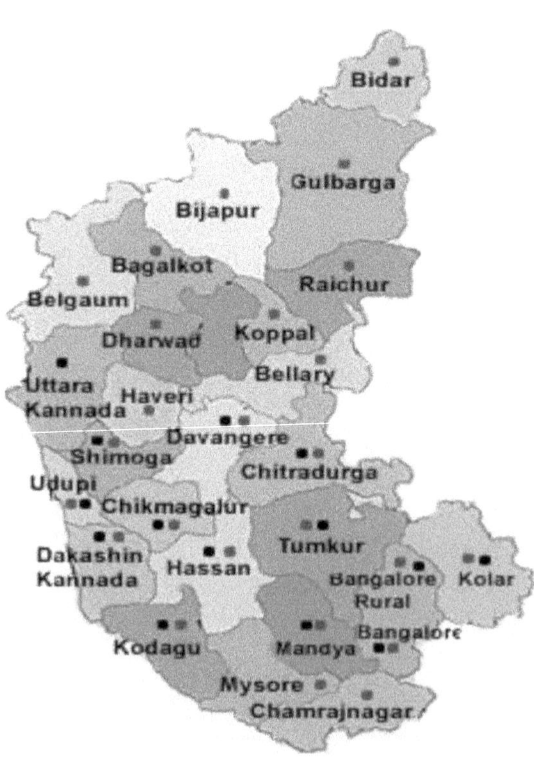

Fig. 8.1. Map of Karnataka (Not to Scale)

The state is bestowed with rich water resources. It has a coastline of 300 km, 27000 km^2 of continental shelf, 2.93 lakh ha of tanks, 2.27 lakh ha of reservoirs, 8000 ha of brackish waters, besides 9000 km of rivers and canals. Total of 74 reservoirs comprising 2.27 lakh ha of water spread area are available in the state for systematic and scientific development. Large reservoirs constitute 80% of the total area, followed by the medium (13%) and small (7%) ones. So far stray efforts were being made to stock available fingerlings in reservoirs here and there but there had been no regular programme to stock the reservoirs with advanced fingerlings on a scientific basis. The traditional fishermen around reservoirs are issued annual licenses to harvest the catch by collecting a nominal license fee.

Table 8.1. Few reservoirs in Karnataka state

Sr.No.	Name of Reservoir	Type of project	Name of river	Year of completion	Gross Command Area (Th ha)
1.	Amaraja	Medium	Amaraja	1973	10.53277
2.	Ambligola	Medium	Endigere	1964	4.34914
3.	Anjanpura	Medium	Kumudavathy	1936	10.52418
4.	Areshankar	Medium	Areshankar	1957	10.18912
5.	Bennithora	Major	Bennithora	NA	24.859
6.	Chittawadagi	Medium	Kadalappana	1971	0.89551
7.	Gundal	Medium	Gundal	1980	NA
8.	Hippargi	Major	Krishna	NA	114.22202
9.	Votehole	Medium	Votehole	NA	9.80846
10.	Vanivilas Sagar	Major	Vedavathy	1908	13.8102
11.	Tungabharda	Major	Tungabharda	NA	NA
12.	Theetha	Medium	Jayamangali	1987	2.06929
13.	Kabini	Major	Kabini	NA	149.65727
14.	Krishnarajasagar	Major	Cauvery	1931	201.15669
15.	Mangla	Medium	Naginihala	1970	2.94549
16.	Kanva	Medium	Kanva	1946	3.44428
17.	Bhadra	Major	Krishna	NA	NA
18.	Bachanki	Medium	Kopanla	1974	NA
19.	Markandeya	Major	Bhima	NA	32.83108
20.	Dudhganga	Major	Dudhganga	NA	24.47475

National Fisheries Development Board-assisted programme scientific development of fisheries of selected reservoirs has been taken up by stocking advanced fingerlings of prime cultivable species. Selected reservoirs are being developed by following public-private partnership and public-co-operative partnership approach. They are strengthen by providing financial assistance for advance fish fingerling stocking programme. It is proposed to enhance the per hectare fish production from the present 20-35 kg/ha to 150 kg/ha by scientific development of the selected reservoirs on a sustainable basis.

National Fisheries Development Board, Hyderabad extended financial assistance to Department of Fisheries for stocking Indian Major Carps fingerlings of 80-100 size for first year stocking programme for the year 2011-12 in 51 reservoirs of the state. For this, Rs. 74.665 lakh were sanctioned towards the cost of 80-100 mm size @ one rupee per fingerling that includes all costs towards production of fingerlings either *in-situ* or *ex-situ* and transportation, for stocking in 51 reservoirs for the year 2011-12. The information on some reservoirs in state is depicted in Table 8.1.

TUNGABHADRA RESERVOIR

This reservoir is the only place in India where reservoir –based pen culture is being carried out from year 1982 onwards for the rearing of carp seed. The reservoir has a water-spread area of 37,814 ha.

Growth and survival of rohu spawn in pens with varying stocking densities commissioned in the periphery of Tungabhadra reservoir was studied by Gireesha *et al* (2003) for 56 days. The pen chambers of 112.5 m^2 each were stocked with 4-day old rohu spawn in numbers based on 4, 5, 6, 7 and 8 million/ha, in two replicates. The soil nutrient characteristics such as NO_3^-N, P_2O_5, organic matter and pH were analysed at the start of the experiment; they were within favourable limits for carp seed rearing (Table 8.2).Weekly samples of water were taken and analysed for physico-chemical characteristics (Table 8.3) and growth of fry in terms of length. All the essential nursery practices were followed. It was observed that the growth was relatively higher at lower stocking densities. While there was no significant difference in growth among the stocking densities of 4, 5 and 6 million spawn/ha, higher average growth was recorded in these three trials compared to those with stocking rate of 7 and 8 million spawn/ha. An attempt was also made to find out the possibility of culturing common carp in pens by stocking six-month old stunted fingerlings (5.8 g) at a stocking density of 8000 nos/ha. Of the total harvested stock, 92% were seen to have attained maturity. The fish attained 95 g in about 2 months, with a survival rate of 88%.

Table 8.2. Soil characteristics recorded at the beginning of the study (Gireesha *et al*, 2003)

Sr.No.	Characteristic	Range
1.	NO_3-N (ppm)	0.132 – 0.382
2.	P_2O_5 (ppm)	1.08 – 2.05
3.	Organic matter (%)	0.23 – -0.46
4.	pH	7.1 – 9.9

Table 8.3. Range of physico-chemical parameters of Tugabhadra reservoir

Date	Days of culture	Air temperature	Water temperature	pH (ppm)	Free CO_2	Dissolved oxygen (ppm)
10/09/96	0	32.3	31.5	6.3 – 6.7	0.02	6.80
17/9/96	7	34.7	32.1	6.8 – 7.1	0.06	7.48
24/9/96	14	33.4	29.6	6.3 – 6.5	0.04	7.92
01/10/96	21	36.1	32.2	7.4 – 7.7	1.20	8.47
08/10/96	28	32.3	29.7	8.1 – 8.4	0.0	10.50
15/10/96	35	29.6	25.1	9.3 – 9.5	0.04	9.56
22/10/96	42	30.3	27.5	7.8 – 8.1	2.12	8.74
29/10/96	49	32.9	27.5	9.4 – 9.5	3.46	7.42
05/10/96	56	33.7	32.2	9.1 – 9.5	3.22	7.12

The zooplankton in the pens mostly consisted of rotifers (*Brachionus, Filinia, Conochilus, Asplanchna*), copepods (*Diaptomus* and *Cyclops* and their laraval stages), cladocerans (*Daphnia, Moina*) and ostracods (*Cypris*). The volume of zooplankton ranged between 1.5 and 3.5 ml/50 liter water during the period of investigation.

The data on growth and survival of rohu seed are given in Table 8.4 and 8.5. At the end of 56 days of rearing in pens, the growth in terms of average length (Table 8.4) and weight (Table 8.5) of rohu fingerlings showed a decreasing trend with increasing stocking density. The lowest stocking density (4 million/ha) recorded the highest growth (6.3 g; 75.22 mm),while the highest density (8 million/ ha) resulted in the lowest growth (4.36 g; 59.32 mm). Similarly, an inverse relationship between stocking density and survival was also observed, with 4 million/ha and 8 million/ha groups recording 29.51% and 20.51% survival respectively .The results of the trial on the growth of common carp are depicted in Table 5.The stunted common carp fingerlings could grow up to 95.0 g (22.5 cm) with a survival rate of 88 % in 56 days. Among them, 62.55% were females and 37.75% were males, of which 92% were mature. The computed production was 625 kg/ha/56 days. The percent increase in growth was maximum (317%) in the first 15 days, after which the increase slowed down.

David *et al.* (1969) have made certain observations on the available forage food of catfishes and several commercial carps in Tungabhadra reservoir. Similarly David and Rajagopal (1974) studied food and feeding relationships of *Puntius kolus, Puntius pulchellus, Labeo fimbriatus, O.vigorsii, Catla catla* and *Tor* spp., from Tungabhadra reservoir. *Puntius kolus* is euryphagic in feeding subsisting on molluscs, insects, ostracods, bacillariophyceae and on grass seeds and decaying plant tissues.

Table 8.4. Length (mm) of rohu recorded during the study (in pen replicates)

Stocking days		Duration (days)								
		0	7	14	21	28	35	42	49	56
4 million/ha	A1	6.92	15.10	24.44	35.53	453.18	52.04	60.53	69.10	77.24
	A2	6.92	14.66	23.93	34.71	44.70	50.08	57.91	65.26	72.90
	Average	6.92	14.88	24.18	35.12	43.94	51.06	59.22	67.18	75.21
5 million/ha	A1	6.92	15.20	23.41	32.48	41.90	49.53	60.40	68.10	76.98
	A2	6.92	14.72	23.91	31.76	40.54	51.52	57.56	64.38	72.86
	Average	6.92	14.96	23.66	32.12	41.22	50.55	58.90	66.24	74.92
6 miilion/ha	A1	6.92	14.95	22.63	30.48	39.27	48.06	58.51	67.26	75.65
	A2	6.92	14.57	22.21	29.82	38.41	49.96	56.79	65.20	72.82
	Average	6.92	14.76	22.42	30.15	38.84	47.51	57.66	66.21	74.26
7 million/ha	A1	6.92	13.65	19.73	27.29	34.76	40.88	48.96	55.52	62.87
	A2	6.92	14.11	20.31	20.31	27.99	35.60	42.24	50.80	66.81
	Average	6.92	13.88	20.02	27.64	35.18	41.56	49.88	57.15	64.84
8 million/ha	A1	6.92	13.21	19.35	25.62	32.55	40.37	46.93	53.23	60.59
	A2	6.92	12.87	18.97	25.04	31.77	39.11	44.99	51.10	58.05
	Average	6.92	13.04	19.16	25.33	32.16	39.74	45.96	52.12	59.32

Table 8.5. Weight (g) and survival (%) of rohu fingerlings recorded at the end of the experiment

		4 million/ha	5 million/ha	6 million/ha	7 million/ha	8 million/ha
Weight (g)	A1	6.93	6.47	6.46	4.57	4.49
	A2	5.67	5.33	5.24	5.31	4.23
	Average	6.30	5.90	5.85	4.94	4.36
Survival (%)	A1	28.23	23.18	23.25	20.62	19.38
	A2	30.79	26.24	24.64	23.84	21.64
	Average	29.51	24.71	23.45	22.23	20.51

Table 8.6. Growth of stunted common carp fingerlings

Days of culture	Length (cm)	% increase in length %	Weight (g)	% increase in weight
0	6.5	—	5.8	—
15	11.7	80.0	24.2	317.0
30	15.3	25.7	48.5	100.0
45	19.3	26.1	72.7	49.8
56	22.5	16.6	95.0	30.6

Puntius pulchellus and *Puntius sarana* are marginal submerged vegetative feeders subsisting on chara, hydrilla and vallisneria.During summer they feed on gastropods by necessity. Feeding habits are correlated with the available food items.*Labeo fimbriatus* is highly adapted to bottom browsing. Its stenophagic feeding on sessile diatoms indicates its selectivity in feeding. *O.vigorsii* is a column feeder preying upon smaller fish and insects. Though euryphagic in habits, it utilizes spirogyra, gastropods and incidentally diatoms, plant matter, ostracods, copepods etc. *Catla catla* is selective in feeding upon copepods,followed by diatoms and insects. *Tor* spp. (*Tor mussullah* and Tor *khudree*) feed upon molluscs, particularly gastropods followed by insects.

NELLIGUDDA RESERVOIR

Nelligudda reservoir was constructed in 1940 below the confluence of two seasonal streams (12°50′N, 77°22′E).Reservoir region, considered a dry zone, is characterized by low rainfall (annual ranges from 679 to 889 mm). The region receives both southwest and northeast showers with more than 50% being received between August and October. The Nelligudda reservoir is sheltered to some extent due to its location in Ramanagaram hill ranges. Direct human impact on the nutrient budget is still trivial. Allochthonous influences on the nutrient budget appear to be significant due to large unintercepted watershed. The morphometric and edaphic characteristics of Nelligudda reservoir are depicted in Table 8.7.

Table 8.7. Morphometric and edaphic factors of Nelligudda reservoir

MORPHOMETRY	
1. Year of construction	1940
2. Elevation (m MSL)	740
3. Maximum depth (m)	10
4. Area at FRL (ha)	80
5. Mean Depth (m)	3
6. Catchmet Area (ha)	7469
7. Shore line development	2.3
EDAPHIC FACTORS	
1. Water temperature (°C)	25.6
2. Transparency (cm)	51.3
3. Alkalinity (mg/l)	142.7
4. Conductivity (u.S.)	373.4
5. Sestonic carbon (mg^4)	5
6. Gross productivity (mg Cnr^2d^1)	2.4

Nelligudda reservoir is a warm dimictic lake with stable thermal stratification for two extended periods (September-October-and February-May) and with an intermittent period characterized by irregular circulation. Continuous mixing occurs during June-August. The water is well buffered, hard, moderately rich in electrolytes and low in transparency. Clinograde distribution of oxygen, high sestonic carbon and chlorophyll 'a' values and high primary production rates suggest that the reservoir is productive (Rao *et al.* 1999).

The reservoir supports fish diversity of 20 species belonging to 6 families, out of which family cyprinidae is dominant with 12 species followed by family bagridae, channidae and cobitidae each with 2 species, while the families cichlidae and siluridae were represented by 1 species each (Table 8.8). The fishery is significantly dominated by *Oreochromis mossambicus*, contributing over 80% to the fish catches. The major carnivores were *Channa gachua, Channa striatus, Ompak bimaculatus, Mystus cavasius* and *Mystus vittatus* that are known to be important predators of fish fry especially that of *Oreochromis mossambicus* may have a regulatory role on the population density of *Oreochrmis mossambicus.*

Tha Catch Per Unit Effort (CPUE) showed near identical seasonal variations, with a primary peak around July and a secondary one around March.CPUE was at its lowest level in winter ad could be due to dispersal of fish in the relatively high volume of water as well as poor activity patterns of fish. The CPUE remained high during major part of 1994 and the highest value of around 50 kg was recorded in June. After September 1994, a sharp decline occurred without further recovery. The CPUE was below 5 kg on most occasion in 1995 and the fishing intensity during this period was only around 0.25 kg/ha as compared to around one craft/ha during June-September 1994.

The monthly fish landings ranged from a low of around 200 kg/ha (May 1995) to a high of 30 t/ha (July 1994) consistently high catches were recorded during June-October 1994 followed by a sharp decline thereafter, the latter due to dwindling of stocks consequent to overexploitation in 1994, using small meshed gill nets. Estimation of annual fish production based on the absolute data on fish landings would be an overestimate due to severe overfishing.

The landing size of *O.mossambicus* was small ad around 40-100 g (7.5 to 15 cm).The size which was around 10-15 cm (mean : 12.7 cm) during June-September 1994 to 7.5 to 12.5 cm (mean: 10.4 cm) after January 1995.

Table 8.8. Fish fauna of Nelligudda Reservoir

Fish fauna
Family : Cyprinidae
1. *Labeo boggut*
2. *Cyprinus carpio*
3. *Amblypharyngodon mola*
4. *Danio aequipinnatua*
5. *Esomus danricus*
6. *Salmostoma belachi*
7. *Puntius dorsalis*
8. *Puntius sarana*
9. *Puntius sophore*
10. *Garra gotyla gotyla*
11. *Parluciosoma daniconius*
12. *Barilius bedalisis*
Family : Bagridae
13. *Mystus vittatus*
14. *Mystus cavasius*
Family Siluridae
15. *Ompak bimaculatus*
Family : Cichlidae
16. *Oreochrmis mossambicus*

Family : channidae
17. *Channa gachua*
18. *Channa striatus*
Family : Cobitidae
19. *Lepidocephalus thermalis*
20. *Nemachielus deninsoni*

The only craft used was coracle. Monofilament surface gill nets, ranging from around 25 to 50 mm mesh bar was the principal gear used for fishing. Each coracle unit consisted two fishermen and 5-7 kg of nets (10-14 pieces).

HEMAVATHY RESERVOIR

The reservoir was created in Hassan district on river Hemavathy,a tributary of the river Cauvery. This is one of the major reservoirs in the state covering a water spread area of 9,162 hectares. The fishermen operating in the reservoir are supposed to have been migrating from the neighbouring states of Maharashtra, Andhra Pradesh and Tamil Nadu. There are five fishing camps located around the Hemavathy reservoir *viz.*, Kerodi, Beejaghatta, Hulugunda, Beekanahally and Shettihally villages. Each camp consists of 20 to 35 families of fishermen and headed by a group leader. The leader normally take care of various activities of the group including obtaining fishing licenses and marketing of fish.Fishermen are usually engaged in fishing activities in groups during the peak season from May to January. During offseason, some of them work as agricultural labourers and a very few of them work in their own agricultural land. The study by Devaraj and Mahadeva (1990) revealed the following status: persons per family 2 to 5;male population among adults 57.14%; active fishermen 47.23%; the youngsters; women and children 42.15%; and persons above 51 years age 40.62%. Majority of them were illiterate and had no exposure to modern fishery practices. Literacy rate was as low as 15.7%. Not more than 10% of children attended schools.

Fisherwomen were found tobe engaged in household works and mending and making fishing nets. Many of them were indebted to money-lending fish merchants. Lack of transport facilities and preservation facilities prevented the fishermen undertaking their own fish sale business. Main reasons for the socio-economic backwardness of fishermen communities are the distribution of the communities in small clusters, inadequate number of nets and lack of proper understanding of the fishery.

The gears used in fishing are gill nets, drag nets, cast nets and long lines. The main craft used for fishing is coracle. The fishermen operated gears for about 265 days in year.

LINGANAMAKKI RESERVOIR

The reservoir was constructed across the river Sharavathi in 1964 for the purpose of hydel-power generation. The reservoir area receives more than 250 cm of rainfall annually. About 80% of this rainfall takes place during the monsoon months of June to August. The reservoir is oligotrophic in nature. The total capacity of the reservoir is 152 TMcft with area of submergence of 326.34 sq.km. at full reservoir level of 554.43 m. Presently about 120 fishermen families are engaged in fishing. All these families are poor, uneducated and the involvement of children in fishing is often noted. The Deputy Director of Fisheries regulates the fishery in Linganamakki reservoir. Obtaining yearly license is compulsory for all the fishermen for fishing in the reservoir. The fee structure for gill net with a length of 500 m is Rs. 1000/- per year; cast net Rs. 300/- per year and hooks Rs. 100/- per year. At present about 200 license holders are in the reservoir area. Thus, the fisheries department earns 1.65 lakh rupees of revenue annually from this source.

In order to avoid overflow and washing off of the seeds, the seed is stocked after the monsoon season. The year wise seed-stocking pattern from 1992-93 to 2001-02 is given in Table 8.9. The quantity of seed stocked depends on availability of the seed.

Table 8.9. Year-wise seed stocking in Linganamakki reservoir, Karnataka

Sr.No.	Year	Seeds in lakh			
		Catla	Rohu	Mrigal	Common carp
1.	1992-93	—	0.42	—	5.4
2.	1993-94	1.2	5.75	—	15.058
3.	1994-95	5.3345	7.0685	—	11.418
4.	1995-96	7.266	11.07	0.35	10.605
5.	1996-97	—	26.362	0.72	9.86
6.	1997-98	3	21.885	6.03	1.25
7.	1998-99	2.448	19.016	1.12	1.4
8.	1999-00	12.115	14.374	—	—
9.	2000-01	—	—	—	5
10.	2001-02	—	—	—	3

The fishermen co-operative society is located in Sagar tehsil of Shimoga district. The society has presently 412 members belonging to fishermen and fish merchants. The fishing activities in the reservoir are more or less under the control of merchants. The society's effort to take control over fish marketing is in vain as the fishermen sell fish to merchants instead of selling to the society. This kind of a system is governed by the mutual dependence of fishermen and merchants. From the beginning of the fishermen and merchants are closely linked due to the flexible credit and services. As fish catch is highly fluctuating over the season, fishermen borrow loans from the merchants for smooth running of their livelihood and the reimbursement is in the form of fish during monsoon season.

The merchants pay on an average of Rs. 22 per kg of fish. As per the estimate yield of the reservoir, the annual revenue obtained by the fishermen on spot becomes Rs.43, 84,990/-. Individually the permanent fishermen earn annually Rs.22042/-.

The reservoir fish production is about 610.76 kg/sq.km at full reservoir level. About 68.23% of the total fish landing is recorded in monsoon. The increase of fish catch during monsoon is due to the tapping of the breeding season. Usually, fishermen shift to shallow areas as the water level rises during the monsoon season. This indicates that they are tapping the breeding grounds of most of the fishes.

Table 8.10. Checklist of fishes from Linganamakki reservoir, Karnataka

Catla catla
Labeo rohita
Labeo fimbriatus
Cirrhinus mrigala
Cirrhinus fulungee
Cirrhinus reba
Cyprinus carpio
Oreochromis mossambicus
Amblypharyngodon mola
Aplochelius lineatus
Barilius canarensis
Chanda nama
Channa marulius
Clarias batrachus
Danio aequipinnatus
Garra gotyla stenorynchus
Glossogobius giuris
Mastacembelus armatus
Lepidocephalichthys thermalis
Mystus cavassius
Mystus keletius
Mystus malabaricus
Nemacheilus ruppelli
Ompak bimaculatus
Pseudoambassis ranga
Pseudeutropius atherenoides
Puntius dorsalis
Puntius fasciatus
Puntius kolus
Puntius narayani
Puntius parrah

Puntius ticto
Rasbora daniconius
Salmostoma boopis
Tor khudree
Tor mussullah
Wallago attu
Xenentedon cancila

Tilapia is slowly gaining its phase in reservoir. As per the local fishermen, the catch of tilapia is increasing over the years, the catch starts at the post monsoon period and during November, it dominates the entire catch. Due to the least demand for this fish in local markets, fishermen treat this fish as an unwanted catch. Scientifically, tilapia is regarded as a hardy, territorial and a powerful competitor in nature. Ecologically; these fishes have adverse effect on the fish diversity. The maximum weight that this fish can attain is 0.5 kg in this reservoir. *Tilapia* is regarded as a highly dominating species in the reservoir, which has the ability to overtake most of the species in competition for food and habitat. The presence of tilapia decreases the population of other fish species. Thus, it is beneficial to minimize its population through selective fishing.

The over exploitation of the reservoir plays a major role in extinction of the fish speies. The department deliberately issues the license to any fishermen. Many fishermen from Tungabhadra reservoir migrate to this reservoir during monsoon season and catch huge quantity of fish. This has adversely affected the livelihood of permanent local fishermen. Overexploitation of the reservoir for fishing has resulted in excessive mortality and reduction in effective population size of the fish.

BHADRA RESERVOIR

The Bhadra reservoir is situated near Lakkavalli village of Tarikere taluka in Chikmangalore district. It is situated at an elevation of 601m above MSL.The reservoir is located at an Latitude 13°42°0N and longitude 75°38°20E. It is a multipurpose project constructed for power generation and irrigation. The average rainfall at reservoir area varies from 117cm to 513 cm.The catchment area of reservoir is about 1968 sq.km. The gears commonly used in reservoir are gill net and cast net.

The reservoir supports fish diversity of 27 species belonging to 5 orders, out of which order cypriniformes is dominant with 12 species followed by order Notopteroidei and siluriformes each with 5 species, while the orders perciformes and cypriniformes were represented by 2 and 1 species respectively (Table 8.11).

Table 8.11. Fish fauna of Bhadra reservoir, Karnataka

Sr.No.	Species
1.	*Catla catla* (Bloch)
2.	*Labeo rohita* (Ham-Buchanan)
3.	*Labeo calbasu* (Ham-Buchanan)
4.	*Cirrhinus mrigala* (Ham-Buchanan)
5.	*Cyprinus carpio* (Linnaeus)
6.	*Gonoproktopterus kolus* (Sykes)
7.	*Cirrhinus fulungee* (Sykes)
8.	*Puntius chola* (Ham-Buchanan)
9.	*Puntius melanostigma* (Day)
10.	*Puntius ophiocephalus* (Raj)
11.	*Puntius phutunio* (Ham-Buchanan)
12.	*Osteobrama cotio cunma* (Day)
13.	*Glyptothorax trewarasae* (Hora)
14.	*Notopterus notopterus* (Pallas)
15.	*Barilius vagra* (Hamilton)
16.	*Salmostoma untrachi* (Day)
17.	*Garra kempi* (Hora)
18.	*Nemacheilus striatus* (Day)
19.	*Nemacheilus corical* (Hamilton)
20.	*Aorichthys aor* (Hamilton)
21.	*Mystus cavasius* (Hamilton)
22.	*Clarias batrachus* (Linnaeus)
23.	*Ambassis kopsii bleeker* (Hamilton)
24.	*Ompak bimaculatus* (Bloch)
25.	*Mastacembelus armatus* (Lacepede)
26.	*Rasbora rasbora* (Hamilton)
27.	*Xenentedon cancila* (Ham.)

Thirumala *et al* (2011) studied water quality of the Bhadra reservoir. The important water quality parameters of reservoir are depicted in the Table 8.12.

Table 8.12. Seasonal variations of physico-chemical parameters of Bhadra reservoir

Parameters	Rainy	Winter	Summer
Water temperature (^{0}C)	26.12	26.10	28.90
pH	7.65	7.47	7.34
Turbidity (NTU)	6.7	7.33	6.00
Electrical conductivity (¼m hos/cm)	71.60	59.38	71.57
Total solids (mg/l)	290.83	337.00	223.00
Total suspended solids (mg/l)	247.15	300.55	170.16
Total dissolved solids (mg/l)	43.67	36.44	39.45
Chloride (mg/l)	12.89	9.05	14.03
Total hardness (mg/l)	33.83	25.50	27.00
Total alkalinity (mg/l)	40.00	32.75	37.25
Nitrite (mg/l)	0.11	0.19	0.37
Ammonia (mg/l)	0.0050	0.0085	0.0230
Dissolved oxygen (mg/l)	7.01	7.18	6.24
Bio-chemical oxygen demand (mg/l)	0.55	2.75	1.00
Phosphate (mg/l)	0.0020	0.0010	0.0025
Calcium (mg/l)	6.86	5.70	5.30
Magnesium (mg/l)	6.36	4.44	6.56
Free carbon-dioxide (mg/l)	7.25	12.25	7.50
Chemical oxygen demand (mg/l)	1.77	2.65	3.32

TIGADI HARINALA

The Tigadi Hainala is a small reservoir in Bailhongi taluka of Belgaum district. The reservoir had a water spread area of 296 ha. The fishery rights of the reservoir were given to Navalgatti village of Bailhongi taluka from 2004-05 for a period of 5 years. The reservoir is regularly stocked with major carp fingerlings. The details of stocking of major carp fingerlings in the reservoir are as follows:

Year	Seed stocking (nos.)	Fish catch (mt)
2004-05	200,000	6.00
2005-06	500,000	10.00
2006-07	2,70,000	18.00

A fishermen's co-operative society was established in 2006 with a view to enhancing the social and economic development of the fisher population of the village and to enable the fisher to get the needed help from the government. Recently a proposal to develop fisheries of reservoir has been submitted to NFDB, Hyderabad.

MADDUR RESERVOIR

Maddur reservoir is situated in southern Karnataka at a height of 649.83 meters above sea level. It lies across 77°10'E Longitude and 12°40'N Latitude .Total water spread area of the reservoir is 4220 hecatres.The maximum depth of the reservoir is 7 meters with a mean depth of 4.90 meters.

Table 8.13. Physico-chemical parameters of the Maddur Reservoir, Karnataka

Sr.No.	Parameters	Range
1.	Air temperature (°C)	22 – 26
2.	Water temperature (°C)	21.20 – 25
3.	Colour	Clear
4.	Transparency (meters)	2.9 – 3.4
5.	Odour	Weedy
6.	Turbidity (NTU)	4.6 – 8.0
7.	pH	7.6 – 8.1
8.	Conductivity (µmhos/cm)	370 – 450
9.	Free Carbon-dioxide (mg/l)	Nil
10.	Dissolved oxygen (mg/l)	6.0 – 8.5
11.	Phenolphthalein alkaliity (mg/l)	25 – 35
12.	Total alkalinity (mg/l)	139 – 161
13.	Total acidity (mg/l)	10 – 18
14.	Total hardness (mg/l)	115 – 123
15.	Calcium (mg/l)	34 – 40
16.	Magnesium (mg/l)	6.06 – 12
17.	Chlorides (mg/l)	18.46 – 23.15
18.	Total dissolved solids (mg/l)	260 – 355
19.	Phosphates (mg/l)	0.016 – 0.021
20.	Nitrates (mg/l)	0.095 – 0.140
21.	Nitrites (mg/l)	0.074 – 0.083
22.	COD (mg/l)	33 – 48

Devaraju *et al.* (2005) studied relationship of physico-chemical conditions and growth of bacillariophyceae during winter in Maddur reservoir. Clear water, weedy odour, and high dissolved oxygen are indicators for abundance of diatoms in the reservoir. The range of physico-chemical parameters of the reservoir are shown in Table 8.13.

The diatoms recorded are depicted in Table 8.14. Altogether 24 diatoms belonging to 11 genera were recorded from this reservoir.

Table 8.14. Diatoms recorded in Maddur Reservoir

Pinnularia acrosphoria, P.cardinaculus, P.dolosa, P.simplex, Navicula vulpina kuetz, N.rhynchocephala kutez, N.viridula kutez, N.venezuelensis, Gomphonema sumatrense firckle, G.lanceolatum, G.gracile, G.andium freugella, Cymbella aspera, C.pusilla grum, C.aspera, C.cumbiiformis, C.osmanabadensis, Stauroneis anceps her, Synedra ulna, Nitzschia obtuse, Eunotia pectinalis, Pleurosigma hippocampus, Cocconeis placentula, Cyclotella catenata.

KELAVARAPALLI RESERVOIR

Fig. 8.2. Kelavarapalli Reservoir

A study by scientists of Central Inland Fisheries Research Institute (CIFRI), Bangalore, says that African catfish – *C. gariepinus* is negatively impacting the commercially important fishes in Kelavarapalli reservoir in Krishnagiri district. Though the Fisheries Department of states like Karnataka has banned the culture of African catfish, considered by many as a menace to the whole freshwater ecosystem, interestingly Tamil Nadu, where the Kelavarapalli reservoir is situated, has not banned it.This fish fetch higher price than tilapia as there is a market for it in certain parts of this State. They enter the reservoir from culture ponds adjacent to river South Pennar during rain and floods. But Nile Tilapia, a dominant invasive fish found in the reservoir, has a positive impact on Indian major carps such as *Catla catla, Labeo rohita* and *Cirrhinus mrigala.* Nile Tilapia, most abundant in the commercial catch and available throughout the year, still remains at limited levels.

The littoral areas of Kelavarapalli reservoir are full of nests of Nile Tilapia and they breed during South-West Monsoon (July-September). There is heavy demand for this fish, especially from poor local people, as it is affordable to the lowest income group in the area.

The scientists modelled the reservoir ecosystem designing its food web and tracing the energy flow incorporating all living organisms and their interactions in the food web. This is the most advanced way of quantifying the flows in ecosystems so that scientific management of reservoir ecosystem can be done for sustainable fisheries management.

The scientists used mass balance model to prove the validity of their findings. The study comes at a time when many in the fisheries sector have expressed their concern over the farming of invasive fish varieties, Nile Tilapia (*Oreochromis niloticus*), Mozambique Tilapia (*O. mossambicus*) and African catfish.

The study finds that this reservoir is relatively at a developmental stage or a young ecosystem which is vulnerable to disturbances (natural or anthropogenic). Though the production is high in young ecosystem, care should be taken not to exploit it indiscriminately. The study points to the need to take up some sort of conservation measure at least during the breeding period of Nile Tilapia.

KHAJI KOTNOOR RESERVOIR

The Khaji Kotnoor reservoir is a perennial reservoir located at near Gulbarga city, which is 22 km away from the Gulbarga University campus. It falls under 17°22'30'N latitude and 76°59'0'E longitude. The total cathment area of Khaji Kotnoor is 265.70 Sq.km and live storage capacity is 5.1784 mm^3 and gross storage of the reservoir is 6.2180 mm^3. This reservoir is used for drinking water and irrigation purpose. The maximum depth of reservoir is 9 meters. Rajashekhar *et al.*(2010) recorded 24 species of zooplankton. Out of which, 10 species belongs to rotifera, 6 species belongs to cladocera, 5 species belongs to copepoda and 3 species of ostracoda. Among zooplankton, particularly rotifera was the dominant group throughout the study period and highest count was recorded in the summer season while low incidence was observed in southwest monsoon season. There was a distinct seasonal fluctuations and composition of the zooplankton in the Khaji Kotnoor reservoir with productive (October to May), retardation (June to August) and recovery (September onwards) periods. The total zooplankton population was dominated by rotifera (41%), cladocera (28%), copepoda (23%) and ostracoda (8%) respectively. Among zooplankton, rotifera was the dominant group. The rotifera group was represented by 10 genera. The most dominant being *Brachionus species,* represented by 4 species *viz., Brachionus angularis, B. candatus, B. calyciflorus and B. rubens.* The others were, *Tricocera cylinderica, T. smiles, Lapadella ovalis, Lecane luna, Keratella tropica,* and *K. cochlearis.*The most common species occurring throughout the year were *Keratella tropica, Keratella cochlearis, Brachinus angularis, Trichocerca similis.* Maximum density of rotifera between 81 ind/l to 329 ind/l were recorded during October 2005 to September 2006. The highest numerical abundance of rotifera population was observed in the month of May (329 ind/l), while low density was observed in the month of

September 2006. The maximum density of rotifera noticed in summer season, while low incidence was recorded in northeast-monsoons season. Among the rotifers *Brachionus angularis, Tricocera cylinderica, Keratella tropica, and K. cochlearis* were dominant species. Cladocera group represented by 6 species *viz., Monia brachiata, Monia macrocopa, Daphnia carinata, Daphnia pluxes, Euryalona orientalis, Alona pulchella.* This group was second dominant group during the study period. The maximum density was observed in southwest monsoon season and northeast monsoon season, while low density was observed in summer season. Among the cladocera *Daphnia Pluex* and *Monia brachiata* were dominant species throughout the study period. The copepoda group is represented by 4 species *Viz., Mesocyclops lukarti, M. hyalinus, Paracyclops fimbriatus,* and *Neodiaptomus strigilipes.* High incidence of copepoda was encountered in southwest monsoon season and northeast monsoon season. The maximum density of cladocera was recorded in the month of October 2005 (223 ind/l), while low density was noticed in summer season in the month of April (39 ind/l). Ostracoda occupied fourth position of zooplankton and represented very low population diversity compared to other groups. This group represented by three species *viz., Hemicypris fossulate, Spirocypris* and *Hyocypris.* This group was also found abundantly in winter season followed by monsoon season during study period. The high number of individuals recorded in the month of December 2005 (58 ind/l) and low number was observed in the month of July 2006 (10 ind/l). The ostracoda was absent in the April and May months.

HARANGI RESERVOIR

It was constructed in year 1984 on river Harangi near Hudgur village of Kodagu district. The Harangi originates in the Pushpagiri Hills of Western Ghats in Kodagu, Karnataka. Heavy rainfall from the south-west monsoon is the source of water in the catchment area of Harangi river which is about 717 km^2. The length of the Harangi from its origin to the confluence with the Kaveri river is 50 km. The river Harangi joins the Kaveri near Kudige in Somwarpet taluka. The reservoir area is 1909 ha with mean depth of 12.6m.

Peak fishing in reservoir was during June, October and February-April. Six tones of fish were caught during March to October 1995 consisting of catla, rohu, mrigal and common carp. Catla grew well in reservoir. Sukumaran and Das (2005) observed thermal stratification during summer months. Strong oxycline was noticed in reservoir was pronounced with anoxic conditions of hypolimnion. The high organic load from the catchments received by Harangi is reflected in the rich soil organic carbon and available nitrogen. Sukumaran and Das (2005) placed the reservoir under productive category.

VANIVILAS SAGAR RESERVOIR

Vanivilas Sagar is a large irrigation reservoir in the Chitradurga district of Karnataka. For the construction of this dam royals of Mysore had to pledge royal jewellery due to shortage of money. Vanivilas was the name of youngest daughter of the then Maharaja of Mysore, that's why it was named as Vanivilas sagara.

The area of the reservoir is 8700 ha, but the actual mean water spread area is only about 2000 ha.The reservoir construction was completed in year 1908 and has been fished since the beginning. Local villagers developed an interest in the fishery from the 1960s onwards, and learned the necessary skills from the migrant fishermen. The dominant fishing unit is a coracle with 3-5 gill nets, operated by two persons. Virtually all fishing units are indebted to particular merchants or commission agents. The amount burrowed by fishing units is reflected in the prices paid by the merchants, who do not formally charge interest.

The reservoir fish resources are divided by their users into "local" fish, comprising mainly predators, and "government", *i.e.*, major carps which are stocked by the government but may also reproduce naturally. On an average, "local" fish dominated by *Notopterus sp, Mystus sp* and *Ompak bimaculatus* account for about 60% of total yield,while the majotr carps *Catla catla, Labeo rohita* and *Cyprinus carpio* contribute about 40%.

Growth of stocked carps (*Catla catla,Labeo rohita* and *Cyprinus carpio*) was extremely good, with estimated asymptotic lengths ranging from 115 to 130 cm.The average gear selection length was 38 cm, corresponding to a weight of about 800 g. The bulk of the catch comprised individuals of 40-80 cm length, estimated to e in their second and third year of life. Total mortality rates were estimated to be 1.2 to 1.6/year.

KARANJA RESERVOIR

The Karanja reservoir a major perennial reservoir of the district and located at Bhalki taluka of the Bidar district at 17°22'30"N Latitude and 76°59'0"E Longitude. It is created due to the construction of dam across the river Karanja, a tributary of Manjra River of Godavari system. It is a medium reservoir having water spread area of 5,673 ha with gross irrigation potential of 1, 62,818 hectares.

Kumar *et al.* (2013) studied ichthyofaunal diversity of Kranja reservoir and reported 64 fish species belonging to 37 genera, 16 families and 5 orders. The order cypriniformes was dominant with 31 fish species followed by order siluriformes 20, perciformes 10, osteoglossiformes 2, and synbranchiformes with one fish species. Although, 64 species were recorded, the cyprinidae was observed as the dominant family with 27 fish species followed by bagridae, 9 fish species.The distribution of fish species is quite variable because of geographical and hydrological conditions. Among the recorded fish species, the high abundance of fish species with maximum availability in number was *Catla catla* followed by, *Labeo rohita,Cirrhinus mrigal, Oreochromis mossambicus, Ctenopharyngodon idella, Cyprinus carpio communis , Cyprinus carpio nudus , Cyprinus carpio specularis, Sperata seenghala, Oreochromis niloticus, Hypophthalmichthys molitrix, Wallago attu, Clarias batrachus, Ompok bimaculatus, Sperata aor* were recorded in all the sites. Fish species such as *Labeo kontius, Osteobrama cotio, Salmostomabacaila, Gagata cenio, Glossogobius giuris, Cirrhinus reba, Garra gotyla, Silonia silonda, Parambassis ranga, Noemacheilus rupelli, Botia almorhae, Rasbora daniconius, Ambassis nama, Anabas testudineus, Puntius chola, Chanda ranga, Labeo gonius* were recorded in lesser number. Fish species *Gagata cenio, Glossogobius giuris, and Osteobrama cotio* reported at only one landing center.

Among cypriniformes, the cyprinidae represented with *Catla catla, Labeo rohita, L. calbasu, L.fimbriatus, L. bata, L. gonius, Cirrhinus mrigal, Cyprinus carpio nudus, C. carpio specularis, C. carpio communis, Ctenopharyngodon idella, Hypophthalmichthys molitrix,Puntius sarana, P. filamentosus, P. dobsoni, P. vittatus, P.ticto, Tor* sp., *Garra gotyla*. The Genus *Labeo* represented by 6 species and *Puntius* with 5 species.

Due to the illegal supply seed of *Clarias gariepinus*,it is recorded in very good numbers and continue to start dominating in the reservoir therefore. Immediate measures should be taken to prevent it from further entering into this system (Kumar *et al.* 2013). Considerable landing of *Oreochromis mossambicus* and *O. niloticus* were recorded in all the fish landing centres. Ecologically, these fishes have adverse effect on the indigenous fish diversity of the reservoir. Fishing rights, including exploitation, stocking and disposal by licensing of Karanja Reservoir, as a rule, is vested with the Karnataka State fisheries Department. In some areas, variety of fishing gear is employed for fishing. Gill nets, both surface and bottom set, are the most common. A variety of traps are employed for catching prawn, air breathing catfishes and murrels, while rod and line are sometimes employed to catch *Wallago attu, Ompak* spp. and *Mastacembelus* spp. Except the issuing of licenses from the State Fisheries Department and stocking of fish seeds, there is no monitoring and regulation of fishing in the Karanja Reservoir due to the inadequate staff and other required facilities (Kumar *et al.* 2013).

ANJANAPURA RESERVOIR

Anjanpura reservoir is situated in Anjanpura village of Shikaripur tehsil in Shimoga district. The reservoir was constructed across the river Kumudvathi in year 1936. It is situated at 14°7′60″N latitude and 75°22′0″E longitude. Water spread area of the reservoir is 698 ha.The climate of the reservoir site is moderately cool.

REFERENCES

- David, A., Ray, P., Govind, B.V., Rajagopal, K.V. and Banerjee, R.K., 1969. Limnology and fisheries of the Tungabhadra reservoir. Bull. No.13,Central Inland Fisheries Research Institute, Barrackpore.
- David, A. and Rajagopal, K.V., 1974. Food and feeding relationships of some commercial fishes of the Tungabhadra Reservoir. *Proceedings of the Indian National Science Academy* 41 : 61-74.
- Devaraju, T.M., Venkatesha, M.G., and Singh Surendra, 2005. Relationship of physico-chemical conditions and growth of bacillariophyceae during winter in Maddur Lake.In:fundamentals of Limnology (Ed. Prof. Arvind Kumar), APH Publishing Corporation, pp. 62-65.
- Gireesha, O.; Basavaraja, N.Tyagi Avinash C., and Shanmukha, S.N., 2003. Tungabhadra reservoir : Rearing of Rohu spawn and common carp fingerlings in Pens. *Fishing Chimes*. 23 (9) : 21-24.
- Kumar Naik A.S., Benakappa S., Somashekara S.R., Anjaneyappa H.N., Jitendra Kumar., Mahesh V., Srinivas H. Hulkoti and Rajanna K.B., 2013. Studies on Ichthyofaunal Diversity of Karanja Reservoir, Karnataka, India. *Int. Res. J. Environment Sci.*. 2(2), 38-43, Paper available online at www.isca.in.

- Lorenzen, K., Mohan, C.V. and Bhatta, R. (in prep.). Assessing the potential for culture based fisheries development : a case study of Vanivilas Sagar reservoir, India. To be submitted to Fisheries Research.
- Rajashekhar M., Vijaykumar K. and Zeba Paerveen,2010.Seasonal variations of zooplankton community in freshwater reservoir Gulbarga District, Karnataka, South India. International *Journal of Systems Biology*. 2(1) : 6-11.
- Rao, Krishan, D.S., Katre, Shakuntala, Jayaraj, E.G. and Reddy, S.R., 1999. Limology and status of the fishery of Nelligudda reservoir (Bangalore: India). *Indian J.Fish.* 46 (3) : 251-264.
- Sreekantha and Ramachandra, T.V., 2005. Fish diversity in Linganamakki reservoir Sharavathi river. *Eco.Env, & Cons.*11 (3-4) : 337-348.
- Sukumaran, P.K. and Das, A.K. 2005. Limnology and fish production efficiencies of selected reservoirs of Karantaka. *Indian J.Fish.* 52(1) : 47-53.
- Thirumala, S., Kiran, B.R. and Kantaraj, G.S., 2011. Fish diversity in relation to physico-chemical characteristics of Bhadra reservoir of Karnataka, India. *Advances in Applied Science Research.* 2 (5) : 34-47.
- Venkateshwarlu, M., Shanmugam, M. and Mallikarjun, H.R., 2002. The fish fauna of Bhadra reservoir, Western ghats. *J. Aqua. Biol.* 17 (1):9-11.
- Veerapa, Govda, H.S., 2009. New initiatives for the development of culture fisheries of Karnataka. *Fishing Chimes.* 29 (1) : 69.

9

KERALA

Fig. 9.1. Map of Kerala (Not to scale)

Kerala lies along the coastline, to the extreme south west of the Indian peninsula, flanked by the Arabian Sea on the west and the mountains of the Western Ghats on the east. This land of Parasurama stretches north-south along a coastline of 580 kms with a varying width of 35 to 120 kms. Cascading delicately down the hills to the coasts covered by verdant coconut groves, the topography and physical characteristics change distinctly from east to west. The nature of the terrain and its physical features, divides an east west cross section of the state into three distinct regions hills and valleys, midland and plains and the coastal region. Located between north latitudes 8018' and 12048′ and east longitudes 74052′ and 72022′, this land of eternal beauty encompasses 1.18 per cent of the country.

The backwaters are a peculiar feature of the state. Canals link the lakes and backwaters to facilitate an uninterrupted inland water navigation system from Thiruvananthapuram to Vadakara, a distance of 450 kms. The Vembanad lake stretching from Alappuzha to Kochi is the biggest water body in the state and is over 200 sq.kms in area. Kuttanad in Alappuzha district alone has more than 20 per cent of India's total length of waterways. The important rivers from north to south are Valapattanam river, Chaliar, Kadalundipuzha, Bharathapuzha, Chalakudy river, Periyar, Pamba, Achancoil and Kalladayar. Other than these, there are 35 more small rivers and rivulets flowing down from the Ghats.

Attempts to develop reservoir fisheries in Kerala started as early as 1960 under the Department of Fisheries. In 1992 the Indo-German Reservoir Fisheries Development Project (IGRFD) started activities with its head quarters at Malampuzha under an agreement between the Government of India and Government of Germany. In 1992 the orientation phase of the project started implementation during which the project focused on the development of Malampuzha, Pothundi, Chulliar, Peechi and Vazhani reservoirs. In 1995, the project started its implementation phase in five additional reservoirs *viz.* Mangalam, Meenkara, Peruvannamuzhi, Kanjirapuzha and Walayar.

The productivity of small reservoirs in Kerala was increased to a level of 53.5 kg/ha/yr, on an average. In individual reservoirs like Chulliar and Meenkara the productivity level has been raised to 227 kg/ha/yr, and 76.5 kg/ha/yr (1992-93 to 1995-96). This level of productivity is among the highest ever reported at the national level. This is an indication that if given proper technical and financial support, the production from small reservoir can be increased substantially so that the fishermen can undertake reservoir fishing as a viable avocation.

Non-availability of reservoirs for conducting fish culture is a major hurdle for development. Recently the Power Department has agreed in principle to let use the hydal power reservoirs for fishing purpose. KSEB have approved in principle to allow fish culture activities in the 11 (eleven) reservoirs under their control in an appropriate way by organizing co-operatives/SHG's of fishermen who are residing in the vicinity of those reservoirs.

Based on the water spread area the reservoirs are classified into 3 categories *viz.* small reservoirs (<1000 ha), medium reservoirs (above 1000 ha to 5000 ha) and large reservoirs (>5000 ha). The district wise important reservoir in Kerala are furnished in Table 9.1 and data on seed stocking in Table 9.2.

Table 9.1. Important Reservoirs in Kerala

Sr.No.	Name of the Reservoir	District	Area (ha)
1	Neyyar	Thiruvanathapuram	1500
2	Peppara	Thiruvanathapuram	582
3	Aruvikkara	Thiruvanathapuram	258
4	Thenmala Dam	Kollam	2590
5	Pamba	Pathanamthitta	570
6	Kakki	Pathanamthitta	1800
7	Maniyar	Pathanamthitta	110
8	Gani and Kallar Dam	Pathanamthitta	25
9	Ponmudi Dam	Idukki	260
10	Anayirankal Dam	Idukki	433
11	Gundala	Idukki	230
12	Mattupetty Dam	Idukki	324
13	Sengulam Dam	Idukki	33
14	Neriyamangalam Dam	Idukki	413
15	Periyar Lake	Idukki	2890
16	Edamalayar Dam	Idukki	350
17	Munnar Head Workers Dam	Idukki	250
18	Kallar Division Dam	Idukki	220
19	Erattayar Dam	Idukki	200
20	Mullaperiyar Dam	Idukki	400
21	Kulamavu Dam	Idukki	6160
22	Cheruthoni Dam	Idukki	—
23	Idukki Arch Dam	Idukki	—
24	Bhoothathakettu	Ernakulum	608
25	Peechi Dam	Ernakulum	1200
26	Vazani	Ernakulum	255
27	Sholayar	Thrissur	870
28	Peringalkuthu	Thrissur	280
29	Chimmini Dam	Thrissur	1000
30	Malampuzha	Palakkad	2313
31	Mangalam	Palakkad	393
32	Meenkara	Palakkad	259

Contd...

Table 9.1: Contd...

33	Chulliyar	Palakkad	159
34	Pothundi	Palakkad	363
35	Walayar	Palakkad	289
36	Parambikulam	Palakkad	2029
37	Thunakkadavu	Palakkad	283
38	Kanjirrampuza	Palakkad	512
39	Cheramangalam	Palakkad	200
40	Tharampilli	Palakkad	244
41	Kuttiyadi	Kozhikode	1052
42	Kakkayam	Kozhikode	1070
43	Peruvannamuzhi	Kozhikode	1050
44	Banasura Sagar	Wayanand	1277
45	Korapuzha	Wayanand	1660
46	Pazhassi	Kannur	648

Table 9.2. Details of seed stocking in Reservoirs of Kerala

Sr.No.	Name of reservoir	Date of stocking	Number of seed stocked		Total
			IMC	Scampi	
1.	Neyyar	09/01/12 27/02/12	—	—	560000
2.	Kullada	31/01/12 16/02/12	971500	560000	971500
3.	Pamba	08/03/12	990000	—	9900000
4.	Ponmudi	27/11/10	195000	—	195000
5.	Neriamangalam	27/11/10	310000	—	310000
6.	Chenkulam	27/11/10	30000	—	30000
7.	Mattuppetty	27/11/10	83000	130000	213000
8.	Kundala	27/11/10	113000	60000	173000
9.	Anayirangal	27/11/10	315000	40000	355000
10.	Idukki	25/02/12	—	2304994	2304994
11.	Periyar	16/03/12	—	460136	460136
12.	Bhoothathankettu	06/12/10 24/12/10	400000	—	400000
13.	Peringalkuthu	20/11/10	353000	—	353000
14.	Sholayar	16/03/12	510190		510190

Contd...

Table 9.2: Contd...

15.	Parambikulam	21/09/12	—	784610	784610
16.	Karappuzha	03/12/10 18/12/12	622250	—	622250
17.	Bhanasurasagar	01/02/11	945000	—	945000
18.	Pazhassi	10/01/12	486000	—	486000

CHULLIAR RESERVOIR

Chulliar reservoir is constructed across Chulliar river, 2 km upstream at is confluence with the Meenkara river, a tributary of Bharathapuzha. The reservoir is situated between Lat. 10°36′N and Long 76°46′E at an altitude of 136.55 m above mean sea level in Palakkad district of Kerala. The construction of dam was started in1961 and completed in 1970. The reservoir has a catchment area of 29.8 km^{2}. The capacity of the reservoir is 13.7 million m^3. The average depth is 8.3m. The shoreline length at FRL is 10,500 m. The waterspread area at FRL is 165 ha. The average water spread area is 159 ha.

With a view to exploiting diverse niches of this reservoir, stocking with quality fish seed was initiated by the state fisheries department as early as 1965-66. The regular fishing in the reservoir was started during 1967-68. Since 1989, stocking as well as fishing is carried out exclusively by SC/ST fishermen co-operative society, with the active support from the department and the Indo-German Reservoir Fisheries Development Project under operation from 1992.

Seed Stocking: The details of fish seed stocking in the reservoir for the period from 1985-86 to 1994-95 are furnished in Table 9.3. Stocking was high at 1580 nos/ha in 1985-86, 1231 nos/ha in 1986-87 and 1554 nos/ha in 1987-88. The species combination included Indian major acrps, common carp and minor carps. Over the period the size of the stocking material variations from 2.6 to 10 cm.

Table 9.3. Fish seed stocking in Chulliar reservoir during 1985-95

Year	Catla	Rohu	Mrigal	*L.calbasu*	*L. fimbriatus*	Common	Total carp	Stocking rate/ha
1985-86	—	15050	229749	—	156	—	244955	1580
1986-87	—	—	133000	—	57800	—	190800	1231
1987-88	—	27	168031	—	72817	—	240875	1554
1988-89	50000	25000	104020	—	13500	50000	242520	1564
1989-90	44000	25000	100000	7920	5580	50000	232500	1500
1990-91	46500	55283	120600	—	—	25000	247383	1596
1991-92	51500	16500	81000	—	—	—	149000	961
1992-93	64240	64000	32000	—	—	—	160240	1034
1993-94	80000	180500	80400	—	—	—	340900	2199
1994-95	11000	155200	20000	—	—	—	186200	1201

Fish Production: The details of fish production during 1985-95 are furnished in Table 9.4. The fish production was maximum during 1991-92 (297.941 kg/ha/yr) and minimum in 1986-87 (8.493 kg/ha/yr).

Table 9.4. Fish production in Chulliar reservoir

Year	Fish production (kg)	Production (kg/ha/yr)
1985-86	2075.150	13.388
1986-87	1316.550	8.493
1987-88	3666.590	23.655
1988-89	7573.800	48.863
1989-90	27207.550	175.532
1990-91	36209	233.609
1991-92	43391	297.941
1992-93	34656.150	223.588
1993-94	38753.550	250.022
1994-95	19869.600	128.190
Total	214719.390	138.528

Generation of Income and Employment: During 1985-95 a total of 214.72 tonnes of fish was harvested from the reservoir. The total revenue generated through the sale of fish at the rate of Rs. 25/kg is estimated to be Rs. 53.68 lakhs. Half of this camoing to Rs. 26.84 lakha was given to the fishing party consisting of 21 fulltime fishermen and was also utilized for other post-harvest operations.

Impact of Fishing on Rural Economy: When 1354 nos of quality carp seed per hectare costing Rs. 333.50 was harvested yielding a gross revenue of Rs.1884.25. The net surplus generated was to the tune of Rs. 1545.75 which appears tobe a fairly good yield.

During 1985-95, the average annual fish production of Chulliar reservoir was 130 kg/ha. On an average 21.47 tonnes of fish was contributed by the reservoir to the annual inland fish production of the state yielding estimated annual revenue of Rs. 5.37 lakhs. Out of this, Rs. 2.68 lakhs (50%) were set apart as share of active fishermen who belonged to the scheduled castes and scheduled tribes living below the poverty line. The average annual gross income for full time fishermen was estimated to be Rs. 1278/-.

Due to increased fish production from the reservoir, more quantity of fresh fish was made available to the rural consumers to meet their nutritional requirements. Surplus fish was transported to the distant urban markets which provided additional employment opportunities to the rural people by way of handling, packaging transport and marketing.

MALAMPUZHA RESERVOIR

Fig. 9.2. A panoramic view of Malampuzha reservoir in Palakkad district

Fig. 8.3. A fisherman with vehicle tyre tube

It is a multipurpose reservoir in Palakkad district. The water spread area of the reservoir is 2313 hectares. It is a multipurpose reservoir constructed across Bharatapuzha river. The local fishermen were organized into local Self-Helping Group (SHG). The reorganization into SHG has generated better cooperation in participatory management and virtually stopped the once rampant poaching in the reservoir. Revenues from prawn/fish catch for year 2005-06 was Rs. 1,475,000 which was almost equal to overall production value of the last 15 years.

The prawn PL stocked during September to January annually. Harvesting was done with large meshed (12,15 and 20 cm) gill nets. The fishermen use an indigenous device, an inflated vehicle tyre tube which is used as a float to sit on and pay out the gill nets. Both prawn and fish are caught by the nets. Fishing continues over the whole year but there is a lull in fishing during the rainy season.

The details of seed stocking and fish production are depicted in Tables 9.5 and 9.6 respectively.

Table 9.5. Seed stocking in Malampuzha Reservoir

Year	Number of seed stocked						
	Catla	Rohu	Mrigal	Common carp	Grass carp	Scampi	Total
2007-08	997250	1483950	539060	270130	—	10,00000	4290390
2008-09	345000	496780	31100	95000	71000	1100000	2138880
2009-10	526069	—	68652	—	—	—	594721
2010-11	300000	900000	1250000	—	177588	—	2627588
2011-12	410500	400000	425000	200000	2500	—	1438000

Table 9.6. Fish production in Malampuzha Reservoir

Year	Fisha production(kg)
2007-08	53093.950
2008-09	Not Available
2009-10	45237.200
2010-11	48056.06
2011-12	55517.500

Braj Mohan (2002) highlighted economic and social conditions of fishermen of Malampuzha reservoir. The fishermen are the members of Malampuzha SC/ ST Reservoir Fisheries Co-Operative Society. About 79.17% of respondents were literate. The scope of gainful employment in fisheries of Malampuzha reservoir is limited to roughly two months a year, *i.e.*, during the monsoon months of June and July. The per capita annual income from fisheries range from a mere Rs. 2100/- to Rs. 47, 250/-. One-fifth of the fishermen reported an income exceeding

Rs. 20, 000/annum. Nearly one-third of the fishermen reported annual earnings of less than Rs. 10, 000/-. This range of variation in earnings indicates unreliability of fishing as a dependable occupation. Fishing is only a minor activity, compared to other economic activities such as agriculture. Fishing in Malampuzha reservoir provide only 28% of the family income. Agriculture labour provides 37% of the family income, followed by general labour (21%) and these together provide more than 50% of the annual income. Incidentally 14% of the annual income comes from services. This 14% of the total income accrues to a handful of people employed in lower grade government service.

The families are extremely conservative in their household expenditure. The maximum household expenditure is Rs. 31,010/-. There were families that survived with an annual expenditure of less than Rs. 10,000. All the families showed a surplus of income over expenditure. Food expenditure constitutes 60 to 80% of the household expenditure.

All the respondents were active in the banking transactions. Marriage, house construction, acquisition of milk animals, cycle and fishing equipments constitute the major purposes for which bank loans are availed. 4.17% of the households colour television, telephone, motorcycle and other two-wheelers, which is an important indication of progressiveness of the fishermen of the Malampuzha reservoir.Fisheries of reservoir mainly comprise tilapia, that multiply in the reservoir, constituting about 75% of the present yield. The fish fauna of reservoir comprises *Oreochromis mossambicus, Etroplus suratensis, Puntius* spp., *Gonoproktopterus curmuca, Parambassis thomassi, Ompak bimaculatus, Catla catla, Cyprinus carpio,*and *Xenentedon cancila.*

MANGALAM RESERVOIR

The reservoir was constructed in Palakkad district with water spread area of 393 ha. The details of seed stocking are depicted in Table 9.7. Seed of *Macrobrachium rosenbergii* was stocked during 2007-08 along with the seed of Indian major carps. Common carp seed was stocked in year 2010-11 only. On an average basis,the reservoir was stocked with 8504 seed/ha/yr. The range of stocking ranged between the minimum of 866 (2011-12) to maximum of 14418 (2008-09) seed/ha/yr.

Table 9.7. Stocking in Mangalam Reservoir of Palakkad district

Year	Number of seed stocked					Total seed stocked	Stocking /ha
	Catla	Rohu	Mrigal	*M.rosen-bergii*	Commom carp		
2007-08	1692000	698900	899500	170000	Nil	3460400	8805
2008-09	1323380	2787944	1555000	Nil	Nil	5666324	14418
2009-10	2775000	949900	937000	Nil	Nil	4661900	11862
2010-11	937560	833910	641110	Nil	169000	2581580	6569
2011-12	300000	300000	Nil	Nil	Nil	600000	866

WALAYAR RESERVOIR

The reservoir was constructed in Palakkad district across the river Walayar, a tributary of Bharathapuzha.The construction of reservoir was started in 1953, partially commissioned in 1956 and completed in 1964. The dam in the river portion is of masonary gravity with a total length of 1478 m and the reservoir has a capacity of 18.40 mm^3 at +203 MSL. There is a road over the dam.

Fig. 9.4.

Table 9.8. Salient features of Walayar Reservoir

1. Type	Straight gravity masonary spillway section with earthen flanking dam
2. Purpose	Irrigation
3. Year of completion	1964
4. Catchment area	106.35 km^2
5. Water spread area	2.59 km^2
6. Gross storage capacity	18.9018 mcm
7. Live storage capacity	18.4 mcm
8. Dead storage capacity	1.35 mcm
9. Spillway length	42 m
10. Spillway gates	3 nos
11. Spillway gate size	12m × 3m

Table 9.9. Seed stocking in Walayar reservoir

Year	Number of seed stocked							Total
	Catla	Rohu	Mrigal	*Scampi*	Commom carp	Tilapia	Grass carp	
2007-08	317379	297171	480892	200000	1000	20,000	Nil	1316642
2008-09	150210	273129	356494	200000	Nil	Nil	25000	1004823
2009-10	497080	389920	284650	Nil	Nil	Nil	Nil	1171650
2010-11	50000	10000	100000	Nil	Nil	Nil	Nil	250000
2011-12	500000	700000	665385	Nil	Nil	Nil	Nil	1865385

IDUKKI RESERVOIR

17 species of fishes, including two exotics are reported from Idukki reservoir (Gopinath and Jayakrishnan, 1984). Presently there is organized fishery in the reservoir (Ansy Mathew and George, 2000). A few local people inhabiting along the reservoir margin are engaged in fishing practices. They use local fishing implements such as thandadivala or kethevala, which are crude gill-nets with mesh size varying from 26 to 180 mm. Gopinath and Jayakrishnan (1984) recorded fish catch of 705.97 kg during 11 months period. Six species that figured in the commercial landings were *Cyprinus carpio, Cirrhinus cirrhosa, Channa gachua, Mastacembelus guentheri, Tor khudree, Oreochromis mossambicus, Ompak bimaculatus* and *Heteropneustes fossilis.*

PEPPARA RESERVOIR

The Peppara reservoir is constructed across the river Kuramana at Latitude 8°30′N and Longitude 77°8′E in the Thiruvananthapuram district. The reservoir has a water spread area of 8.50 km^2 at full reservoir level at an elevation of 112.50 m above mean sea level. The catchment area spreads about 86 km^2 with an average annual rainfall of 481 cm. Besides the main river, three other perennial streams (Todai-ar, Attai-ar and Kavi-ar) empty into the reservoir.

The limnological studies by Thomas and Azis (2000) in the Peppara reservoir revealed that the reservoir water is an ideal site for aquaculture and fishery development.

The air temperature fluctuated from 21.5°C in July to 31.5°C in January. The maximum water temperature recorded was 31.5°C (March) for the surface and bottom waters. The low water temperature noticed during the monsoon months was due to the monsoon rain, cloudy sky and cold weather. The water temperature exhibited spatial and temporal variations.

pH in the reservoir varied from 5.10 to 7.53. The surface water pH was generally higher than that of bottom water at all the stations during the monsoon period. The higher concentration of carbon dioxide might have increased the hydrogen ion concentration which in turn lowers the pH of bottom layer. pH was acidic during the monsoon and the post-monsoon periods. This can be attributed to heavy rainfall and the consequent runoff experienced in the reservoir catchment area.

Higher concentrations of free carbon dioxide were invariably noticed in the bottom water. Free carbon dioxide in the surface water ranged from 4 to 20 mg/l, whereas those of the bottom water varied from 4 to 44 mg/l.

Total alkalinity showed only negligible seasonal and vertical variations. Values for the surface water ranged from 8 to 16 mg/l and those of the bottom water ranged from 6.14 to 18 mg/l.

The range of dissolved oxygen for the surface water was 5.84 to 9.56 mg/l and that of bottom water was 2.05 to 9.62 mg/l. Dissolved oxygen showed high concentrations in the surface waters. Depletion of oxygen was noticed in the bottom layer presumably due to the decomposition of submerged vegetation in the impounded area.

AKKULAM-VELI LAKE

The Akkulum-Veli Lake is situated about 5 kms north west of Thiruvananthapuram between Latitudes 8.25′ and 8.35′ and Longitudes 76.50′ ad 76.58′ E, the lake is having an area of <1 km surrounded by lateritic hillocks. Serious environmental degradation is being experienced by this system due to municipal waste disposal, eutrophication, excessive tourism load, effluent discharge, developmental activities etc. Two canals, viz. the Kulathur canal and Parvathy puthenar join the veli lake in the northern side. The Channankara canal connects the veli lakr with Kadinamkulam kayal in the north. Seepage of sewage from Muttathara sewage makes the water extremely polluted. Kannamoola canal joins the eastern part of the Akkulam lake. Sewage from the Thiruvananthapuram city and drainage from the suburbans are brought into the lake through the Kannammoola canal. In the vicinity of the lake there are two factories. The English Clays Limited and Travancore Titanium Products Limited. The clay factory discharge and its effluents directly to the lake, while the effluent discharged from Travancore Titanium Products Limited to the sea find its way to the lake when the river mouth remains open.

Pushpangadan and Nair (2012) reported bacterial load and species composition of bacteria in Akkulam Veli lake. Both the species composition and total bacterial load was found to be considerably affected by the type of pollution.

SASTHAMCOTTA LAKE

Sasthamcotta Lake is the biggest freshwater lake of Kerala which extends to about 3.42 sq.km. It forms the major source of drinking water for people of Kollam Municipal Area and its surrounding villages under the control of Kerala Water Authority. The lake is famous for fishes like *Etroplus suratensis, Etroplus maculates, Ambassis dayi* and *Horabagarus brachysoma*. Recently, the quality of lake water is threatened by the activities of local inhabitants by way of waste discharge, bathing, washing and agricultural practices.

Table 9.10. Water quality of Sasthomcotta Lake in relation to indicator parameters of soap-detergent-fertilizer pollution and coliforms during September 1990 to February 1991

1. pH	6.6 to 6.8
2. Total alkalinity ($CaCO_3$ mg/l)	11.0 to 16
3. Total Hardness (mg/l $CaCO_3$)	5.8 to 9.0
4. Calcium Hardness (mg/l $CaCO_3$)	4.2 to 6.3
5. Total phosphorus (mg/l PO_4/l)	0.32 to 0.96
6. Total Organic Nitrogen (mg/l)	0.22 to 0.32
7. Nitrate (mg/l)	0.11 to 0.14
8. Sulphate (mg/l)	0.09 to 0.12
9. Sodium (μg/l)	0.60 to 0.84
10. Potassoium (μg/l)	<1.5
11. Colforms (MPN/100 ml)	1882 to 2090
12. Faecal colifroms (MPN/100 ml)	1370 to 1750

Prakasam and Joseph (2000) studied water quality of Sasthamcotta lake in relation to primary productivity and pollution from anthropogenic sources (Table 9.10). The surface water temperature varied from 27.8°C to 31.2°C indicating narrow fluctuation (Prakasam and Joseph, 1991). The average transparency values during rainy and post rainy seasons were 134.4 and 139.7 cm respectively. Water pH was acidic and the total alkalinity ranged from 10.1 to 14.55 $CaCO_3$ mg/l. Surface water contained low amounts of free CO_2 but dissolved oxygen was moderately high throughout the investigation. Coductivity was also low indicating the scarce amounts of dissolved salts. Phosphate content was slightly high and comparatively greater during rainy season than in the post rainy season. Nitrates were absent from March to April, but showed high values during June to September. The gross primary productivity of the lake was 0.31 0.99 gc/m^3/day. During the rainy season there was a slight increase in productivity. The total hardness of the water was low and may be classified as soft water (Praksam and Joseph, 2000). The total phosphorus concentration was not high as could be expected of a lake subjected to detergent pollution. Similarly sulphate concentration and level of potassium were also low. Pollutants from bathing, washing and cultivation were not making any significant impact on the water quality of the lake.Microbial analysis showed that coliforms, especially faecal ones were present to a level of about 1500 (MPN) per 100 ml of water sample, which indicated faecal contamination of water.

Thus, it can be concluded that the water characteristics of the Sasthamcotta lake were not favourable for high productivity and that the lake water was not polluted except for faecal contanmination.The lake carried better quality of drinking water.

PERIYAR LAKE

The Periyar River which originates from Chokkampetty with its manifold tributaries drains the Sanctuary area The river Mullayar which originates at Kottamalai, joins Periyar at Mullakudy and forms the Mullaperiyar River. The Periyar Lake had been formed by constructing a dam across the Mullaperiyar River by the erstwhile Madras Government in 1895. The lake is at an elevation of 900 m. The water spread area of the lake is 26 km^2 with depth ranging from 46 m to 32 m. Portions on the western side of the plateau are drained by rivers Pamha and Azhutha, which form the boundary of the reserve. Rocks are crystalline and of plutonic origin. The average rainfall of the area is 2500 mm. The area receives both South-West and North-East monsoons. The maximum precipitation recorded was in the month of July and minimum in January.

Minimol (2001) documented fishery management of Periyar lake.There were fortyone species of fishes recorded in the Periyar Lake during the study. These belonged to 5 different orders, 11 families and 24 genera. Of these 2 are endemic to the Periyar. *Oreochromis mossambica, Hypselobarbits kurali, Cyprinus carpio* and *Tor khudree* were the four important species of fishes caught from the Periyar lake. Other species which were caught in smaller numbers along with these included *Hypselobarbus periyarensis, Heteropneustes fossilis, Mastacembelus armatus* and *Labeo fimbriatus.*

The fishermen of Periyar Lake, comprise two tribes, 'The Mannans' and 'The Paliyans'. Of these, the Mannans are traditionally fishermen while the Paliyans collected honey and other minor forest produce. The Forest Department has permitted only the tribes for fishing in the lake. A co-operative Society for fish marketing was established at Kumily in 1984 and each tribal fisherman is a member. In the beginning, the Forest Department granted permission to only those members of the society who had identity cards. On every Friday, the Wildlife Preservation Officer attested their pass to catch fish. The system had been continuing up to 1986. However the Forest Department gradually lost interest and the process of issuing passes came to a stand still alter 1986. This has resulted in the intrusion of non tribals into fishing in the lake.

The nets used are Kooral net (mesh size 4 cm × 5 cm), Vaika net (mesh size 11 cm × 11 cm), 1/2 Number (Mesh size 6 cm × 6 cm), 1 Number (Meshsize 10 cm × 10 cm), (5) 2 Number (Mesh size 14 cm × 14 cm), 3 Number (16 cm × 16 cm), and (7) Cast net (Mesh size 2 cm × 2 cm). ½ number net is for catching small sized fishes mostly *Oreochromis mossambica.* Kooral net and 1 number are for Hypselobarbus kurali and *Oreochromis mossambica.*, 2 number is for *Cyprinus carpio* and 3 number is for big fishes, mostly *Tor khudree.* According to the tribals the best tackle for *Tor khudree* is hook and line. Besides the cast nets, the gill nets are also used.

REFERENCES

- Braj Mohan, 2002. Economic status of the fishermen of Malampuzha reservoir. In: Riverine and Reservoir Fisheries of India (Proc. of the Nat. Seminar on Riverine and Reservoir Fisheries-Challenges and Strategies). Editors: Boopendranath, M.R.,

Meenakumari, B., Joseph, J., Sanakr, T.V., Pravin, P. and Edwin, L.), p. 421-425. Society of Fisheries Technologists (India), Cochin.

- Gopinath, P. and Jayakrishnan, T.N., 1984. *Fishery Technology*. 21(2), 131.
- Mathew, Ansy N.P. and George, J.P. 2002. Reservoir Fisheries: Post-independent scenario and strategies for development. In: Riverine and Reservoir Fisheries of India (Proc. of the Nat. Seminar on Riverine and Reservoir Fisheries-Challenges and Strategies). Editors: Boopendranath, M.R., Meenakumari, B., Joseph, J., Sanakr, T.V., Pravin, P. and Edwin, L.), pp. 415-.420. Society of Fisheries Technologists (India), Cochin.
- Minimol, K.C., 2001. Fishery Management of Periyar Lake, Thesis submitted to Mahatma Gandhi University in partial fulfillment of the requirements for the degree of Ph.D. in Zoology.
- Prakasam, V.R. and Joseph, M.L., 1991. Oxygen, carbon dioxide and temperature cycle in lake Sasthamcotta (Kerala) with a note on biological productivity. In: Current Trends in Fish, Fishery Biology and Aquatic Ecology (Eds. A.K. Yousuf, M. Raia and M.Y.Quadry), University of Kashmir, Srinagar, India.
- Prakasam, V.R. and Joseph, M.L., 2000. Water quality of Sasthamcotta lake, Kerala (India) in relation to primary productivity and pollution from anthropogenic sources. *J. Environ. Biol.* 21(4):305-307.
- Pushpangandan Sini Bhadrasenan and C. Radhakrishnan Nair., 2012. Impact of environmental pollutants on microbial load (pathogens) in Akkulam–Veli Lake. *Poll. Res.* 31(1) : 107-109.
- Thomas, Sabu and Azis Abdul, P.K., 2000. Physico-chemical limnology of a tropical reservoir in Kerala, S. India. *Ecol. Env. & Cons.* 6 (2) : 159-162.
- Velayudhan, T.D. and Sidharthan, T., 1998. Influence of stocking on fish yield in Chulliar reservoir and its impact on rural economy. *Fishing Chimes*. 18 (4) : 17-20.

10

MADHYA PRADESH

In Madhya Pradesh there are many freshwater wetland areas in the form of lakes and man-made reservoirs. The reservoirs are constructed primarily for flood control, conservation of rainwater, irrigation, power generation and water supply to cities and industries. Fishing development in these water bodies is considered as a secondary activity. The responsibility of fishery development in reservoir has been assigned to the state fisheries departments. The state has 3 lakh ha water spread area in the form of reservoirs, and ponds, out of which 2.50 lakh ha area is I the form of reservoir, and 0.50 lakh ha in in the form of village ponds. At present fish culture activities are carried out in 24048 reservoirs/ village ponds in 2.90 lakh g area. Five major river systems namely Narmada, Tapti, Ganga, Mahi and Godavari also pass from the state and have a network of 17088 km.The average fish production from reservoirs of the state is 53 kg/ ha. Production from village ponds is reported at 1375 hg/ha. Total fish production in the state from all sources is 62060 tonnes/yr (44787 tonnes from village ponds, 10517 tons from reservoirs and 6756 tonnes from rivers).

About 75% of the total production from the reservoir fishery of state is marketed outside the state. Within the state the sale of fish is confined to the near by areas of reservoirs and cities like Bhopal and Jabalpur. The profit margin from local sales is more than that of outstation sales. This is because the marketing of the catch in distant cities like Delhi and Kolkata involves additional packing and transportation charges. Besidies the prices in out station markets are fully controlled by intermediaries like the wholesalers/commission agents over when the producer (Federations) has no control. Hence there is a need to develop a strong chain of retail outlets for fish sale inside the state. This will enhance the returns of fishermen and will provide the population a cheap source of protein at affordable prices. For this an initial survey has to be undertaken in the state of Madhya Pradesh to assess the consumer preference and local demand.

Fig. 10.1. Map of Madhya Pradesh (Not to Scale)

TAWA RESERVOIR

The Tawa is a multipurpose man-made reservoir used for irrigation, hydel power generation and for fishing activity. It is situated on the reservoir Tawa in Husangabad district. Tawa river, a left bank tributary originates from Mahadeo hills in Chindwara district flows through Betul and joins Narmada in Husangabad district. It is the longest tributary of the river Narmada on left bank. The reservoir is located near Ranipur village, 35 km from Itarsi railway junction. The dam is positioned at Latitude of 22°30′40″ N a Longitude of 77°56′30″ E. The catchment ara of the project is 5982.9 km^2. The construction of dam was initiated in 1956 and completed in 1974. The reservoir was filled to its full capacity, only in 1979. It was transferred to Madhya Pradesh State Fisheries Development Corporation, which was continued till 1994.

The population around the reservoir mainly consists tribals (Gondas and Korus). Apart from tribals traditional fishermen communities such as Kahars and Dimmers are also common. The people who came to work as labourer during construction of dam, and later on settled around Tawa reservoir. The villages around the reservoir are involved mainly in fishing activities.

Table 10.1. Basic Features of Tawa Reservoir

Reservoir	Tawa
River	Tawa on The Narmada
District(s)	Hoshangabad
Number of Displaced Villages	44
Reservoir Area in ha (At full tank level)	20 050
Reservoir Area in ha (At minimum level)	4 240
Average Reservoir Area in ha	12 145
First year of Fishing	1979
Management regimes of Fishing	Fisheries Department (1975-79) MPFDC (1979-94) Contractor (1994-95) Free Fishing (1995-96) Cooperative Federation (1996 onwards)
Average Productivity* (1990 to 1995)	10.60

After a long struggle, the displaced tribal of Tawa reservoir got the fishing rights in the reservoir on co-operative basis. Thirty-three primary fishermen of co-operative societies of displaced tribals, five primary societies of traditional fishermen and their federation (Tawa Displaced Tribal Fish Production and Marketing Co-operative Federation, briefly known as Tawa Matsya Sangh). TMS registered under Madhya Pradesh Co-operative Act 1960. In the initial agreement (1996) TMS got exclusive fishing rights for five years, which was further extended in 2001 for another five years. But the fishing rights of the local tribal and the TMS regime is presently under threat because of project tiger, where there are talks to amalgamate the surrounding Bori and Panchmarhi sanctuaries and the Satpura National Park. Presently the TMS has 34 primary co-operatives and 6 affiliated co-operativies under it. The membership of the fishermen co-operative society is restricted to the project affected people and those residing within a radius of 3 km from the periphery of the reservoir. The responsibility to manage the fish resources including procurement of catch, transportation, marketing and stocking of fingerlings in the reservoir lies with TMS. TMS through co-operation of the villages are managing the fish resource and preventing illegal fishing. TMC successfully marketed the fish at local level as well as in distant markets of Howrah, Delhi, Lukhnow, Jabalpur, Bhopal etc. Along with increasing fish production, fish seed were also stocked and fish conservation rules for the reservoir were also observed properly. The reservoir fall will below the stocking target. This is mainly because of the non-availability of sufficient number of good quality fish seed at the right time. The average stocking density of the reservoir is 215 no/ha.

Indian major carp seed is stocked in the reservoir. The present regime in Tawa is stocking fingerlings at an average rate of 45.33% of catla, 28.83% of rohu and about 25.64% of mrigal. In other terms, it amounts to an average of 28 lakh fingerlings stocking/year, and an average of 236 fingerlings/hectare (Table 10.2).

Table 10.2. Seed stocking in Tawa Reservoir

Year	Catla%	Rohu%	Mrigal%	Total	Stocking/ha
1997-98	52.75	18.4	28.85	2,613,865	215
1998-99	42.9	28.86	28.24	2,7,90,460	230
1999-2000	45.73	33.04	21.23	2,9,47800	242
2000-01	41.01	35.23	23.76	3,219,800	265
2001-02	54.12	26.73	19.15	3,111,320	256
2002-03	39.25	33.13	27.61	2,734,270	225
2003-04	42.98	26.4	30.62	2,654,700	219
Mean	45.53	28.83	25.64	2,867,459	236

The fishermen are provided with fishing crafts under government schemes like IRDP and webbing under no profit no loss basis. The catch per unit effort (CPUE) is calculated as kg of fish caught/fishing unit/year. In 1996-97 the fishing in reservoir was only for 85 days, hence the sudden decline in CPUE (Table 10.3).

Table 10.3. Catch per unit Effort (CPUE) in Tawa reservoir (1990-91 to 1997-98)

Year	CPUE
1990-91	2178
1991-92	2655
1992-93	1529
1993-94	1431
1994-95	1602
1995-96	—
1996-97	49
1997-98	1248

The fishermen of Tawa reservoir mostly use gill net, Gol net, hook and lines etc. Some fishermen have learnt new fishing techniques like the shore seine. The mesh size regulation as well as the regulation in the size of the catch is also enforced strictly. This is reflected in the catch composition also.

The major groups of the fishes in the landings of the Tawa reservoir are major carps, local majors, local minors, minnows and miscellaneous (Table 10.4). The common fishes found in the Tawa reservoir are *Catla catla, Labeo rohita, Cirrhinus mrigala, Labeo calbasu, Hypophthalmichthys molitrix, Ctenopharyngodon idella, and Mystus seenghala.* Out of these *Catla catla* and *Cirrhinus mrigala* dominate the catch. All the predatory fishes found in the reservoir fetches higher prices in the market, compared to the introduced species in the reservoir. The exotic fishes such as *H. molitrix* and *C. idella* are very rarely caught.

Table 10.4. Group-wise composition of landings (in %) of the total catch of the Tawa reservoir

Year	1990-91	1991-92	1992-93	1993-94	1994-95	1995-96	1996-97	1997-98
Major Carps	82.94	85.14	85.86	87.33	7.36	–	80.14	82.50
Local Majors	2.63	3.02	2.30	2.43	7.45	–	14.24	9.65
Local Minors	6.74	9.08	4.69	4.82	10.14	–	5.60	7.85
Minnows	6.99	2.25	6.28	4.17	8.72	–	–	–
Miscell-aneous	0.7	0.51	0.87	1.25	0.09	–	–	–

The fish production in Tawa reservoir does not show any regular pattern. The maximum fish production was recorded during 1999-2000 *i.e.,* 393.169 tonnees.

Table 10.5. Fish production In Tawa reservoir

Year	Fish Production (Tonnees)	Yield Rate (Kg/Ha/Yr)
1990-91	130.69	10.75
1991-92	141.01	12.05
1992-93	88.67	07.30
1993-94	84.42	06.95
1994-95	176.18	15.90
1995-96	–	–
1996-97	93.23	07.68
1997-98	245.81	20.24
1999-2000	393.169	32.34
2003-04	189.25	15.56
2004-05	382.12	31.43

Table 10.6. Earning of the fishermen engaged in fishing

Average No. of Fishermen	393
Average earning/Day (Rs.)	29.01
Total No. of Fishing Days/Yr	267
Earnings/Year (Rs.)	7746
Fishing System	Co-operative mg

GANDHISAGAR RESERVOIR

Gandhisagar is the second largest reservoir in country, next only to Hirakud reservoir in Orissa. It is situated in Mandsaur district of Madhya Pradesh. The water spread area of reservoir is 66000 ha. The average fish yield in Gandhisagar reservoir is about 37.3 kg/ha/yr. Details of fish production and fish yield are depicted in Table 10.7. The average fish yield in reservoir was computed at 37.3 kg/ha/yr.

Table 10.7. Fish production and yield rate in Gandhisagar reservoir (1990-91 to 1997-98)

Year	Fish production (tonnes)	Yield rate (kg/ha/yr)
1990-91	738.43	17.87
1991-92	679.25	16.44
1992-93	406.01	9.82
1993-94	880.60	21.31
1994-95	3424.33	82.87
1995-96	2986.5	72.27
1996-97	1393.7	33.73
1997-98	1823.09	44.12

The major groups of fishes in the landings of the reservoir are major carps, local majors, local minors, minnows and miscellaneous (Table 10.8).

Table 10.8. Group composition of landings (in percentage) of the total catch in the Gandhisagar reservoir

Year	Major carps	Local majors	Local minors	Minnows	Miscellaneous
1990-91	85.53	7.06	6.37	—	1.04
1991-92	89.17	3.92	5.56	—	1.35
1992-93	81.17	8.16	8.24	—	2.43
1993-94	89.14	3.31	5.31	—	2.24
1994-95	81.02	11.75	6.97	—	0.26
1995-96	84.01	10.02	5.65	—	0.32
1996-97	87.92	5.58	6.15	—	0.35
1997-98	91.64	3.12	3.72	—	1.52

In Gandhisagar reservoir fishery is well managed ,major carps and local majors constitutes almost 90% of the total landings followed by local minors and minnows. This shows that the major share of the catch comprises quality fish, which fetch good market price. The stocking of the fish seed in Gandhisagar reservoir is under the control of Madhya Pradesh Fisheries Development Corporation (MPFDC).

The fishermen of Gandhisagar can be grouped in the lower middle class whose income range is between Rs. 15000 to 45000/-. The CPUE is highest in Gandhisagar reservoir, when compared to other reservoirs in state. Table 10.9 provides CPUE in Gandhisagar reservoir from 1990-91 to 1997-98.

Table 10.9. CPUE (kg of fish caught/fishing unit/year) in Gandhisagar reservoir

Year	CPUE
1990-91	—
1991-92	2717
1992-93	1376
1993-94	2985
1994-95	5775
1995-96	4864
1996-97	4016
1997-98	6605

In Gandhisagar reservoir the mesh size regulation as well as the regulation in size of the catch is also enforced strictly. This is reflected in the catch composition also. Minnows were absent in the fish landings of reservoir. Poaching is a major problem in Gandhisagar reservoir. Lack or manpower, heavy costs, vastness of the area, hostile terrain and absence of vehicles makes the task difficult for the concerned fishery managers to take effective step against this menace.

In Gandhisagar reservoir, motorboats carrying ice are deployed for the quick transportation of the catch.

Downgrading is a major problem encountered during the disposal of the catch. Fish, which do not meet the standards of the intended markets, sold at a lower price due to down grading. Downgrading is usually done by the commission agents/wholesalers/contractors who lift the catch. The catch data of Gandhisagar fishery shows that about 6% of the total landings in 1997-98 are classified as downgraded fish, which formed 3.75% of the sales turnover. This downgraded catch will fetch only half the value of the rest of the catch.

Studies on phytoplanktonic dynamics and productivity fluctuations in Gandhisagar reservoir were conducted by Rao and Choubey (1990). Annual records of the phytoplanktonic populations on a qualitative basis are given in Table 10.10. These were represented by all five major taxa of algae, *viz.*, cyanophyceae (12 species), chlorophyceae (22 species), bacillariophyceae (21 species), dinophyceae (3 species) and euglenophyceae (1 species). A total of 59 phytoplanktonic species were recorded.

Zitzschia palea was the most dominant form among all plankton.It showed a continuous upward trend through December to September (273 to 700 units/l). The vertical distribution of *N.palea* showed its densest population at a depth of 50 to 60 feet.

Table 10.10. List of phytoplankton in Gandhisagar reservoir

Cyanophyceae: *Microcystis* sp., *Merismopedia* sp., *Oscillatoria* spp., *Anabaena* sp., *Lyngbya* sp., *Nostoc* sp., *Spirulina* sp., *Phormidium* sp., *Arthrospira* sp. **Chlorophyceae**: *Eudorina* sp., *Pandorina* sp., *Volvox* sp., *Scenedesmus* sp., *Cosmarium* sp., *Ulotrix* sp., *Zygnema* sp., *Microspora* sp., *Dispora* sp., *Pediastrum* sp., *Spirogyra communis, Oedogonium* sp., *Pachycladon* sp., *Actinastrum* sp., *Chlorella vulgaris, Closteridium* sp., *Stichococcus* sp., *Staurastrum* sp., *Cladophora glometra.*
Bacillariophyceae: *Pinnularia* sp., *Bacillaria* sp., *Navicula* spp. *Amphipleura* sp., *Fragillaria* sp., *Nitzschia palea, Melosira* sp., *Cyclotella* sp., *Cocconeis* sp., *Cymbella* sp., *Gomphonema* sp., *Diatomella* sp., *Pinnularia interupta, Diatoma* sp., *Synedra ulna, Cylindrotheca* sp., *Acicular* sp., *Asterionella sp.*
Dinophyceae: *Ceratium* sp., *Gymnodium* sp., *Gonyaulax* sp.
Euglenophyceae: *Euglena viridis.*

Microcystis sp. Formed the second most dominant phytoplankter in the reservoir. Highest density of this species was recorded in summer season both at surface and bottom. It showed continuous increase through December to June (76.4 to 512 units/l). After June, a steep decline in this population was observed and this continued up to December. *Micrcystis* sp. showed the characteristics unimodal pattern in its annual density. The population of this species was dense up to 18 feet depth after which it gradually declined.

The primary productivity in Gandhisagar showed moderate level. Productivity at surface waters was maximum in April followed by May and June. From January to March the productivity values at surface and at a depth of 2.5 m were equal. Higher primary productivity values were recorded at a depth of 2.5 m in April.

Table 10.11. Fish fauna of Gandhisagar reservoir

Carps: *Catla catla, Labeo rohita, Labeo calbasu, Labeo gonius, Labeo boggut, Cirrhinus reba, Cirrhinus mrigala.*
Catfishes: *Mystus seenghala, Mystus aor, Wallago attu, Silonia silondia, Ompak Bimaculatus.*
Forage fishes: *Danio devario, Rasbora daniconius, Ambassis ranga, Ambassis nama.*
Murrels: *Channa striatus, Channa marulius, Channa punctatus.*
Mullets: *Liza corsula.*
Featherback: *Notopterus notopterus.*
Miscellaneous fishes: *Glossogobius giuris, Xenentodon cancila, Nandus nandus.*

Attempts are made in Gandhisagar reservoir to introduce single and two-boat trawling for eradication of predators and trash fishes and for capturing major carps and other commercially important fishes with a view to studying its feasibility in Indian reservoirs. The average catch/hr of fish computed for single boat trawling at two speed ranges of 2.00-2.50 and 2.50-3.00 knots and two boat bottom trawling at 3.00-4.00 knots are 21.57 kg, 31.30 kg, 35.83 kg and 31.71 kg respectively. About 92.17% of the total catch consisted of economic varieties like Catla, Rohu, Murrels, Mullets and Featherbacks in two boat midwater trawling, in contrast to 91%, 75.45%, and 64.20% of non-commercial varieties of fish obtained in single and two boat bottom trawling. Thirteen additional species, such as four species of forage fish, three species of miscellaneous fishes, three species of murrels,one species each of spiny eel, mullet and featherback could also be obtained from different methods of trawling in annual landing 1985-86,apart from the 12 species of fish normally exploited y using gillnets in Gandhisagar reservoir. Two boat midwater trawling at 3-4 knots is recommended fore the exploitation of commercial varieties of fish and single and two-boat bottom trawling at 2-3 knots are suitable for eradication of uneconomic species of fishes in reservoir.

HALALI RESERVOIR

Halali reservoir was constructed in Raisen district about 40 km from Bhopal.The water spread area of the reservoir is 7712 ha. The morphometry of reservoir is depicted in Table 10.12. Soil reaction was neutral to moderately alkaline. Available nitrogen (mg/100 g of soil) was 57.83. pH of water in the reservoir varied from 8.3 to 8.4. Alkalinity ranged from 92 to 142 mg/l. The range of dissolved oxygen was 6.9 to 8.0 mg/l. The phosphate and silicate concentration ranged from 71 to 180 µg/l and 0.4 to 1.7 mg/l respectively. The nitrate values in reservoir water varied within a range of 220 to 4715 µg/l (Das *et al.* 2008). The fish production and yield rate in the reservoir is given in Table 10.13.

Table 10.12. Morphometry of Halali Reservoir in Madhya Pardesh

1. Location	Raisen/Vidisa
2. River	Halali
3. Year of construction	1973
4. Latitude	23°30′
5. Longitude	77°30′
6. FRL (m)	458.40
7. DSL (m)	448.95
8. Maximum depth (m)	29.5
9. Minimum depth (m)	5.3
10. Water spread area at FRL (ha)	7712
11. Catchment area (km^2)	699
12. Productive area (ha)	4795
13. Shore line (km)	65
14. Average annual rainfall	1108
15. Purpose of dam	Irrigation

Table 10.13. Fish production and yield rate in Halai reservoir, Madhya Pradesh

Year	Fish production (tonnes)	Yield rate (kg/ha/yr)
1990-91	142.56	29.73
1991-92	114.7	23.92
1992-93	73.47	15.32
1993-94	162.79	33.95
1994-95	224.75	46.87
1995-96	NA	NA
1996-97	243.51	50.78
1997-98	238.84	49.81

The major groups of fishes in the landings of the reservoir are major carps, local majors, local minors, minnows and miscellaneous species (Table 10.14).

Table 10.14. GroupWise compositions of landings (in percentage) of the total catch in the Halali reservoir, Madhya Pradesh

Fish groups	90.91	91.92	92.93	93.94	94.95	95.96	96.97	97.98
Major carps	16.99	21.64	33.45	23.39	25.1	—	20.76	13.39
Local majors	24.74	29.57	26.8	31.91	28.91	—	27.35	40.23
Local minors	58.27	48.79	39.7	44.59	36.84	—	40.75	29.41
Min-nows	—	—	—	—	8.87	—	11.14	16.97
Misce-llaneous	—	—	0.05	0.11	0.28	—	—	—

The group wise composition of landings in recent years show that about 50% of the catch comprise of local minors and minnows. These two groups mainly constitute small sized fishes (<1 kg) and they are of low economic value. The average market price of these groups is less than half of the value of major carps and local majors. This results in low profit margins from the fisheries. Hence ,even though the yield rate of this reservoir is fairly high, the profit margins from the fishery is low when compared to large reservoirs.

About 300 fishermen associated with more than 12 fishermen co-opertive societies and Self Helping Groups are earning their livelihood from the reservoir.

The stocking of fish seed in Halai is under the control of MPFEDC. The average water spread area of the reservoir is 4795 ha and the average seed stocking density is 213 no./ha. The reservoir fall well below the stocking rate as recommended by Madhya Pradesh State Fisheries Department (500 nos./ha). This is mainly because of the non-availability of sufficient number of good quality fish seed at the right time.

The earnings of the fishermen engaged in fishery and the CPUE in the reservoir are depicted in Tables 10.15 and 10.16 respectively. The CPUE is calculated as kg of fish caught/fishing unit/year.

Table 10.15. Earnings of the fishermen engaged in Halali reservoir fishing (1997-98)

Average no. fishermen	Average earnings/day (Rs.)	Total no. of fishing days/year	Earnings/year (Rs.)	Fishing system
130	5124	251	12862	contract

Table 10.16. Catch per unit effort (CPUE) in the Halai reservoir

1990-91	91-92	92-93	93-94	94-95	95-96	96-97	97-98
—	1712	1097	1714	1873	—	3745	3675

Tomot *et al.* (2008) carried out water quality monitoring of Halali reservoir with reference to cage aquaculture. The range of different limnochemical parameters were observed such as dissolved oxygen 6 to 9.2 mg/l, free carbon dioxide nil-8 mg/l, pH 7.8-8.8, transparency 52-140 cms, chloride 18-60 mg/lit, total alkalinity 90-160 mg/lit and total hardness 80-150 mg.lit. Two units of 8 cages were installed in the reservoir. The size of each cage was 3x3x3 meters. The cage was made of High Density Polyethylene (HDPE) square mesh netting of 6 mm sizes used for raising advanced fry to fingerling and 15 mm size for raising table size fish from advanced fingerlings. For proper growth specially formulated feed were given by using floating tray in cages. Indian major carps were used as candidate species for growth and survival in cage experiment. It was observed that growth rate varied from 0.54 to 2.03 mm/day and survival rate from 57.9 to 80.6%. Fingerlings reared above 100 mm size in cages and then released into the reserfvoir for further growth which supports the high rate of survival.

SAMPNA RESERVOIR

This is an irrigation project, located in Betul district Khargone of Madhya Pradesh. It is an earthen dam constructed on Sampana river, a tributary of Narmada river. The construction was started in the year 1952 and completed in the year 1958. Its live storage capacity is 14.31 MCM. The detail morphometric features of reservoir are depicted in Table 10.17.

Table 10.17. Morphometric features of Sampna Reservoir in Madhya Pardesh

1. Location	Betul
2. River	Sampna
3. Year of construction	1958
4. Latitude	22°8′25″
5. Longitude	77°11′
6. FRL(m)	694.10
7. DSL (M)	686.10
8. Maximu depth (m)	11.4
9. Minimum depth (m)	6.4
10. Water spread area at FRL (ha)	262
11. Catchment area (km^2)	44.75
12. Productive area (ha)	175
13. Shore line (km)	8.0
14. Avergae rainfall (mm)	1203

The soil reaction was slightly acidic *i.e.*, 6.5 (Das *et al.* 2008). Free CO_2 remained absent at the surface water but definitely its presence was noticed at the bottom layer of water to the extent of 3.0 to 10 mg/l, indicating accelerated decomposition of bottom organic matter. Dissolved oxygen varied between 7 to 10 mg/l and alkalinity showed a range from 82 to 120 mg/l. Chlorides showed a variation between 14.2 to 17 mg/l. The nitrate values in reservoir water varied within a range of 8 to 48 μg/l. The phosphate concentration recorded from 8 to 26 μg/l. Reservoir water showed silicate concentration between 1.6 to 2.9 mg/l.

The occurrence of *Oreochromis mossambicus* and *Aristichthys nobilis* was observed in Sampna reservoir. The growth of catla was adversely affected on account of big head of this reservoir which needs to be taken care of. The intensive fishing round the year and heavy stocking of major carps may bring down the population of tilapia to provide a foundation for development of major carps in Sampna reservoir.

Mesh regulation is not enforced properly. Strict mesh regulation should be imposed to prevent the exploitation of small-sized stocked species. Small-meshed nets could be operated in certain seasons and areas to catch minor carps and minnows under supervision. The best season for such operation would be the period of summer when water level is low. The regulation on capture size with catla minimum 2 kg and rohu/mrigal minimum 1 kg should be strictly followed.

DAHOD RESERVOIR

Reservoir has good concentration of hydrophytes.The introduction of grass carp may pay dividend in this reservoir. A higher rate of stocking of grass carp

(40%) is required in Dahod reservoir. The morphometric features of the reservoir are depicted in Table 10.18. Basin soil texture of the reservoir is predominantly sandy-loam in nature. Soil reaction acidic (5.8) and available P (mg /100 gm of soil) recorded at 0.59.

Table 10.18. Morphometric features of Dahod Reservoir

1. Location	Raisen
2. River	Bangnar
3. Year of construction	1958
4. Latitude(N)	23°2′
5. Longitude (E)	77°29′
6. FRL (m)	459.94
7. DSL (m)	453.85
8. Maximum depth (m)	9.2
9. Minimum depth (m)	3.4
10. Water spread area at FRL	820
11. Catchment Area (km^2)	51.79
12. Productive Area (ha)	460
13. Gross capacity ($10^6 x m^3$)	27.75
14. Average Annual inflow($10^6 x m^3$)	32.54
15. Shore line (km)	37.5
16. Shore development	3.7
17. Average Annual rainfall (mm)	1145
18. Purpose	Irrigation

Table 10.19. Salient features of sediment and water in Dahod Reservoir

Sediment	
1. pH	5.7 – 6
2. Organic C (%)	1.1 – 1.5
3. Total Nitrogen (%)	0.054 - 0.074
4. C/N ratio	18 – 22
Water	
1. Temperature (°C)	22 – 30
2. pH	8.4 – 9.04
3. Specific conductivity (µS/cm)	112 – 132
4. Dissolved oxygen (mg/l)	8.2 – 12
5. Alkalinity (mg/l)	82 – 114

Contd...

Table 10.19: Contd...

6. Chloride (mg/l)	14 – 15
7. Nitrate (μg/l)	4 – 40
8. Phosphate – P (μg/l)	4 – 8
9. Silicate-Si (mg/l)	0.4 – 1.2

Table 10.20. Primary production and fish production potential of Dahod Reservoir

1. GPP ($MgC/m^3/h$)	107.7
2. NPP ($MgC/m^3/h$)	73.2
3. Respiration ($MgC/m^3/h$)	35.5
4. Assimilation Efficiency (%)	66.3
5. Fish Yield (kg/ha/yr)	37

KISHANPURA LAKE

Kishanpura lake is located in west-south direction near Chota betma village on Indore-Dhar Road (22°30"N Latitude and 75°39"Longitude) at 22 km from Indore. It is situated in a less populated area of the Indore township. The lake basin has shrubs on two sides while on other two sides it is open agriculture land. The lake receives considerable amount of domestic sewage from the city and is also a bathing place. The lake is mainly fed by surface water runoff in the rainy season. It is a shallow lake. Due to suitability of habitat, lake shore is densely populated by benthic invertebrate communities and fish species. Sharma *et al* (2007) studied biodiversity of benthic macroinvetebrates and fish species communities of Kishapura lake near Indore. In Kishanpura Lake the bethic communities were represented by species belonging to sponges, oligocheates, Hirdudinea,molluscs, chirnomids and chaoborus. The oligocheate population were represented by *Dero diginata, Dero dorsalis, Dero cocperi, Tubifex tubifex, Stylaria fossularis, Branchiodrilus, Sempri, Nais communis* and *Chaetogaster* sp. The most abundant oligocheate diversity was observed in summer season while minimum in rainy season. The Hirudinea group is represented by 2 species *viz. Glossiphonia sp.* and *Hemiclepsis sp.* The most dominate species among Hirudinea was Glossiphonia. The molluscs were very high towards the shore line and included *Lymnea acuminate,Bellamya benglensses, Digoniostoma pulchella, Melanoids tubercularis, Thira scabra, Gyrallus sp., Pissidium clakeanum* and *Vivipara bengalensis.* Generally the abundance of molluscs was found to be associated with the aquatic vegetation.Macrophytes found in the lake included species such as *Hydrilla, Vallisneria, Nymphea, Lemna, Ceratophyllum, Pistia* and *Chara.*

The fish fauna of Kishanpura Lake was represented by 29 species (Sharma *et al.* 2007).

Table 10.21. Fish fauna of Kishanpura Lake of Indore

1.	*Rasbora daniconius*
2.	*Puntius ticto*
3.	*Puntius sarana*
4.	*Catla catla*
5.	*Cirrhina mrigala*
6.	*Labeo rohita*
7.	*Labeo calbasu*
8.	*Labeo gonius*
9.	*Labeo bata*
10.	*Cyprinus carpio*
11.	*Notopterus notopterus*
12.	*Notopterus chitala*
13.	*Mystus seenghala*
14.	*Mystus bleekeri*
15.	*Mystus cavasius*
16.	*Mystus tengra*
17.	*Channa striatus*
18.	*Channa punctatus*
19.	*Channa gachua*
20.	*Mastacembelus armatus*
21.	*Noemachilus aureus*
22.	*Lepidocephalus guntea*
23.	*Wallago attu*
24.	*Xenentedon cancila*
25.	*Heteropneustes fossilis*
26.	*Ompak bimaculatus*
27.	*Gudusia chapra*
28.	*Danio malavaricus*
29.	*Clarias batrachus*

KOLAR RESERVOIR

Basin soil texture in the reservoir is predominantly sandy loam in nature. Soil reaction in Kolar reservoir is neutral to moderately alkaline. Organic carbon (%) was 0.28 in Kolar reservoir. Free carbon dioxide remained absent at surface water of Kolar reservoir. In Kolar reservoir, the main catla fishery is on declining trend.Hence,artificial recruitment needs to be generated through stocking fingerlings of desired species. A size of 100 mm and above is ideal for quick growth and better survival. The stocking should be done during post-monsoon, the period of water level stabilization and spurt of zooplankton.

Table 10.22. Morphometry of Kolar Reservoir

1. Location	Sehore
2. River	Kolar
3. Year of construction	1988
4. Latitude (N)	22°58′
5. Longitude (E)	77°21′
6. FRL (m)	462.20
7. DSL (m)	453.60
8. Maxuimum depth (m)	42
9. Minimum depth (m)	11.30
10. Water spread area at FRL (ha)	2380
11. Catchment Area (km^2)	508
12. Productive Area (ha)	1928
13. Gross capacity ($10^6 x m^3$)	270
14. Average Annual inflow ($10^6 x\ m^3$)	194.56
15. Shore line (km)	29.5
16. Shore development	1.7
17. Annual Average rainfall (mm)	1230
18. Purpose of dam	Irrigation

Table 10.23. Salient features of sediment and water in Kolar Reservoir

Sediment	
1. pH	7.0 – 7.6
2. Organic C (%)	0.2 – 0.4
3. Total Nitrogen (%)	0.014 – 0.020
4. C/N ratio	14 – 19
Water	
1. Temperature (°C)	18.2 – 30
2. pH	7.5 – 7.9
3. Sepcific conductivity (µS/cm)	310 – 380
4. Dissolved oxygen (mg/l)	7.3 – 8.6
5. Alkalinity (mg/l)	160 – 180
6. Chloride (mg/l)	19.9 – 22.2
7. Nitrate (µg/l)	20 – 38
8. Phosphate – P (µg/l)	4 – 12
9. Silicate-Si (mg/l)	2.5 – 3.8

RAMPUR RESERVOIR

The Rampur reservoir is one of the artificial water bodies of Guna district. The reservoir is situated at a distance of 34 miles from Guna City. Rampur is the nearest village. The direction of the reservoir is north west of Guna and it lies between 24°47′N and 77°10′E. The reservoir has a catchment area of approximately 102 sq miles and it is a part of Chambal Jamuna basin. The reservoir serves the purpose of irrigation for the nearby villagers. It is also used for pisciculture of major and minor carps and locally available fishes.The physico-chemical studies of the reservoir were under taken to throughout the year to find out the seasonal variations in the physico-chemical factors of the reservoir. The purpose was to gain a basic knowledge of the reservoir as it would not only enable us to enhance the limnological knowledge but also explore possibilities for better management, development and augmentation of pisciculture.Monthwise variations in physico-chemical characteristics of Rampur reservoir are depicted in Tables 10.24 and 10.25.

Table 10.24. Observations at station 'A' of Rampur reservoir of Guna district, Madhya Pradesh

Months	Temp-erature (°C)	Transp-arency (cm)	pH	Free CO_2 (ppm)	Dissolved oxygen (ppm)	Alkalinity (ppm)	Nitrates (ppm)	Chlorides (ppm)
Jan-04	19.2	30.2	8.0	4.1	6.2	91	0.099	5.2
Feb-04	19.1	32.0	8.1	4.2	6.4	89	0.089	5.8
Mar-04	21.6	35.0	7.9	3.9	6.5	96	0.082	6.0
Apr-04	24.0	35.6	8.1	4.1	8.0	97	0.091	6.1
May-04	28.9	40.0	8.2	3.2	8.9	105	0.097	6.5
Jun-04	31.4	42.0	8.5	2.5	9.0	110	0.127	8.0
July-04	30.8	30.2	8.6	6.5	3.8	87	0.211	8.5
Aug-04	28.2	13.2	8.1	8.0	3.2	67	0.233	7.1
Sept-04	27.0	12.9	7.9	10.2	3.0	65	0.233	6.9
Oct-04	26.2	24.0	7.7	6.7	4.9	82	0.189	6.2
Nov-04	21.2	29.0	7.8	5.6	5.0	90	0.123	6.0
Dec-04	20.0	29.5	7.9	5.4	5.6	94	0.110	5.4

The results show that the water temperature varies with the atmospheric temperature *i.e.,* being maximum during summer (31.4°C at station A and 32°C at station B June 2004) and minimum during winter (19.2°C at station A and 19.5°C at station B).

Maximum transparency was found during winter and minimum during rainy season at both stations. The minimum transparency during rainy season was due to high turbidity of water caused by suspended slit and organic debris.

Table 10.25. Observations at station 'B' of Rampur reservoir of Guna district, Madhya Pradesh

Months	Temperature (°C)	Transparency (cm)	pH	Free CO_2 (ppm)	Dissolved oxygen (ppm)	Alkalinity (ppm)	Nitrates (ppm)	Chlorides (ppm)
Jan-04	19.5	30.1	8.0	4.0	6.5	95	0.091	5.6
Feb-04	19.2	31.9	8.0	4.1	6.6	86	0.089	58.0
Mar-04	21.0	35.0	7.9	3.9	6.5	96	0.085	6.0
Apr-04	24.3	35.8	8.1	4.2	8.1	96	0.088	6.5
May-04	28.9	39.0	8.1	3.1	8.9	110	0.099	7.0
Jun-04	32.0	42.2	8.6	2.5	9.2	115	0.133	8.2
July-04	31.0	33.1	8.6	5.6	4.0	90	0.221	8.8
Aug-04	28.2	14.2	8.2	7.3	3.6	76	0.230	7.1
Sept-04	27.5	13.5	8.0	10.2	3.5	68	0.241	6.8
Oct-04	26.3	23.6	7.9	6.9	4.8	82	0.156	6.5
Nov-04	22.0	28.4	7.8	5.8	5.2	95	0.123	6.6
Dec-04	21.1	29.5	7.9	5.5	5.6	94	0.111	5.4

As pH is concerned, no significant variation was observed throughout the year. Though slightly lower values were found during winter and slightly higher values during summer, but throughout the year it remained alkaline.

Free carbon dioxide varies considerably throughout the year. The presence of free CO_2 is due to incomplete utilization in photosynthesis. During the monsoon, the rate of photosynthesis being low due to cloudy weather, free CO_2 values are high.

Variations were observed in the concentration of dissolved oxygen. Low values of oxygen were recorded during rainy season. The concentration of oxygen then increases and reaches maximum during summer. The high values during summer are associated with the rise in phytoplankton, while during rainy season low values may be due heavy influx of the water resulting in poor photosynthesis.

Alkalinity of the water is its capacity to neutralize a strong acid and is characterized by the presence of all hydroxyl ions capable of combing with the hydrogen ions. Total alkalinity in the present study shows seasonal variations. The values were high during summer and low during monsoon. The steep fall during monsoon may be due to dilution of the water.

The concentration of the nitrates was found maximum during rainy season. The concentration then decreases during winter and again increases during summer. The high values of nitrates during rainy season may be due to addition of nitrates into the water by run-off water.

The chloride concentration exhibited seasonal variations. It was high during summer and low during winter. The high values during summer may be attributed to low water levels.

MANSAROVAR RESERVOIR

The Mansarovar reservoir of Bhopal is also known as Chunabhatti. It is under environmental stress due to siltation, human encroachment, high macrophytic populations and sewage input from various sources. Since the water quality affected production of fish food organisms, it also affects fish production. The reservoir with 73 hectares of productive area. It is situated in the southern part of Bhopal. Its main source of water is from the catchment area of 8.29 sq.kms. Its storage capacity is 2.29 m.cu.m. ad water spread at full reservoir level is 0.797 m.sq.m.

Kulshrestha *et al.* (1992) studied the several variations in the limnochemical characteristics of the Mansarovar reservoir. The surface temperature ranged between 18.6 to 28.5°C in winter and summer respectively with a difference ranging from 1 to 3°C without any thermal stratification.

Transparency values ranged between 13.6 to 32 cm with high turbidity in summer which may be due to the degeneration of blue green algae reducing the light penetration.

Total suspended solids ranged from 8 to 718 mg/l. The values in column and bottom were higher than at surface.

The conductivity ranged from 750 μ mhos cm^{-1} during postmonsoon to 1600 during summer. Such variation may be due to the concentration and dilution of solutes in water.

The pH was alkaline rangig between 7.2 to 9.5. The values increased in summer and decresased in monsoon and winter. The decrease in pH during winter may be due to the decrease in photosynthesis while during monsoon it may be due to greater inflow of water. The pH exhibited higher values at surface and gradual decrease at column ad bottom respectively. Maximum values during summer may be due to increased photosynthesis of the algal blooms resulting into the precipitation of carbonates of calcium and magnesium from bicarbonates causing higher alkalinity.

The free carbon dioxide was rarely observed at surface. It was observed during monsoon in column ad maximum values of 8 mg/l were recorded at the bottom during summer.The absence of free carbon dioxide may be due to its complete utilization in photosynthetic activity or its inhibition by th presence of appreciable amount of carbonates in water. Higher values of free carbon dioxide at the polluted station may be due to depletion of oxygen contents and subsequent increase in the build up of free carbon dioxide by the process of anaerobic digestion of sewage wastes.

Total alkalinity ranged between 121 to 294 mgl. A decline in the alkalinity was observed during the monsoon which may be due to dilution of water. High alkaline conditions may be due to high photosynthetic activity of the phytoplankton, sewage input and water temperature.

The values of carbonates ranged between nil to 88 mg/l with maximum in winter.These values showed decreasing trend in column ad bottom. Higher values may be due to increased photosynthesis. The carbonates showed a significant position relation with pH, sodium,potassium, calcium and negative relationship with free carbon dioxide. Phytoplankton take up bicarbonate ions during photosynthesis and reak up its carbonate and OH ions. The concentration of these ions in water depends on the photosynthetic activity. The bicarbonates raged between 116 to 294 mg/l with maximum values in winter.

The dissolved oxygen ranged between 4 to 9.9 mg/l in surface samples. The lower levels of dissolved oxygen at bottom and column may be due to lack of water and air action and decomposition of organic matter. The dissolved oxygen showed direct relationship with bicarbonates, magnesium, total hardness, carbonates and total kjeldhal nitrogen which indicate high photosynthetic activity related to increase in nutrients and dissolved oxygen levels.

The biochemical oxygen demand (BOD) values ranged from 26.3 to 60. The chemical oxygen demand (COD) values ranged from 78 to 147.9 mg/l with maxima in summer in surface and column samples. In bottom, the maxima was recorded in post-monsoon. Higher values of the COD in summer may be due to high temperature.

The chloride values ranged from 49.6 to 98.3 mg/l. Its maxima was recorded in post-monsoon period. The lower values in winter and monsoon may be due to looking up in sediments due to low temperature or input of water in rainy season.

The values of total hardness ranged between 121.5 to 240.6 mg/l with maxima in winter. The total hardness increased at surface and column and decreased in bottom.

The values of calcium ranged between 81.8 to 193.2. Higher values of calcium may be due to addition of detergents used for cloth washing.

The magnesium values ranged between 5 to 16.5 mg/l with maxima in summer or winter and minima during monsoon. The high temperature and evaporation may result in high concentration, whereas the dilution caused by rain could result in lowering of these values.

The sodium values ranged from 69.8 to 139.6 mg/l. Its maxima and minima were recorded during winter and summer respectively. An increasing trend was visible from surface to bottom.

Potassium ranged between 19 to 48.2 mg/l with maxima in summer and winter. Sewage receiving area showed higher values of potassium.

Orthophosphates ranged from 0.010 to 0.067 mg/l with decreasing trend from surface to bottom. The maximum values were obtained at places which receive dtergets and sewage pollution.

RAMSAGAR RESERVOIR

The reservoir is located approximately 80 kms south of Gwalior (Fig. 10.2). It is a small man-made reservoir with 140 ha water spread area. It was built over a Nichroli Nallah in the basin of Sindh river. The reservoir is located approximately 8 km north-west of Datia city. Geographically, it lies between 25°40′ N Latitude and 78°23′ E Longitude. There is an earthen dam on the Northern side of reservoir. The length of the dam is 474 m. The maximum height of the dam above foundation is 13.41 m. The gross storage capacity of the reservoir is 5.866MCM. The reservoir's command area constitutes an area of more than 1050 hectares. Reservoir is used for different purposes like drinking water supply, irrigation, fisheries. The reservoir is surrounded by hills on the eastern and western sides, which drain their runoff in the reservoir through various nullahs prominent among these being the Pateria from eastern side, Badi nadi from southern side and Patparu from the western side.

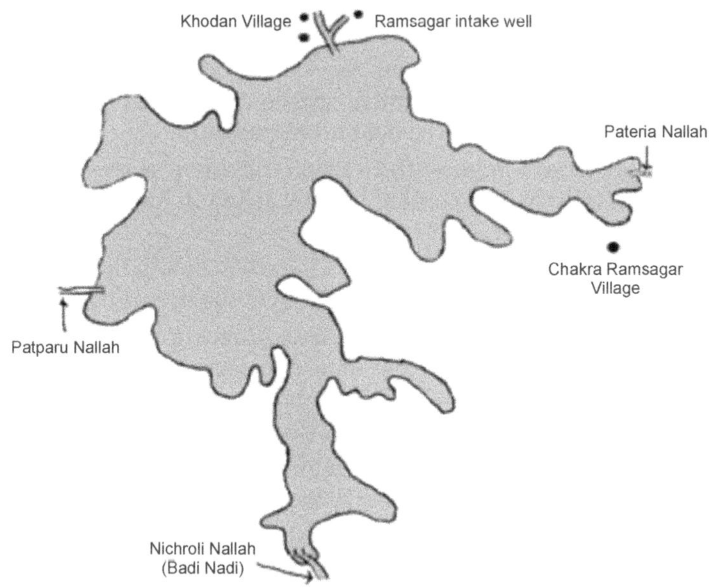

Fig. 10.2. Map of Ramsagar reservoir

Rao and Garg (2013) carried out field surveys to prepare biomap of Ramsagar reservoir by using Global Position System and reported 13 species of aquatic vegetation, 14 species of molluscs, 5 species of aquatic insects, 26 species of fish both natural as well as released by the Fisheries Department, 2 species of aquatic reptiles and 15 species of migratory birds in the reservoir.

The Ramsagar reservoir is a mesotrophic water body with moderate quantity of nutrients (*Garg et al*, 2009). The fishermen cooperatives have rights on the fishing activities. Killing of migratory birds is also reported. Locals are not aware of the importance of migratory birds and the legislation. Due to mixing of pesticides through water runoff from the agriculture fields the fishes from Ramsagar reservoir have been infected.

In Ramsagar reservoir, the water temperature increased during warmer months and decreased during colder months. In Ramsagar reservoir, maximum water level was recorded in post-monsoon period while also minimum water level was recorded in summer season.

Water was turbid in monsoon season with yellow brown colour, while green colour in winter and transparent green colour was observed in summer season. The transparency in Ramsagar reservoir ranged from 66.59-116.00 cm with low value during monsoon season.

The electrical conductivity was found to fluctuate between 108.00 $GScm^{-1}$ to 246.30 GS cm^{-1} and falls within the range observed for Indian waters. Maximum turbidity was recorded as 16.72 NTU during monsoon season, while minimum value was recorded as 2.17 NTU during winter season. During monsoon season silt, clay and other suspended particles contribute to the turbidity values while during winter season settlement of silt, clay resulting low turbidity.

TDS showed highest value of 239.00 mg/l in the month of June-2004. The variation in pH in Ramsagar reservoir was between 7.41 to 8.95.

The dissolved oxygen was found to vary from 6.78 to 11.59 mg/l with maximum concentration during winter season. In summer season, dissolved oxygen decreased due to increased water temperature.

In Ramsagar reservoir, highest free carbon dioxide was recorded as 6.32 mg/ l. However, its absence or low content was recorded in most of the times due to alkaline nature of reservoir.

The total alkalinity ranged from 64.25 to 146.25 mg/l, which makes the reservoir as nutrient rich and highly productive water body.

Chloride concentration varied from 13.13 to 22.36 mg/l with the concentration higher in summer season and lower in monsoon season. In aquatic environment calcium serves as one of the micronutrients for most of the organisms. The calcium contents in Ramsagar reservoir varied from 11.21 to 33.81 mg/l.

The rich concentration of nutrient such as sulphates (1.50-8.87 mg/l), nitrates (0.011-0.033 mg/l), nitrite (0.004-0.029 mg/l), phosphates (0.013-0.054 mg/1) and silicates (0.65-8.42 mg/l) has been observed in Ramsagar with greater concentration during summer season due to evaporation. The ammonia content varied from its absence in August-2004 to 0.85 mg/l in June-2003 with higher values in summer season and lower value in monsoon season.

BOD in Ramsagar reservoir was recorded in the range of 0.93 mg l-1(September-2004) to 4.68 mg/l with low values in monsoon and high values in summer season. COD ranged from 3.60 to 17.40 mg/l with higher values during summer season.

In Ramsagar reservoir, the magnesium content up to 5.60 mg/1 was observed with higher concentration during winter season and lower concentration in monsoon season.

Sodium content fluctuated between 16.75 to 34.30 mg/l. A higher value of sodium during summer season and lower value in monsoon season was recorded. Evaporation of water is a significant factor in increasing sodium level during summer season.

Potassium content varied from 1.97 to 4.86 mg/l (June 2004). Higher concentration of potassium was recorded in summer season during second year of the study and lower concentration was recorded in monsoon season during first year of the study period while second year of the study period, lower potassium content was recorded in winter season.

Altogether 13 species of macrophytes were recorded during study period in reservoir (*Garg et al.,* 2009). They are *Nymphaea nouchali, Trapa bispinosa, Ipomoea aquatica, Ipomoea carnea, Ceratophyllum demersum, Hydrilla verticillata, Vellisneria spiralis, Lemna minor, Wolffia arrhiza, Potamogeton crispus, Najas minor, Aponogeton natans* and *Cyperus articulatus*. The dominancy of macrophytes was observed in winter and summer season during both years of study period while in monsoon season three macrophytes species *i.e., Nymphaea nouchali, Trapa bispinosa* and *Lemna minor* were not observed. In order to have proper management and best possible use of the reservoir, the macrophytes will have to be controlled (*Garg et al.*, 2009). This can be achieved by mechanical removal or by biological means using grass carp. The nallahs, streams and rivulets joining the reservoir should be obstructed by constructing stop and check dams. This will not allow the siltation in reservoir

Garg et al. (2013) reported eighteen species of birds belonging to 5 different orders and 11 families were recorded. Out of these, family ardeidae with 4 species was dominant followed by charadridae, anatidae, rallidae, palacrocoracidae, sturnidae, muscicapidae, alcedinidae, dacelonidae, cerylidae and meropidae. Coot (*Fulica atra*, Linnaeus) have been the most common and abundant species of family rallidae.

UADISAGAR LAKE

Udaisagar Lake is the main source of water supply to the zinc smelter ,a factory of Hindusthan Zinc Limited. The lake is getting polluted by drainage of river Ahar. River Ahar carries the domestic, sewage and industrial wastes. Udaisagar is highly eutrophic lake (Vijayvergia, 2008). Fish mortality is also reported from the lake. The most serious aspect of the eutrophication of Udaisagar lake is water bloom formation by *Microcystis aeruginosa* in summer and rainy months. Higher values of nitrate ad phosphate in the rainy season might be responsible for the development of water bloom by *Microcystis aeruginosa* and may have accelerated the process of eutrophication. The occurrence of *Microcystis aeruginosa* might have adverse effect with regard to clogging of gills and toxic effect on fish life. Vijayvergia (2008) studied the seasonal variations in the water characteristics of lake Udaisagar for the year 1986 and 2006. A close perusal of water characteristics and consequences reveals that the lake is towards an increasing state of eutrophy and of much more deteriorated water quality.

TIGHRA RESERVOIR

The Tighra freshwater reservoir sitated about 23 km west of Gwalior city at an altitude of 218.58 m from mean sea level near SADA magnet city. The Tighra reservoir lies on 26-12'0"E Longitude. The reservoir was surrounded by hills from three sides. There water spread area of 2112 hectares. The average depth of the reservoir 24 meter. The reservoir is rain fed during monsoon periods. The reservoir is also fed by Sank river.

Uchchariya and Saksena (2012) recorded 51 species of zooplankton from Tighra reservoir. Out of 51,29 species belong to rotifera, 10 species to cladocera, 5 species to protozoa, 4 species to copepoda and 3 species to ostracoda. Brachionus has been found to be most abundant genus among rotifera, while mesocyclops among copepoda, and moina among cladocera were dominant genera of zooplankton.

Table 10.26. Fish Diversity of Tighra Reservoir in Gwalior

Name of fish
1. *Catla catla*
2. *Cirrhinus mrigala*
3. *Cirrhinus reba*
4. *Labeo rohita*
5. *Labeo calbasu*
6. *Labeo bata*
7. *Labeo gonius*
8. *Labeo fimbriatus*
9. *Ctenopharyngodon idella*
10. *Cyprinus carpio*
11. *Rasbora daniconius*
12. *Mystus seenghala*
13. *Mystus bleekeri*
14. *Mystus cavasius*
15. *Mystus vitatus*
16. *Puntius chola*
17. *Puntius sarana*
18. *Puntius sophor*
19. *Puntius ticto*
20. *Xenentedon cancila*
21. *Channa marulius*
22. *Channa Puncatus*
23. *Channa starius*
24. *Clarius batrachus*
25. *Heteropneustes fossilis*
26. *Notopterus notopterus*
27. *Notopterus chitala*
28. *Ompak bimaculatus*
29. *Ompak pabo*
30. *Rita rita*
31. *Wallago attu*

Mahor (2011) made survey on fish fauna of the Tighra reservoir and reported fish diversity of 33 species. out of the 33 species, one species of Catla (*Catla catla*), two species of Cirrhinus (*Cirrhinus mrigala* and *Cirrhinus reba*), five species of Labeo (*Labeo rohita, Labeo calbasu, Labeo fimbriatus, Labeo bata, Labeo gonius*), one species of Rasbora (*Rasbora daniconius*), four species of Mystus (*Mystus seenghala, Mystus bleekeri, Mystus cavasius, Mystus vitatus*), four species of Puntius (*Puntius chola, Puntius ticto, Puntius sarana, Puntius sophore*), one species of Xenentedon (*Xenentedon cancila*), three species of Channa (*Channa maralius, Channa punctatus, Channa striatus*), one species of clarius (*Clarius batrachus*), one species of Heteropneustes (*Heteropneustes fossilis*) two species of Notopterus (*Notopterus chitala, Notopterus notopterus*),two species of Ompak (*Ompak bimaculatus, Ompak pabda*), one species of Rita (*Rita rita*), one species of Wallago (*Wallago attu*), one species of Ctenopharyngodon (*Ctenopharyngodon idella*), one species of Cyprinus (*Cyprinus carpio*), one species of Hypophthalmichthys (*Hypophthalmichthys molitrix*), one species of Mastacembelus (*Mastacembelus armatus*). A total thirty two species encountered in the Tighra reservoir, there belonging to 5 order, 9 family and 33 genera. (Table 10.26). Mahor (2011) also accounted bait and hooks and gill net as the common gears used in fishing of Tighra reservoir.

BANSAGAR RESERVOIR

It is a newly constructed reservoir across river Sone. Its water spread area is 26511 hectares. In the year 2000, fishing was started with a nominal contribution of 7.99 tonnes. Maximum production was noted in the year 2005-06 *i.e.*, 279.375 tonnes and the catch fell down in the year 2006-07 to 193.426 tonnes.

SANJAY SAGAR RESERVOIR

Sanjay Sagar Reservoir is located in the Guna district, about 50 km from Guna township. The average rainfall of Guna is 910mm. Sanjay Sagar reservoir is a medium irrigation reservoir constructed in 1982 with the help of World Bank Funding.It lies between latitude 24-22-30" and longitude 77-14-30" at an altitude of 480.05 meters above mean sea level. The dam is constructed for drinking water and fish culture purposes. It is constructed on river Gomukh which is tributary of river Parvati. It is 630 meters long and has the capacity of 37.51 million cubic meters while live capacity is 35.16 million cubic meter. Maximum water level is 451.34 meter while river bed level is 421.88 meter. The reservoir has a net cultivable command area of 6100 hectares while gross command area is 6973 hectares. About 46 villages are benefited from the reservoir. This reservoir is an important unit for the irrigation in both district Guna and Ashoknagar.

Solnaki *et al.* (2011) reported 16 fish species from Sanjay Sagar reservoir. They also classified fishes according to their economic importance.

Table 10.27. Fish fauna of Sanjay Sagar Reservoir

1. *Catla catla*
2. *Cirrhina mrigala*
3. *Cirrhina reba*
4. *Cyprinus carpio*
5. *Labeo bata*
6. *Labeo calbasu*
7. *Labeo fimbriatus*
8. *Labeo gonius*
9. *Labeo rohita*
10. *Mystus aor*
11. *Mystus seenghala*
12. *Clarias batrachus*
13. *Wallago attu*
14. *Mastacembelus armatus*
15. *Ophiocepohlaus marulius*
16. *Ophiocephalus striatus*

INDIRASAGAR RESERVOIR

Indira Sagar Project (ISP) situated on River Narmada, 10 Km from village Punasa in Khandwa district of Madhya Pradesh. ISP is a multipurpose Project with an installed capacity of 1000 MW, with annual energy generation of 2698 Million Units in Stage-I, 1850 Million Units in Stage-II and 1515 Million Units in Stage-III and annual irrigation of 2.65 lac. ha on a cultivable command area (CCA) of 1.23 lac. ha. Total catchment area at the Dam site is 61642 sq.km. ISP is the mother project for the downstream projects on Narmada Basin with largest reservoir in India, having 12.22 Bm^3 storage capacity.It is the newly constructed reservoir across river Narmada with a water spread area of 49,855 hectares. In this reservoir fishing is being done for the last three years. Starting in the year of 2003-2004. Statistics show an increasing catch trend in this reservoir. In the year 2004-05, the fish production was 71.5 tonnes, it was 104.793 tonnes in the year 2005-2006 and 204.533 tonnes in 2006-2007. Matsya Maha Sangh has formed fishermen's societies for catch improvement of Indirasagar reservoir.

Table 10.28. Morphometry of Indirasagar Reservoir in Madhya Pradesh

Catchment Area (Sq. km)	61,642
Rainfall (*a*) Maximum (mm) (*b*) Minimum (mm) (*c*) Average (mm)	 1879 603 1288
Top Bund Level (TBL) (in meter)	267.00
Max Water Level (MWL) (in meter)	263.35
Full Reservoir Level (FRL) (in meter)	262.13
Minimum Draw Down Level (MDDL)(in meter)	243.23
Crest Level (Spillway) (in meter)	245.13
Water Spread Area at FRL (Sq. kms)	913.48
Gross storage capacaity Bm3 (Billion Cumec) M.A.F. (Million Acre Feet)	 12.22 9.90
Live storage capacaity Bm^3 (Billion Cumec) M.A.F. (Million Acre Feet)	 9.75 7.90
Dead storage capacaity Bm^3 (Billion Cumec) M.A.F. (Million Acre Feet)	 2.47 2.00
CONCRETE GRAVITY DAM (IN METER) (*i*) Total Length (in meter) (*ii*) Non- overflow (in meter) (*iii*) Overflow portion Ogee shaped spillway (in meter) (*iv*) Qty. of concrete (CUM) Maximum Height (in meter)	 653 158 495 13,92.000 92

RADIAL CREST GATES	
Number	20
Length (in meter)	20
Height (in meter)	17
POWER HOUSE	
Type of Power House Surface	Surface
Installed Capacity (in MW)	1000
Type of Turbine Francis	Francis
SUBMERGENCE	
Villages affected due to submergence at FRL	249
Population affected	129000
LAND	
Cultivated Area (in ha)	44.363
Other area (in ha)	5.565
Forest Area (in ha)	41.420

Vyas *et al.* (2009) studied the spatial and temporal variation of fish fauna in the stretch of Indira Sagar submergence area. This study attempts to document the fish biodiversity of Indira Sagar reservoir submergence area and tributaries of river Narmada *viz.* Ganjal and Machak, temporal variation therein and also assess the impact of formation of reservoir on the riverine fish fauna. Altogether fifty two species belonging to 28 genera, 13 families and 7 orders were encountered. Twenty species were recorded in Machak river whereas twenty nine species were recorded in Ganjal River, nineteen species were recorded in Narmada river near Nemavar village and thirteen species were reported at the tail end of the reservoir near Purni. The study indicates that this region is still a hotspot of fish biodiversity and conservation measures should be followed to sustain fish biodiversity of tributaries.

BARGI RESERVOIR

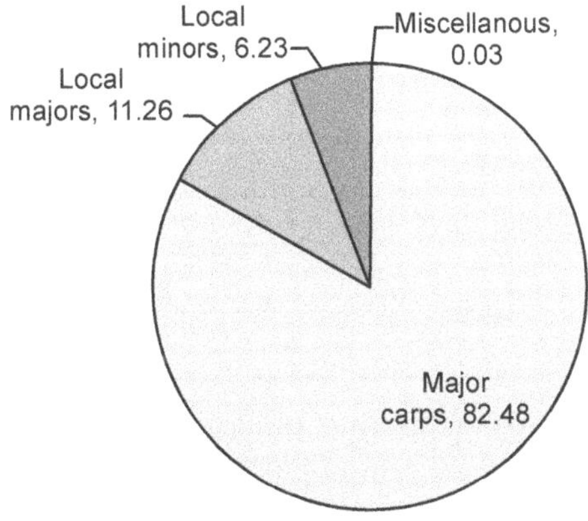

Fig. 10.3. Group wise composition of landings (in %) of the total catch in Bargi reservoir (year 1997-98)

The apex federation of fishermen co-operatives is managing the fishery of Bargi reservoir. The reservoir is in Jabalpur district with water spread area of 27296 ha. Table 10.29 shows the details of fish production from Bargi reservoir. The group wise composition of landings for year 1997-98 (in percentage) are depicted in Fig. 10.3.

Table 10.29. Details of fish production in Bargi Reservoir

Year	Fish production in tonnes	Yield rate (kg/ha/yr)
1990-91	41.96	2.52
1991-92	94.57	5.66
1992-93	175.658	10.55
1993-94	536.04	33.39
1994-95	574.62	34.51
1995-96	582.08	34.96
1996-97	406.02	24.39
1997-98	422.11	25.35

In Bargi reservoir the fishery is well managed, major carps and local majors constitutes almost 90% of the total landings followed by local minors. This shows that the major share of the catch comprises quality fish, which fetch good market price.

In Bargi the seed stocking is carried out by the apex federation of the fishermen co-operatives. The stocking rate proposed by the Madhya Pradesh State Fisheries Department is 250 nos/ha for reservoirs of the size category above 12000 ha.The Bargi reservoir has average water spread area of 16649 ha and the average stocking density is 186 nos./ha. The reservoir fall well below this stocking target (Ninan, 2002). This is mainly because of the non-availability of sufficient number of good quality fish seed at the right time.

Table 10.30. Earnings of the fishermen engaged in Bargi Reservoir (1997-98)

1. Average number fishermen	880
2. Average earnings/day (Rs.)	22.49
3. Total number of fishing days/year	293
4. Earnings/year	6590
5. Fishing system	Co-operative managemet

The catch per unit effort (CPUE) is lowest in Bargi when compared to major reservoirs of Madhya Pradesh. Ninan (2002) calculated CPUE at 959 kg of fish caught/fishery unit/year. In Bargi the closed fishing season is declared during rainy season. The mesh size regulation as well as the regulation in the size of the catch is also enforced strictly. This is reflected in the catch composition also. Minnows and other under sized species are almost absent in the fish landings.

In Bargi poaching is a major problem. Lack of manpower, heavy costs, vastness of the area, hostile terrain and absence of vehicles makes the task difficult for the concerned fishery managers to take effective step against this menace.

BARNA RESERVOIR

In Barna there is co-operative society which has the fishing rights. The reservoir is in the Raisen district with water spread area of 7705 ha. The details of the fish production in reservoir and yield /ha/yr is given in Table 10.31. Group wise composition of landings (in percentage) of the total catch for year 1997-98 is shown in Fig. 10.4.

Table 10.31. Details of fish production in Barna Reservoir

Year	Fish production in tonnes	Yield rate (kg/ha/yr)
1990-91	32.89	6.86
1991-92	79.97	16.7
1992-93	103.26	21.55
1993-94	152.18	31.76
1994-95	102.08	31.3
1995-96	Not Available	Not Available
1996-97	151.43	31.61
1997-98	140.94	29.42

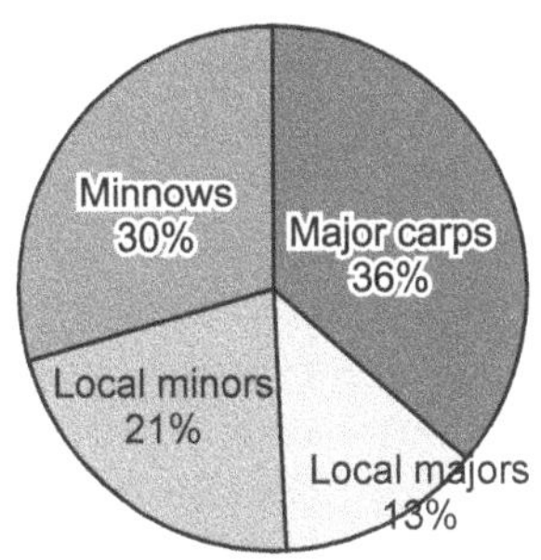

Fig. 10.4. Groupwise composition of landings (in %) of the total catch in Barna Reservoir (Year 1997-98)

In case of Barna, group wise composition of landings show that about 50% of the catch comprise of local minors and minnows. These two groups mainly constitute small sized fishes *i.e.*, < 1 kg and they are of low economic value. The average market price of these groups is less than half of the value of major carps and local majors.This result in low profit margins from the fishery. The significant landings of low value, under sized fishes from Barna reservoir shows that the fishing system is not properly managed leading to the depletion of the stock of quality fish groups. Conservation methods like mesh size regulation is not strictly followed which is evident from the landings of under sized fishes (Ninan, 2002).

The stocking if fish seed is under the control of Madhya Pradesh Fisheries Development Corporation (MPFDC). The stocking rate proposed by the Madhya Pradesh State Fisheries Department is 500 nos./ha for the reservoirs of the size category of 4001 to 12000 ha. The Barna reservoir is with average water spread area of 4791 and the average stocking density is 196 nos./ha. The reservoir fall well below this stocking target. This is mainly due to non-availability of sufficient number of good quality fish seed at the right time.

Ninan (2002) calculated catch per unit effort (CPUE) in Barna reservoir at 2237.CPUE calculated as kg of fish caught/fishing unit/year.

Table 10.32. Earnings of the fishermen engaged in Barna Reservoir (1997-98)

1. Average number fishermen	125
2. Average earnings/day (Rs.)	39.93
3. Total number of fishing days/year	240
4. Earnings/year	9584
5. Fishing system	Contract

GOHAD RESERVOIR

Mishra and Sakesna (2012) studied gonadosomatic index and fecundity of *Labeo calbasu* from Gohad reservoir in Gohad town of Madhya Pradesh. Ovarian weight of the fish ranged from 140 ± 320 g with a mean value of 201 ± 16.56 g. The ovarian weight is almost 20% of the body weight in fully mature fishes. Mean gonadosomatic index increased gradually from May and reach to peak in July and then decreased in August. The gonadosomatic index of *Labeo calbasu* was ranging from 18.22 to 22.10 with a mean of 19.87 ± 0.383. The high value of gonadosomatic index is indicative of mature stage of gonads in the fish. The mature ova are round and heavily laden with yolk.

Fecundity of fish increase with the increase in size, weight of fish and weight of gonad. The average weight and length of fish were 1006 ± 75.22 g and 38.85 ± 1.29 cm respectively. The average fecundity was measured as 402217 ± 30661 ova per fish with an average fecundity 400.5±8.40 ova per gram of body weight and 10241 ± 401.8 ova per cm of fish length.

REFERENCES

- Das, A.K., Shrivastava, NP., Vass, K.K. and Pandey, B.L., 2008. Managemant strategies for enhancing fish production in Madhya Pradesh reservoirs. In : Proceedings of TAAL-2007 : The 12th World Lake Conference (Editors: Sengupta, M. and Dalwani, R.), pp. 1295-1300.

- Garg, R.K, Rao, R.J and Saksena, D.N., 2009. Water quality and conservation management of Ramsagar reservoir, Datia, Madhya Pradesh. *Journal of Environmental Biology,* 30(5) 909-916.
- Garg, R.K., Rao, R.J. and Saksena, D.N. 2013. Spatial relations of migratory birds and water quality management of Ramsagar reservoir, Datia, Madhya Pradesh, India, *Journal of Ecology and the Natural Environment,* 5(10) : 335-339.
- Kartha, K.N. and Rao. K.S., 1990. Experimental trawling in Gandhisagar reservoir. pp.144-149. In : Jhingran, Arun G. and V.K. Unnithan (eds.). Reservoir Fisheries in India. Proceedings of the National Workshop on Reservoir Fisheries, 3-4 January, 1990. Special Publication 3, Asian Fisheries Society, Indian Branch, Mangalore, India.
- Kulshrestha, S.K., M.P. George, Rashmi Saxena, Malini Johri and Manish Shrivastava, 1992. Seasonal variations in the limnochemical characteristics of Mansarovar reservoir of Bhopal. In S.R. Mishra and D.N. Saksena (eds), Aquatic Ecology, Ashish Publishing House, New Delhi, pp. 275-292.
- Mahor, R.K., 2011. A survey of fish and fisheries of the fresh water reservoir Tighra, Gwalior, Madhya Pradesh. *Shod Samiksha aur Mulyanka.* 2(25) : 49-50.
- Mishra, Shailja and Saksena, D.N.2012. Gonadosomatic index and fecundity of an Indian major carp, *Labeo calbasu* in Gohad reservoir. *The Bioscan,* 7(1) : 43-46.
- Ninan, George, 2002. Fishery management practices of major reservoirs in Madhya Pradesh. pp.363-369. In : M.R. Boopendranath, B.Meenakumari, Jose Joseph, T.V. Sankar, P.Pravin and Leela Edwin (eds.). Riverine and Reservoir Fisheries of India. Proceedings of the National Seminar on Riverine and Reservoir Fisheries-Challenges and Strategies, 23-24 May 2001. Society of Fisheries Technologists (India), Cochin, India.
- Rao, K.S. and Choubey Usha, 1990. Studies on phytoplankton dynamics and productivity fluctuation in Gandhisagar reservoir. pp. 106-112. In : Jhingran, Arun G. and V.K. Unnithan (eds.). Reservoir Fisheries in India. Proceedings of the National Workshop on Reservoir Fisheries, 3-4 January, 1990. Special Publication 3, Asian Fisheries Society. Indian Branch, Manglaore, India.
- Rao, R.J. and Garg, R.K. 2013. Conservation of wetlands in Madhya Pradesh: A case study of Ramsagar Reservoir in datia District, North Madhya Pradesh using GIS, paper available on www.gisdevelopment.net.
- Singhai, S; G.M.A. Ramani and U.S.Gupta, 1990. Seasonal variations and relationship of different Physico-chemical characteristics in Newly made Tawa Reservoir. *Limnologica* (Berlin)., 21(1) : 263-301
- Sharma, M., 1992. Hydrobiological studies of Halali Reservoir with reference to zooplanktons and fishery prospects. Ph.D. thesis, Baraktulla University, Bhopal M.P.
- Sharma Dushyant and Renu Jain, 2000. Physico-chemical analysis of Gopalpura Tank of Guna District (M.P.). *Ecol.Env. & Cons,* 6(4) : 441-445
- Sharma, Shailendra, Joshi, Vibha, Kurde, Sushma and Singhvi, M.S., 2007. Biodiversity of benthic macroinvertebrates and fish species communities of Kishanpura Lake, Indore, Madhya Pradesh. *J. Aqua. Biol.* 22(1) : 21-24.
- Solanki, Pradeep, Shiv Singh, I.V. Sharma and R. Mathur, 2011.Fish fauna of Sanjay Sagar reservoir of district Guna (M.P.). *Biological Forum.* 3(1) : 44-45.
- Tomot, Praveen, Mishra Rajeev and Somdutt, 2008. Water quality monitoring of Halali reservoir with reference to cage aquaculture as a modern tool for obtaining

enhanced fish production. In : Sengupta ,M. and Dalwani, R. (Editors). Proceedings of TAAL-2007, the 12th World Lake Conference : 318-324.

- Uchchariya, D.K. and Saksena, D.N., 2012. Zooplankton diversity of Tighra reservoir of Gwalior, Madhya Pradesh. *Flora and Fauna,* 18(2) : 233-242.
- Vijayvergia, R.P., 2008. Eutrophication: A case study of highly eutrophicated lake Udaisagar, Udaipur (Rajasthan), India with regards to its nutrient enrichment and emerging consequences. In : Sengupta, M. and Dalwani, R. (Editors). Proceedings of TAAL-2007, the 12th World Lake Conference : 1557-1560.
- Vyas, Vipin; Parashar, Vivek; Damde, Dinesh ; Singh Satyendra. 2009. Fish biodiversity of Narmada in submergence area of Indira Sagar reservoir. *Journal of the Inland Fisheries Society of India.* 41(2) : 18-25.

11

MAHARASHTRA

Maharashtra is the third largest state in India with an area of 308 lakh hectares. It is bordered by the states of Madhya Pradesh to the north, Chhattisgarh to the east, Andhra Pradesh to the southeast, Karnataka to the south, and Goa to the southwest. The state of Gujarat lies to the northwest, with the Union territory of Dadra and Nagar Haveli sandwiched in between. The Arabian Sea makes up Maharashtra's west coast. The Western Ghats or Sahyadri ranges run parallel to the coast, at an average elevation of 1,200 metres (4,000 ft). To the west of these hills lie the Konkan coastal plains, 50–80 kilometres in width and to the east of the Ghats lies the flat Deccan Plateau. The Western Ghats form one of the three important watersheds of India, from which many South Indian rivers originate, like Godavari, Bhima, Koyna and Krishna. It is the second most populous after Uttar Pradesh and third largest state by area in India.

The geographical area of Maharashtra state is 308 lakh ha and its cultivable area is 225 lakh ha. Out of this, 40% of the area is drought prone. About 7% of the area is flood prone. The highly variable rainfall in Maharashtra ranges from 400 to 6000 mm and occurs in a four month period between June-September with the number of rainy days varying between 40 and 100. The estimated average-annual availability of water resources consist of 164 km^3 of surface water and 20.5 km^3 of subsurface water. In Maharashtra, of the 5 river basin systems, 55% of the dependable yield is available in the four river basins (Krishna, Godavari, Tapi and Narmada) east of the Western Ghats. These four river basins comprise 92% of the cultivable land and more than 60% of the population in rural areas. About 45% of state's water resources are from West Flowing Rivers which are mainly monsoon specific rivers emanating from the Ghats and draining into the Arabian Sea.

Fig. 11.1. Map of Maharashtra (Not to scale)

Table 11.1. Inland Fisheries Resources of Maharashtra

Details	Number	Area (ha)
Irrigation tanks above 200 ha	192	1,85,887
Irrigation tanks below 200 ha	2065	1,01,896
Zilla Parishad and other tanks	31425	90,122
Total number of tanks	33740	3,77,905
Riverine length	19456 km	—

Table 11.2. Details of selected reservoirs in Maharashtra State

Sr.No.	Reservoir	Area (Ha)	District
1.	Dongargaon	120	Nanded
2.	Nagzari	128	Nanded
3.	Loni	143	Nanded
4.	Shirpur	117	Nanded
5.	Kedarnath	150	Nanded
6.	Sudha	103	Nanded

Contd...

Table 11.2: Contd...

7.	Mahaligi	129	Nanded
8.	Pethwadaj	192	Nanded
9.	Jamkhed	163	Nanded
10.	Chandola	114	Nanded
11.	Kundrala	144	Nanded
12.	Karadkhed	147	Nanded
13.	Yedur	174	Nanded
14.	Mannar	1560	Nanded
15.	Talni	217	Nanded
16.	Urdhuv Mannar	750	Nanded
17.	Yeldari	9472	Parbhani
18.	Karpara	551	Parbhani
19.	Masoli	373	Parbhani
20.	Dhom	1394	Satara
21.	Kanher	1123	Satara
22.	Rajewadi	812	Satara
23.	Yeralwadi	546	Satara
24.	Urmodi	813	Satara
25.	Tarli	288	Satara
26.	Ekburgii	218	Washim
27.	Adol	182	Washim
28.	Mohagavan	120	Washim
29.	Urdhuv morna	118	Washim
30.	Kolhi	112	Washim
31.	Chakteerth	122	Washim
32.	Sonala	240	Washim
33.	Adan	740	Washim
34.	Manyad	360	Jalgaon
35.	Bori	630	Jalgaon
36.	Hiwra	576	Jalgaon
37.	Dhaigaon	239	Jalgaon
38.	Bahula	374	Jalgaon
39.	Hathnur	5100	Jalgaon

Contd...

Table 11.2: Contd...

40.	Waghur	2372	Jalgaon
41.	Anjani	603	Jalgaon
42.	Manyarkheda	102	Jalgaon
43.	Sur	157	Jalgaon
44.	Mahsla	104	Jalgaon
45.	Bambrud	100	Jalgaon
46.	Kalamsara	162	Jalgaon
47.	Mahsva	125	Jalgaon
48.	Kankraj	103	Jalgaon
49.	Bhokerwadi	118	Jalgaon
50.	Nisardi	153	Jalgaon
51.	Abhora	113	Jalgaon
52.	Gul	153	Jalgaon
53.	Terna	497	Osmanabad
54.	Rui	275	Osmanabad
55.	Wagholi	160	Osmanabad
56.	Wadala	142	Osmanabad
57.	Bori	411	Osmanabad
58.	Harni	239	Osmanabad
59.	Palas-Nilegaon	119	Osmanabad
60.	Kalegaon	102	Osmanabad
61.	Bikhar Sangvi	120	Osmanabad
62.	Borgaon	116	Osmanabad
63.	Kothalwadi	104	Osmanabad
64.	Ramganga	174	Osmanabad
65.	Sangmeshwar	448	Osmanabad
66.	Arsoli	115	Osmanabad
67.	Khandeshwarwadi	159	Osmanabad
68.	Khasapur	330	Osmanabad
69.	Chandani	340	Osmanabad
70.	Sakat	246	Osmanabad
71.	Sina-Kolegaon	2228	Osmanabad
72.	Turori	144	Osmanabad

Contd...

Table 11.2: Contd...

73.	Koregaon	178	Osmanabad
74.	Kolsur	162	Osmanabad
75.	Sarwadi Naichakur	110	Osmanabad
76.	Murali	105	Osmanabad
77.	Dagaddhanora	142	Osmanabad
78.	Jakatpur	251	Osmanabad
79.	Benitura	306	Osmanabad
80.	Morna	203	Sangli
81.	Sankh	414	Sangli
82.	Basappawadi	165	Sangli
83.	Shegaon	113	Sangli
84.	Dodnalla	135	Sangli
85.	Atpadi	100	Sangli
86.	Nimbwade	140	Sangli
87.	Bhivargi	151	Sangli
88.	Bhakuchiwadi	130	Sangli
89.	Hingangaon	153	Sangli
90.	Siddhewadi	113	Sangli
91.	Ankalagai	103	Sangli
92.	Dnyaneshwar Sagar	3511	Ahemadnagar
93.	Ghatshi Paragon	188	Ahemadnagar
94.	Visapur	163	Ahemadnagar
95.	Ghod	2132	Ahemadnagar
96.	Mand ohal	154	Ahemadnagar
97.	Sina	949	Ahemadnagar
98.	Khairi	260	Ahemadnagar
99.	Adhala	135	Ahemadnagar
100.	Manar	1560	Ahemadnagar
101.	Urdhuv Manar	750	Ahemadnagar
102.	Wan		Beed
103.	Erai	4250	Chandrapur
104.	Dina	827	Gadchiroli
105.	Pujari tola	1831	Gondia

Contd...

Table 11.2: Contd...

106.	Bodalkasa	305	Gondia
107.	Pothara	450	Wardha
108.	Pen Takli	592	Buldhana
109.	Nalganaga	1100	Buldhana
110.	Tulsi	348	Kolhapur
111.	Vadagon	120	Kolhapur
112.	Katepurna	805	Akola
113.	Dagadparva	288	Akola
114.	Uma	217	Akola
115.	Vishwamitra	122	Akola
116.	Nirguna	275	Akola
117.	Wan	461	Akola
118.	Chandpur	328	Bhandara
119.	Shivnibandh	295	Bhandara
120.	Rawanvadi	170	Bhandara
121.	Sorna	116	Bhandara
122.	Nira Devghar	783	Pune
123.	Varasgaon	1178	Pune
124.	Madanwadi	170	Pune
125.	Shetphal Havelli	196	Pune
126.	Andhra khore	154	Pune
127.	Niradevghar	783	Pune
128.	Wadiwale	230	Pune
129.	Panshet	784	Pune
130.	Nazre	221	Pune
131.	Wadaj	352	Pune
132.	Thetewadi	217	Pune
133.	Bhama Askhed	884	Pune
134.	Pauna	1594	Pune
135.	Varasgaon	1178	Pune
136.	Veer	1776	Pune
137.	Chasakman	1821	Pune
138.	Dimbhe	1278	Pune

Contd...

Table 11.2: Contd...

139.	Manikdoah	1088	Pune
140.	Yedgaon	1225	Pune
141.	Pimpalgaon Jogga	2000	Pune
142.	Bhatghar	1650	Pune
143.	Kasar sai	336	Pune
144.	Waitarna	2702	Nashik
145.	Mukne	1067	Nashik
146.	Kadwa	409	Nashik
147.	Gangapur	1156	Nashik
148.	Waldevi	207	Nashik
149.	Kashpi	205	Nashik
150.	Alandi	252	Nashik
151.	Waghad	556	Nashik
152.	Palkhed	340	Nashik
153.	Karanjwan	698	Nashik
154.	Punegaon	211	Nashik
155.	Wazarkhed	465	Nashik
156.	Chankapur	664	Nashik
157.	Haranbari	297	Nashik
158.	Girna	3290	Nashik
159.	Talegaon	123	Nashik
160.	Manikpunj	119	Nashik
161.	Bhawli	182	Nashik
162.	Tisgaon	180	Nashik
163.	Bhojapur	167	Nashik
164.	Nagasakya	185	Nashik
165.	Darna	1778	Nashik
166.	Bendsura	—	Beed
167.	Manjara	—	Beed
168.	Kundlika	—	Beed
169.	Mehakari	—	Beed
170.	Borna	120	Beed
171.	Rampur	120	Solapur

Contd...

Table 11.2: Contd...

172.	Hotgi	190	Solapur
173.	Mamdapur	100	Solapur
174.	Pathri	123	Solapur
175.	Bhabulgon	130	Solapur
176.	Nerle	130	Solapur
177.	Vadshivne	107	Solapur
178.	Ekruk	1842	Solapur
179.	Ujani	39000	Solapur
180.	Dubdubi	285	Solapur
181.	Kurnur	366	Solapur
182.	Jawalgaon	654	Solapur
183.	Hingni (Pangaon)	563	Solapur
184.	Ashti	720	Solapur
185.	Tisangi (S)	1156	Solapur
186.	Mangi	221	Solapur
187.	Buddehal	346	Solapur
188.	Nimgaon (M)	324	Solapur
189.	Dhubdhubi	285	Solapur
190.	Galhati	343	Aurangabad
191.	Ambatanda	105	Aurangabad
192.	Siddeshwar	2574	Hingoli
193.	Kalamnuri	134	Hingoli
194.	Jui	132	Jalna
195.	Nagthus	186	Jalna
196.	Mandla	153	Jalna
197.	Devidhaegaon	132	Jalna
198.	Ghalati	142	Jalna
199.	Nirkheda	113	Jalna
200.	Somthana (V)	184	Jalna
201.	Rajewadi	183	Jalna
202.	Banegaon	116	Jalna
203.	Hatadi	100	Jalna
204.	Hastur Tanda	154	Jalna

Contd...

Table 11.2: Contd...

205.	Musa Bhadrayan	174	Jalna
206.	Manepuri	129	Jalna
207.	Khadkeshwar	123	Jalna
208.	Pirkalyan	343	Jalna
209.	Kalyan Girja	302	Jalna
210.	Upper Dudhna	245	Jalna
211.	Dhamana	237	Jalna
212.	Kasura	222	Jalna
213.	Dhangar pimpri	263	Jalna
214.	Nimna Dudhna	4549	Jalna
215.	Waki	164	Jalna

Table 11.3. Profile of some reservoirs in Maharashtra state

	Terna	Kundlika	Benitura	Chandani	Mehakari	Kurnur	Borna
Location	Dhokhi	Majalgaon	Omerga	Paranda	Ashti	Naldurg	Parli
Year of completion	1970	1986	1994	1965	1966	1968	1983
River	Terna	Kundlika	Benitura	Chandani	Mehakari	Bori	Borna
Type of dam	Earthfill	Earthfill	Earthfill	Earthfill	Earthfill	Earthfill	Earthfill
Height (m)	15	28.45	13.48	17.18	27.63	23.7	22.3
Length (m)	2651	1403	1780	1920	1308	1206	866
Gross storage capacity ($10^3 m^3$)	22910	46350	12810	20700	16130	35240	10908
Reservoir area ($10^3 m^2$)	380	6850	Not-Available	813	38	570	2191
Effective storage capacity ($10^3 m^3$)	18630	NA	NA	15220	13000	32670	9060
Volume content ($10^3 m^3$)	186	NA	NA	289	163.5	45	460
Purpose of dam	Irrigation	Irrigation	Irrigation	Irrigation	Irrigation	Irrigation	Irrigation

Table 11.3. (Contd.) Profile of some reservoirs in Maharashtra state

	Manikdoh	Khairi	Sakol	Devargan	Siddeshwar	Kundrala	Radhanagri
Location	Junner	Jamkhed	Udgir	Udgir	Siddeshwar	Mukhed	Radhanagri
Year of completion	1984	1989	1992	1993	1968	1969	1954
River	Kukadi	Kar	Local	Devari	Purna	Local	Bhogawati
Type of dam	Gravity	Earthfill	Earthfill	Earthfill	Earthfill	Earthfill	Gravity
Height (m)	51.8	18.91	17.65	15.68	38.26	18.5	42.68
Length (m)	930	1210	1425	1715	6353.2	999	1143
Gross storage capacity ($10^3 m^3$)	308060	15110	12689	13410	250850	14680	236810
Reservoir area ($10^3 m^2$)	18434	492	4256	4010	40580	253	18218
Effective storage capacity ($10^3 m^3$)	283070	13743	10950	10670	80940	12990	33320
Volume content ($10^3 m^3$)	596	54	371	NA	907.2	370	NA
Purpose of dam	Irrigation/ Hydro-electricity	Irrigation	Irrigation	Irrigation	Irrigation	Irrigation	Irrigation

Table 11.3. (Contd.) Profile of some reservoirs in Maharashtra state

	Khadakwasla	Vaitarna	Galhati	Sukhana	Girna	Itiadoh
Location	Haveli	Mumbai	Ambad	Aurangabad	Nandgaon	Arjuni
Year of completion	1880	1954	1966	1968	1969	1970
River	Muta	Vaitarna	Galhati	Sukhana	Girna	Garvi
Type of dam	Earthfill Gravity	Gravity	Earthfill	Earthfill	Earthfill	Earthfill Gravity
Height (m)	32.9	82	13.3	16.92	54.56	29.85
Length (m)	1539	567.07	2987	446	963.17	505
Gross storage capacity ($10^3 m^3$)	86000	204980	13840	21340	608980	288830
Reservoir area ($10^3 m^2$)	14800	8.39	NA	6782	60040	46910

Contd...

Table 11.3: Contd...

Effective storage capacity (10^3m^3)	56000	174790	NA	18480	525920	225120
Volume content (10^3m^3)	1170	0.06	NA	68	2042	911
Purpose of dam	Irrigation Water supply	Water supply	Irrigation	Irrigation	Irrigation	Irrigation

Table 11.3. (Contd.) Profile of some reservoirs in Maharashtra state

	Koradi	**Masoli**	**Tawarja**	**Sina Kolegaon**	**Purna**
Location	Mehkar	Gangakhed	Latur	Paranda	Amravati
Year of completion	1979	1981	1982	2007	NA
River	Koradi	Masoli	Tawarja	Sina	Purna
Type of dam	Earthfill	Earthfill	Earthfill	Earthfill	Earthfill
Height (m)	19.31	24.84	14.3	36.6	38
Length (m)	900	1086	2222	NA	3120
Gross storage capacity (10^3m^3)	22500	34080	20520	150490	41759
Reservoir area (10^3m^2)	6465	6970	741	1529	5880
Effective storage capacity (10^3m^3)	15120	27390	16950	89340	35370
Volume content (10^3m^3)	1193	626	361	234	1277
Purpose of dam	Irrigation		Irrigation	Irrigation	Irrigation/ Hydropower

ISAPUR RESERVOIR

The Isapur reservoir is constructed across river Penganga at village Isapur in Pusad taluka of Yeotmal district. The length of reservoir is 3730 meters with gated spillway on the right side. The maximum height of the dam is 48 meters.

It is situated within the Latitudes 19°16′30″N to 20°-30′N. Pulle *et al.* (2003) studied the annual heat budget and summer heat income of the reservoir for year 1997-99. The annual heat budget was found to be 72.6125 gm/cal/sq.ft. at S_3 during theyear 1997-98, while 67.3666 gm/cal/sq.ft. at S_1, 11.9263 gm/calsq.ft at S_2 and 54.076 gm/cal/sq.ft at S_3 during the year 1998-99. The summer heat income recorded was 743.7875 gm/cal/sq.ft at S1, 291.87 gm/cal/sqft at S_2 and 383.91 gm/cal/sq.ft at S_3 during the year 1997-98,while 753.664 gm/cal/sq.ft at S_1,293.897 gm/cal/sq.ft at S_2 and 453.895 gm/cal/sq.ft. at S_3 during the year 1998-99.

KANDHAR RESERVOIR

Kandhar town in Nanded district is bounded by 18°15′ and 19°55′N latitude and between 77°40′ and 78°15′E longitude. The general elevation above mean sea level (MSL) is 1830. Kandhar reservoir near Kandhar town is also known as Jaitung sagar. Surve *et al.* (2004) studied population dynamics of microzootic fauna of Kandhar dam. They observed monthly variations of microzootic fauna such as protozoans, rotifers, helminth eggs and arthropods.The minimum population was observed during summer while maximum during monsoon season. Rotifer was the most dominant group followed by arthropods, helminth eggs and protozoans. About 19 zooplankton species were observed by Surve *et al.* (2004).

About 3 protozoan species namely *Balantidium coli, Entamoeba* and *Giardia lambila* were observed. *Ascaris lumbricoides, Enterobius vermacularis, Fasciola hepatica, Hymenolepis nana* and *Trichuris trichura* were observed. The eggs of *Ascaris lumbricoides* were found most prevalently followed by *Fasciola hepatica, Enterobius vermacularis, Hymenolepsia nana* and *Trichuris trichura*. Six rotifer species namely *Brachionus, Filina, Chromatogaster, Keratella, Epiphanes* and *Monostylla* were observed. Among the zooplankton components, the rotifers were abundant. The dominance of rotifer may be attributed to their dependence on abundant particulate organic matter.

PETHWADAS RESERVOIR

The dam is across the confluence of three nalas, namely Wartala ala, digrus nala and Anamand nala near village Kallali in Kandhar taluka of Nanded district.The dam was constructed in 1973. It is situated between the latitude 18-47′–0″ and longitude 77–18′–0″.The main purpose of the dam is irrigation. The dam water is also used for fish culture, drinking and domestic purposes. The dam submergence of total area is 297 hecatres. The catchment area of dam is 39.2 sq.miles.The morphometry of reservoir is depicted in Table 11.4.

Water temperature ranged between 20.5 to 36°C. Secchi disc transparency ranged from 27.2 to 81.2 cm. The total solids were I the range of 250 to 410 mg/l. The total dissolved solids were recorded in the range of 186 to 310 mg/l. The total suspended solids were in the range of 56 to 154 mg/l. The dissolved oxygen was found to be in the range of 2.8 to 9.6 mg/l. Carbon dioxide was in the range of 7 to 8.3 mg/l.pH raged from 7.02 to 7.85. The total alkalinity was recorded in the range of 88 to 212 mg/l. total hardness of water varied between 128 to 168 mg/l (Table 11.5). Pawar *et al.* (2006) recorded 21 fish species from Pethwadas dam (Table 11.6).

Table 11.4. Morphometry of Pethwadaj Reservoir

1. Year of construction	1973
2. Type of dam	Earthfill
3. Height of dam (m)	19.5
4. Length of dam (m)	1260
5. Gross storage capacity (10^3m^3)	11600
6. Reservoir area (10^3m^2)	2970
7. Effective storage capacity (10^3m^3)	9040
8. Designed spillway capacity (m^3/sec)	1185
9. Volume content (10^3m^3)	495
10. Purpose	Irrigation

Table 11.5. Water quality of Pethwadas dam in Nanded district

Parameters	Range
1. Water temperature (°C)	20.5 to 36
2. Water transparency (cm)	27.2 to 81.2
3. Total solids (mg/l)	250 to 410
4. Total dissolved solids (mg/l)	186 to 310
5. Total suspended solids (mg/l)	56 to 154
6. Dissolved oxygen (mg/l)	2.8 to 9.6
7. Free Carbon dioxide (mg/l)	7 to 8.3
8. pH	7.02 to 7.85
9. Total alkalinity (mg/l)	88 to 212
10. Total hardness (mg/l)	128 to 168
11. Calcium (mg/l)	64 to 100
12. Magnesium (mg/l)	9.76 to 22.936

Table 11.6. Fishes recorded in Pethwadas dam in Nanded district

Species	Status
1. *Catla buchanani*	A
2. *Catla catla*	A
3. *Labeo rohita*	A
4. *Labeo calbasu*	A
5. *Cirrhinus mrigala*	A
6. *Barbus ticto*	A
7. *Barilius bendelis*	A
8. *Cyprinus carpio*	M
9. *Channa marulius*	A
10. *Channa striatus*	M
11. *Channa punctatus*	M
12. *Channa gachua*	M
13. *Clarias batrachus*	R
14. *Mastacembelus armatus*	R
15. *Nemacheilus botia*	R
16. *Notopterus chitala*	A
17. *Notopterus kapirat*	A
18. *Wallago attu*	A
19. *Mystus seenghala*	A
20. *Rasbora daniconius*	A
21. *Anabus* sp.	A

A = Abundance, R = Rare and M = Moderate

SHIKACHIWADI RESERVOIR

The reservoir is about 20 km from Nanded city. It was constructed near village Shikachiwadi.The reservoir water is used for irrigation and fish culture activities.The reservoir is infested with aquatic weeds like *Ipomoea carnea, Cyperus* sp., *Hydrilla* sp and *Chara* sp. The waterbody is rich with avifauna. Kulkarni *et al.* (2006) reported 93 species of birds belonging to 39 families and 16 orders. The identified birds include 51 species of residient, 32 species are resident migrant, 11 species winter migrant, 7 species migrant, 1 species breeding migrant and 2 species passage migrant.

The 15 species of birds namely red wattle lapwing, little brown dove, red-vented bulbul, white breasted kingfisher, indian myna, grey heron, common coot, white throated munia, blue rock pigeon, indian pond heron, spot bill duck, eurasian collard dove, little egret, small bee-eater and rose-ringed parakeet were common in all the seasons. Further, common swallow, black drango, indian robin were placed in the category of breeding migrant uncommon.

The little grebe, grey wagtail, common sand piper, balck winged stilt, little cormorant and little ringed plover were observed in the month of October. Babbier, suift, open bill, white ibis, pintail, large grey babbler, marsh sandpiper, grey shrike, greater covcal were recorded in the moth of November. The shrike, black necked stork, ucrlew sandpiper, sand grouse were recorded in February, while bar headed goose, paradise flycatcher, rufous tailed finch and lark were recorded in March. Brain fever bird, sunbird, ashy prinia wabbler, pied wagtail, tree pipit were recorded in the month of April and May. Flocks rosy starling was found after winter.The bird demoiselle crane was observed only once. While bellied Heron and asian paradise flycatcher were oberrved in less number. Vultures and flamingos were not recorded. Similarly comb ducks and pheasant tailed jacana were also not recorded. Black stork (a threatened species) were observed by Kulkarni *et al.* (2006).

DAHIPAL RESERVOIR

Gaike *et al.*(2011) studied hydrobiology of Dahipal dam of Jalna district. At the catchment area, the annual rainfall varied from zero to 600 mm in theyear of 2008-09. The maximum of 144 mm was recorded in the month of July and there was no rain in December, January, February and March. Atmospheric temperature varied from 20°C to 33°C. The maximum of 33°C was recorded in the month of May and minimum was 20°C in the month of December. The water temperature varied from 19°C to 30°C. The maximum of 30°C was recorded in May; the minimum of 19°C was recorded in December during 2008-09. The dissolved oxygen varied from 9.86 to 13.67 mg/l. The maximum of 13.6 mg/l was recorded in the month of March and the minimum of 10.02 mg/l was recorded in the month of September. The low dissolved oxygen values coincided with high temperature during the summer months. The free carbon dioxide varied from 4.2 to 11.5 mg/l. The maximum carbon dioxide was recorded in the month of May and minimum in the month of October. The calcium varied from 16.2 to 31.1 mg/l. The maximum of 31.1 mg/l was recorded in the month of July and minimum of 17.3 mg/l was recorded in the month of January. Chlorides in Dahipal dam varied from 30.6 to 59.3 mg/l. The maximum of 59.3 mg/l was recorded in the month of January and the minimum of 30.6 mg/l was recorded in the month of April. The water sample meets the WHO limits for the trace metals and the physico-chemical properties. The water is not polluted.

DUDHNA RESERVOIR

Dudhna reservoir was constructed on river dudhna near village Somthana in Badnapur taluka of Jalna district in Marathwada region of Maharashtra.The reservoir site is located at 75041′ to 75042′E and is about 29 kms towards west from Jalna town.The water spread area of the reservoir is about 243 ha. The height and length of reservoir is 16.5 m and 2.46 km respectively. The reservoir water irrigates 3400 ha of agricultural land.

Seed of *Catla catla, Labeo rohita, Cirrhinus mrigala* and *Cyprinus carpio* is regularly stocked in the reservoir. Shaikh *et al.* (2011) reported 27 fish species under 7 orders from this reservoir. Gill net, cast net and drag net are the common gears used in fishing.

EKBURJII RESERVOIR

Ekburjii is one of the water bodies in Washim district. The total water spread area of water body is 218 hectare. Though this waterbody was created mainly for irrigation purpose, now its water is used for domestic supply of Washim town. It is also used for fish cultivation.The salient features of the reservoir are depicted in Table 11.7.

Table 11.7. Salient features of Ekburjii Reservoir in Washim district

1. Year of construction	1964
2. River	Chandrabhaga
3. Type	Earthfill
4. Height of dam (m)	23.7
5. Length of dam (m)	830
6. Gross storage capacity (10^3m^3)	14100
7. Reservoir area (10^3m^2)	218
8. Effective storage capacity (10^3m^3)	11960
9. Volume content (10^3m^3)	566
10. Purpose	Irrigation

The reservoir is stocked with *Catla catla, Labeo rohita, Cirrhinus mrigala* and exotic carps *viz., Hypophthalmichthys moiltrix, Ctenopharyngodon idella,* and *Cyprinus carpio.*Sone and Malu (2000) reported eight predatory ad four weed fishes from this reservoir (Table 11.8). To avoid predation of stocked major carp by preadatory fishes only advance fingerlings or yearlings must be stocked in Ekburjii reservoir.The fish production from this reservoir is 1 ton/ha/yr (Sone and Malu, 2000).

Table 11.8. Fish fauna of Ekburjii Reservoir in Washim district

PREDATORY FISHES *Ophiocephalus marulius, Ophiocephalus punctatus, Ophiocephalus striatus, Ophicephalus gachua, Ophiocephalus stewartii, Notopterus chitala, Clarias batrachus, Wallago attu.*
WEED FISHES *Ambassis ranga, Puntius sarana, Puntius ticto, Nemacheilus zenatus.*

PUJARI TOLA

It is an earthfill and gravity dam built on Bagh river in year 1970 near Amgaon in Gondia district of Vidhrbha region. The height of the dam above lowest foundation is 19.2 m while the length is 2661 m. The volume content is 664 km^3 and gross storage capacity is 65110 km^3. The surface area of reservoir is 17650 km^2 with capacity of 48690 km^3. Project has a spillway of ogee type.

Length of the spillway is 187.76 m, spillway consists of 13 radial type os spillway gates. The dam has catchment area of 69.93 thousand hectares. In year 2012 NFDB sactioned funds to MFDC for stocking of the reservoir with quality seed of carps.

ADAN RESERVOIR

River Adan, a principal tributary of the Painganga lies between Long. 770.22′ Lat. 200.17′ to long 780.21′ Lat 190.9′ in Washim district of Maharashtra. This 209.21 km long river meets Painganga River in Yavatmal district. Three dams have been built on Adan; one at its origin near Sonala village other near Karanja (Lad) city and third, Gokhi dam is built on a small tributary. The river flows through scrubland and degraded type of dry deciduous forest with extensive agriculture.

Dhamani village is situated at the bank of river Adan in Washim district. This village has 25 families of fishermen and a population of 200 people completely dependent on fish of river Adan. Bhois are adept in various river fishing methods and are regularly employed for fishing in larger tanks. Their livelihood is closely intertwined with the river: they grow Singade (water chestnuts) in tanks and melons, cucumbers and vegetables on sandy stretches along the banks of streams.

In 1977 Adan dam was built in the middle stretch of river Adan for irrigation and for providing drinking water for Karanja Lad city. Height of the dam is 30.13 m and length 755 m (2,477 ft), with gross storage capacity of 18,789.97 m3.The water spread area of reservoir is about 740 ha. The dam is located at 20°24′28.90″N, 77°32′42.69″E. It is an earthfill and rockfill dam.

Dr. Nilesh Heda of Samwardhan (NGO in Washim district) has been working closely with river communities for over a decade. He studied the impact of Adan dam on riverine ecosystem. Adan Dam came up on the Adan River in 1977. The dam changed the entire structure of river and its flora and fauna. It transformed fishermen from being the kings of the river to being the slaves of the dam. Habitat transformation is the principal effect of Adan dam. Bhois of Adan now fish in stretch of 23 kms upstream of the dam, as the water levels are too low in the downstream river for most of the year. Fish in the river congregate in river pools. Each and every pool has a name and a story behind it. Fish are found in these pools throughout the year, even when the river goes dry. However, according to local fishermen, after dam construction, river flow reduced drastically, pools were filled with silt and fishing became more difficult. Accumulated silt led to unchecked growth of hydrophytes which not only resulted in increased evapotranspiration losses, but also caused intense itching during fishing.Habitat transformation and sedimentation led the Bhois to abandon traditional fishing practises like Isor, which is installing a mesh like structure of branches at riffles for fishing. Traditionally place of each Isor was fixed for a particular family and its ownership was inherited. Such traditional practises are not only intuitive, but also sustainable. Bundless agriculture, deforestation in the basin area, unchecked water extraction contributed to low flows and resultant siltation in pools.

Both diversity and abundance of fish has changed dramatically in the post dam scenario. Round the year availability of fishes, is now confined to few monsoon months. The fish like *Eutropiichthys vacha, Anguilla bengalensis, Barilius sps, Tor khudree, Tor mussullah, Gonoproktopterus kolus,* fish of family Balitoridae (Loaches) have been completely wiped out from the upside dam area of river. At the same time, the abundance of the fishes like *Garra mullya, Rasbora daniconius, Amblypharyngodon mola,* fishes of family bagridae, *Ophiocephalus* spp has reduced drastically. These species of fishes are popular among the fish consumers of the area and fishermen get good market price for them. At the same time, species which tolerate water fluctuation, but are not prized as edible fish are increasing, *e.g., Puntius sophore, Puntius ticto, Chanda nama, Parambassis ranga, Osteobrama cotio cotio, Osteobrama cotio peninsularis* are increasing. Volume of catch before and after dam is also a matter of concern. According to the locals, fish catch ranged from 10 to 20 kg per head per day and after the dam, this has come down to an average of 1 to 2 kg per head per day, which provides merely Rs 100 per day. The consequences of this phenomenon are impinging directly on the economics, health and well-being of the family.

According to knowledgeable elders of the community, Otters, which were abundant in the river, stand exterminated. Along with aquatic animals like black prawns, crabs, tortoises were also consumed and sold. The abundance of these species has fallen sharply in the post dam scenario due to changes in flows and silt regime. Siltation of the dam has resulted in choking of crab burrows and affected tortoises. Riparian vegetation is affected by stagnant water in the upstream and reduced water in the downstream of the dam. The medicinal plants which were once abundant along and in the river like *Pan Kanda,* Denali grass,trees of *Ajani, Khadkya Raj, Sanjivani, Amarvel, Pen Ghagra, Mirchi kand, Nark Kand* are now depleted. Denali grass sticks were used by the local fishermen to anchor fishing nets, now these are replaced by thermocol. Shifting the ownership in 1977, when work on Adan dam was completed, local fishermen were assured fishing rights in the reservoir. They registered a cooperative society in anticipation. However, local leaders, wealthy businessmen and government officials created a situation where year after year fishing contracts went to politically strong mafia. This contractor now hires local fishermen and provides wages per kg. Local people do not have right to fish in the reservoir and few kmabove it. 'Exotic' threat Every year, Adan reservoir is stocked with culture fishes, especially major carps. However, the fish seeds imported from the Bengal are often contaminated with seeds of invasive alien species like *Oreochromis mossambicus,* African magur (*Clarias gariepinus*), grass carp (*Ctenopharyngodon idella*), silver carp (*Hypophthalmichthys molitrix*), common carp (*Cyprinus carpio*) and so on. These fish are not consumed and crowd out the local species.

The data obtained from district fisheries departemt of Washim is depicted in Table 11.9. The data on fish seed stocking and fish catch revealed that the fish catch varied from 81.08 kg/ha/yr to 1116.21 kg/hr/yr.

Table 11.9. Fish seed stocking and catch in Adan reservoir

Year	Fish production	Kg/ha/yr	Stocking (in lakh)	Stocking (nos/ha/yr)
2008-09	60000	81.08	6.80	919
2009-10	120000	162.16	45.00	6081
2010-11	273000	368.9	100.00	13514
2011-12	826000	1116.21	50.00	6757

The information on fisheries of Katepurna reservoir (805 ha) in Akola district is given in Table 11.10. The seed stocking varied from 6,72,500 to 700,000 with an average of 849 nos/ha/yr. The fish production for the period of 2007-08 to 2009-10 ranged from 124.22 kg/ha/yr to 223.60 kg/ha/yr.

Table 11.10. Deatails of seed stocking and fish production from Katepurna Reservoir

Year	Seed Stocked (Lakh)	Fish Production (MT)	Seed Stocked (ha/yr)	Fish Production (ha/yr)
2007-08	6.725	100	746	124.22
2008-09	7.0	150	870	186.33
2009-10	7.50	180	932	223.60

The detail information on seed stocking and fish production from Kunghala reservoir (77 ha) and Dina reservoir (827 ha) of Gadchiroli district in depicted in Tables 11.11 and 11.12.

Table 11.11. Seed stcoking and fish production in Kunghala reservoir

Year	Seed Stocked (Lakh)	Fish Production (MT)
2007-08	1.00.000	0.8
2008-09	7500	1.0
2009-10	NA	NA
2010-11	NA	NA
2011-12	60,000	1.3

Table 11.12. Fisheries of Dina Reservoir

Year	Seed Stocked (Lakh)	Seed Stocked (Nos./ha/yr)	Fish Production (MT)	Fish Production (kg/ha/yr)
2007-08	5.60	677	12	14.51
2008-09	5.20	629	1.60	1.93
2009-10	4.40	532	1.55	1.87
2010-11	6.30	762	1.00	1.20
2011-12	2.50	302	45.00	54.41

The information available on Chandpur reservoir (328 ha), Shrnibandh Reservoir (295 ha) and Sorna reservoir of Bhandara district is presented in Tables 11.13, 11.14 and 11.15 respectively.

Table 11.13. Seed stocking and fish yield in Chandpur Reservoir

Year	Seed Stocked (Nos.)	Seed Stocked (Nos./ha/yr)	Fish Production (kg)	Fish Production (kg/ha/yr)
2007-08	4,00,000	1220	3800	11.58
2008-09	3,75,000	1143	1300	3.96
2009-10	4,25,000	1296	1100	3.35
2010-11	4,10,000	1250	1250	3.81
2011-12	3,00,000	915	5000	15.24

Table 11.14. Fisheries of Shrnibandh Reservoir in Bhandara district

Year	Seed Stocked (Nos.)	Seed Stocked (Nos./ha/yr)	Fish Production (kg)	Fish Production (kg/ha/yr)
2007-08	4,25,000	1441	19000	64.40
2008-09	3,50,000	1186	23000	77.96
2009-10	3,75,000	1271	17000	57.62
2010-11	4,90,000	1661	23000	77.96
2011-12	5,90,000	2000	14000	47.75

Table 11.15. Fish seed stocking and fish production in Sorna Reservoir

Year	Seed Stocked (Nos.)	Seed Stocked (Nos./ha/yr)	Fish Production (kg)	Fish Production (kg/ha/yr)
2007-08	2,00,000	1724	2100	18.10
2008-09	1.90,000	1638	2200	18.96
2009-10	1,80,000	1552	2300	19.82
2010-11	1,70,000	1466	3100	26.72
2011-12	2,00,000	1724	3100	26.72

CHHATRI LAKE

The Chhatri lake is situated on the East of the Amravati city. It is situated at above 450 meters above the sea level and located at 77°46′E longitude and 20°53.8′N latitude. It was constructed during 1880. The details of morphometry of reservoir are depicted in Table 11.16. The lake area is surrounded by weeds and grassy hilly area from North East and East sides on which the domestic animal graze adding the excreta in the area which gets decomposed and during rainy season it comes under water spread area.

Deshmukh (2001) conducted ecological investigation on Chhatri lake. The water temperature ranged between 15.2°C to 31.8°C. Temperature exhibited significant negative correlation with dissolved oxygen and free carbon dioxide and significant positive correlation with silicates and phosphate content of the lake. TDS fluctuated between 450 mg/l (December 1997) to 100 mg/l (August 1997). TDS showed moderately significant positive correlation with nitrate contents. Turbidity exhibited two peaks in a year one in the month of May 1998 and other in August 1997.Conductivity ranged between 127.67μ mhos/l and 236.16 μ mhos/l. It showed positive correlation with silicates and sulphates. The pH of water ranged from 8.2 to 9.2. Maximum dissolved oxygen was recorded in winter season. Dissolved oxygen showed moderately significant correlation with temperature. Free carbon dioxide was found mainly during April, May and June only, which may be due to depletion of dissolved oxygen and its presence can also be due to anaerobic digestion of dead aquatic plants during the periods. The magnesium hardness was more. The high quantities of magnesium could interfere in the fish metabolism and hence the lake water is not suitable for intensive fish culture (Authors of thesis). Chloride ranged between 21.65 mg/l to 77.11 mg/l with maxima in rainy season. The high chlorides in lake water are due to organic wastes of animals and nearby industries. A sudden increase in sulphate was observed from January reaching to maximum in April. The higher values of sulphate during summer can be attributed to the evaporation of water while the higher values in rainy season could be due to the entry of sulphates from catchment area of the lake due to run off.

Table 11.16. Morhometric features of Chhatri lake of Amravati

1. Average water spread area (at full supply level)	2.9600 sq.miles
2. Average water spread area (at high flod level)	3.2685 sq.miles
3. Gross storagecapacity	325 lac gallons
4. Maximum depth at high flood level	15feet
5. Mean depth at high flood level	13.2 feet
6. Catchment area	3.6659 sq.miles

Deshmukh (2001) observed 28 species of chlorophyceae, 18 species of bacillariophyceae,13 species of myxophyceae and 3 species of euglenophyceae.

Deshmukh (2001) reported 62 species of zooplankton from the lake. Summer months exhibited higher population of zooplankton and during rainy season least population of zooplankton was observed. Rotifers dominated the zooplankton population (42.30%) followed by cladocerans (22.73%), copepods (17.22%), protozoans (15.23%) and ostracods (2.86%). The group rotifera was represented by 28 species, followed by twelve species of protozoans, nine species of cladocerans, eight species of copepods and five species of ostracods.

Deshmukh (2001) observed 28 species of insects, out of which 5 were coleopteran, 7 were hemiptera, 5 were odonata, 2 were trichoptera, 5 from dipteral and 4 from ephemeroptera.

Deshmukh (2001) recorded 32 fish species from Chhatri lake. 17 species from order cypriniformes, 5 species from order channiformes, 4 species each from orders perciformes and siluriformes and one species each from order osteoglossiformes and mastacembeliformes.

RAWANWADI RESERVOIR

The reservoir is located near Rawanwadi village about 20 km from Bhandara.The average water spread area of reservoir is 170 ha. Recently, during 2011-12, the society purchased 2,00,000 seed and fish production in the same year was 10,000 kg *i.e.*, 58.82 kg/ha/yr.

Kalbande *et al.* (2012) reported 29 species of fishes from Rawanwadi lake,out of which 19 were of order cypriniformes, 2 of perciformes, 1 of clupeiformes, 4 of ophicephaliformes, 2 of siluriformes and 1 of synbranchiformes (Table 11.17). Orderwise distribution showed dominance of cypriniformes with 12 species followed by ophiocephaliformes with 4, siluridae and perciformes with 2 and clupeiformes and synbranchiformes each with one.

Table 11.17. Fishes of Rawanwadi Reservoir in Bhandara district

ORDER CYPRINIFORMES	*Mystus vittatus, Mystus seenghala,Mystus cavassius, Lepiodcephalus guntea, Ompak pabda, Clarias batrachus, Oxygaster bacaila, Glossogobius giuris, Chela bacila, Catla catla, Labeo rohita, Cirrhinus mrigala, Cyprinus carpio, Rasbora daniconius, Rasbora rasbora, Puntius amphibious, Puntius ticto, Puntius sophore, Heteropneustes fossilis*
ORDER CLUPEIFORMES	*Notopterus notopterus*
ORDER OPHIOCEPHALIFORMES	*Ophiocephalus punctata, Ophiocephalus striatus, Ophiocephalus maruliuslu, Ophicephalus oreientalis*
ORDER SILURIDAE	*Wallago attu, Bagarius bagarius*
ORDER PERCIFORMES	*Nandus nandus, Chanda ranga*
ORDER SYNBRANCHIFORMES	*Mastacembelus armatus*

PENTAKLI RESERVOIR

It is a medium irrigation project constructed on river Painganga near Pentakli village in Mehakar taluka of Buldhana district. The water storage capacity of dam is 67.355 mcum. The average rainfall at dam site is 761.32 mm. The dam water is used for drinking, irrigation, fishing and domestic use. Ubharhande and Sonawane (2012) reported 21 fish species belonging to 10 families (Table 11.18). Cyprinidae was dominant with 10 species followed by channidae and mastacembelidae with 2 species, balitoridae, bagridae, claridae, belontidae, notopteridae, cichlidae and poecilidae 1 species each. Ubharhande and Sonawane (2012) also reported that the water is suitable for fisheries.

Table 11.18. Fish fauna of Pentakli Reservoir in Buldhana district

FAMILY CYPRINIDAE	*Labeo rohita,Puntius sarana,Puntius ticto, Salmostoma phulo, Garra lamta, Rasbora daniconius, Catla catla, Cirrhinus mrigala, Cyprinus carpio, Hypopthalmichthys molitrix*
FAMILY BALITORIDAE	*Neamacheilus beavani*
FAMILY MASTACEMBELIDAE	*Masatcembelus armatus, Mastacembelus pancalus*
FAMILY BAGRIDAE	*Mystus bleekeri*
FAMILY CLARIDAE	*Clarias batrachus*
FAMILY NOTOPTERIDAE	*Notopterus chitala*
FAMILY CHANNIDAE	*Channa punctatus, Channa orentalis*
FAMILY CICHLIDAE	*Oreochromis mossambicus*
FAMILY BELONIDAE	*Xenentedon cancila*
FAMILY POECILIDAE	*Poecilia reticulate*

NALGANAGA RESERVOIR

Nalganga reservoir (1100 ha) is one of the most important water bodies in the Buldana District, of Vidharbha region providing drinking water to villages of the Buldana district. The reservoir was constructed across Nalganga River and is located between Longitudes 7°15′E – 7°32′E and Latitudes 4°32′ – 4°37′N in the western part of the Vidharbha region. The area has the normal vegetation comprising of trees, various plants and aquatic macrophytes. It is characterized by high ambient temperature usually about 35.5°C and above. Relative humidity fluctuates between 60 and 95% and rainfall averaging about 1500 to 2000 mm (Bharambe, 2013). This high rainfall often increases the volume of water in the reservoir hence providing good fishing opportunity for the residents. Fishing is one of the major activities in Nalganga reservoir.

The seasonal variation of the physical and chemical parameters is given in Tables 11.19, 11.20 and 11.21. Dry season mean values of the physical and chemical parameters are shown in Table 11.19; while the wet season values are on Table

11.20. The combined seasonal parameters are shown in Table 11.21. The dry season ambient temperature range from 31.76 – 32.74°C (32.27 ± 0.40°C) while the wet season range was 29.17 – 30.14°C (± 29.58 ± 0.41°C (28.80 ± 15°C) while the wet season values were between 26.45 and 26.90°C (26.70 ± 0.19°C). The pH recorded in the dry season was between 6.01 and 6.13 (6.08 ± 0.05) while the wet season values were 6.55 – 6.87 (6.74 ± 0.14). The total alkalinity in the dry season range from 9.50 to 9.72 mg/L with a mean of 9.59 ± 0.9 mg/L. This was higher than the wet season which was between 6.73 and 9.34 mg/L (7.90 ± 1.08 mg/L). The electrical conductivity in the dry season was within the range of 13.39 and 22.22/cm (16.08 ± 4.48 while the wet season values range from 11.68 to 15.46/cm (13.44 ± 1.55). The turbidity values in the dry season were between 2.44 and 2.56 NTU (3.40 ± 1.31). This was lower than the wet season values which were between 5.25 and 5.79 NTU (5.61 ± 0.25 NTU). The dry season dissolved oxygen values ranged between 5.17 and 5.83 mg/L (5.63 ± 0.33 mg/L). This was lower than the wet season values which range between 5.89 and 6.43 mg/L (6.14 ± 0.22 mg/L). The salinity in the wet season was zero for all the stations. However, station 3 recorded salinity of 0 to 0.09‰ (0.03 ± 0.04‰). The phosphate values in the dry season range between 0.05 and 0.14 mg/L (0.11 ± 0.04 mg/L) while the wet season range was between 0.05 and 0.09 mg/L (0.06 ± 0.02 mg/L). The nitrate values in the dry season range between 0.33 to 0.56 mg/L (0.42 ± 0.10 mg/L) which was higher than the wet season values which range 0.11-0.34 mg/L (0.21 ± 0.10 mg/L).

Table 11.19. Dry season mean values and standard deviation of physico-chemical parameters of water at the various stations in Nalganga reservoir

Physico-Chemical					
Parameters	Station1	Station 2	Station 3	Mean	SD
Ambient temperature (°C)	32.31 ± 0.70	31.76 ± 1.14	32.74 ± 1.26	32.27	0.40
Water temperature (°C)	29.02 ± 2.22	8.73 ± 1.02	28.66 ± 1.88	28.80	0.15
pH	6.13 ± 0.39	6.01 ± 0.42	6.09 ± 0.36	6.08	0.05
Alkalinity (mg/L)	9.50 ± 0.94	9.56 ± 0.93	9.72 ± 1.50	9.59	0.09
Conductivity (s/cm)	13.39 ± 2.33	22.22 ± 2.04	14.33 ± 3.56	16.08	4.48
Turbidity (NTU)	2.56 ± 0.83	2.50 ± 0.53	2.44 ± 6.37	3.40	1.31
Dissolved oxygen (mg/L)	5.49 ± 0.80	5.17 ± 1.02	5.83 ± 0.98	5.63	0.33

Contd...

Table 11.19: Contd...

Salinity (%)	0.00	0.00	0.09 ± 0.07	0.03	0.04
Phosphate (PO4) (mg/L)	0.05 ± 0.03	0.14 ± 0.23	0.14 ± 0.18	0.11	0.04
Nitrates (NO_3) (mg/L)	0.33 ± 0.40	0.38 ± 0.31	0.56 ± 0.99	0.42	0.10

Table 11.20. Wet season mean values and standard deviation of physical and chemical parameters at the various station in Nalganga reservoir

Physico-Chemical					
Parameters	**Station1**	**Station 2**	**Station 3**	**Mean**	**SD**
Ambient temperature (°C)	29.17 ± 0.92	29.42 ± 1.23	30.14 ± 1.77	29.58	0.41
Water temperature (°C)	26.74 ± 0.55	26.90 ± 0.88	26.45 ± 0.76	26.70	0.19
pH	6.80 ± 0.39	6.55 ± 0.57	6.87 ± 0.44	6.74	0.14
Alkalinity (mg/L)	7.62 ± 2.44	9.34 ± 2.48	6.73 ± 1.42	7.90	1.08
Conductivity (s/cm)	11.68 ± 1.68	13.19 ± 3.30	15.46 ± 4.27	13.44	1.55
Turbidity (NTU)	5.25 ± 1.92	5.79 ± 2.27	5.79 ± 2.30	5.61	0.25
Dissolved oxygen (mg/L)	5.89 ± 0.07	6.09 ± 0.51	6.43 ± 0.89	6.14	0.22
Salinity (%)	0.00	0.00	0.00	0.00	0.00
Phosphate (PO_4) (mg/L)	0.09 ± 0.14	6.05 ± 0.02	0.05 ± 0.04	0.06	0.02
Nitrates (NO_3) (mg/L)	0.11 ± 0.19	0.18 ± 0.28	0.34 ± 0.54	0.21	0.10

(Source: Bharambe and Patil, 2013)

Table 11.21. Combined seasonal mean values, standard deviation and size range of dry and wet season physico-chemical parameters in water in Nalganga reservoir

	Dry season		Wet season	
Parameter	Mean ± SD	Range	Mean ± SD	Range
Ambient temperature (°C)	32.27 ± 0.40	31.76 – 32.74	29.58 ± 0.41	29.17-30.14
Water temperature (°C)	28.80 ± 0.15	26.66 – 29.02	26.70 ± 0.19	26.45 – 26.90
pH	6.08 ± 0.05	6.01 – 613	6.74 ± 0.14	6.55 – 6.87
Alkalinity (mg/L)	9.59 ± 0.09	9.50 – 9.72	7.90 ± 1.08	6.73 – 9.34
Conductivity (s/cm)	16.08 ± 4.45	13.39 – 22.22	13.44 ± 1.55	11.68 – 15.46
Turbidity (NTU)	3.40 ± 1.31	2.44 – 2.56	5.61 ± 0.25	5.25 – 5.79
Dissolved oxygen (mg/L)	5.63 ± 0.33	5.17 – 5.83	6.14 ± 0.22	5.89 – 6.43
Salinity (%)	0.03 – 0.04	0.00 – 0.09	0.00 – 0.00	0.00 – 0.00
Phosphate (PO_4) (mg/L)	0.11 ± 0.04	0.05 – 0.14	0.06 ± 0.02	0.05 – 0.09
Nitrates (NO_3) (mg/L)	0.42 ± 0.10	0.33 – 0.56	0.21 ± 0.10	0.11 – 0.34

The seasonal variation of the physical and chemical parameters is given in Tables 11.19, 11.20 and 11.21. Dry season mean values of the physical and chemical parameters are shown in Table 11.19; while the wet season values are on Table 11.20. The combined seasonal parameters are shown in Table 11.21. The dry season ambient temperature range from 31.76 – 32.74°C (32.27 ± 0.40°C) while the wet season range was 29.17 – 30.14°C (± 29.58 ± 0.41°C (28.80 ± 15°C) while the wet season values were between 26.45 and 26.90°C (26.70 ± 0.19°C). The pH recorded in the dry season was between 6.01 and 6.13 (6.08 ± 0.05) while the wet season values were 6.55 – 6.87 (6.74 ± 0.14). The total alkalinity in the dry season range from 9.50 to 9.72 mg/L with a mean of 9.59 ± 0.9 mg/L. This was higher than the wet season which was between 6.73 and 9.34 mg/L (7.90 ± 1.08 mg/ L). The electrical conductivity in the dry season was within the range of 13.39 and 22.22 s/cm (16.08 ± 4.48 while the wet season values range from 11.68 to 15.46 :/cm (13.44 ± 1.55). The turbidity values in the dry season were between 2.44 and 2.56 NTU (3.40 ± 1.31). This was lower than the wet season values which were between 5.25 and 5.79 NTU (5.61 ± 0.25 NTU). The dry season dissolved oxygen values ranged between 5.17 and 5.83 mg/L (5.63 ± 0.33 mg/L). This was lower than the wet season values which range between 5.89 and 6.43 mg/L (6.14 ± 0.22 mg/L). The salinity in the wet season was zero for all the stations. However, station 3 recorded salinity of 0 to 0.09% (0.03 ± 0.04%). The phosphate values in the dry season range between 0.05 and 0.14 mg/L (0.11 ± 0.04 mg/L) while the wet season range was between 0.05 and 0.09 mg/L (0.06 ± 0.02 mg/L). The nitrate values in the dry season range between 0.33 to 0.56 mg/L (0.42 ± 0.10 mg/L) which was higher than the wet season values which range 0.11 – 0.34 mg/L (0.21 ± 0.10 mg/L).

PAROLA RESERVOIR

Parola reservoir is one of the minor irrigation projects in Marathwada region.Its construction was started in 1964-65 and completed in 1971. The reservoir is very close to village Parola in Hingoli district.Reservoir was constructed across a tributary of Kayadhu river which itself is a tributary of river Godavari. Reservoir site is about 2 miles away from Hingoli-Akola state highway. The reservoir is situated in the Latitude 19°47′ and Longitude 77°09′. The reservoir submergences the total area of 177.80 acres.The gross commanding area of dam is 1559 acres,while the catchments is spread across 6.76 sq.miles. The reservoir water is mainly used for irrigation and fishing purpose.

Jayabhaye and Madlapure (2006) reported 28 species of zooplankton from Parola reservoir, out of which 14 rotifers, 5 copepods, 3 ostracods and 6 were species were cladocerans.The *Brachionus* spp. dominated the zooplankton in Parola reservoir (Table 11.22).

Table 11.22. Zooplankton diversity in Parola Reservoir

Rotifera	*Asplanchana sp., Asplanchana intermedia, Brachionus durgae, B.forficula, B.pallas, B.angularis, B.calycifloris, B.calyciflorus vandoreas, B.rubens, Filinia bory, F.terminales, F.longiseta, Keratella bory, Keratella quadrata*
Copepoda	*Argulus foliceous, Cyclops, Mesocyclops* sp., *Microcyclops* sp., *Phyllodiaptomus* sp.
Ostracoda	*Cypris, Stenocypris, Strandesia*
Cladocera	*Alona rectangular richardi sars, Ceriodaphnia laticaudata, Ceriodaphnia cornuta, Moina, Moina branchiate jurine, Moina micrura.*

Jayabhaye *et al.* (2007) recorded 43 phytoplankton species belonging to 4 algal groups *i.e.*, chlorophyceae, bacillariophyceae, cyanophyceae and euglenophyceae from Parola reservoir. The maximum phytoplankton population was during summer ad minimum during rainy season. Among the phytoplankton species members of chlorophyceae were found to be dominant. The chlorophyceae were represented by 18 species with dominance of *Chlorella* sp., *Pediastrum* sp., and *Scendesmus* sp. The maximum polulation of mebers of chlorophyceae was recorded in May. Bacillariophyceae were represented by 10 species with dominance of *Navicula* sp., *Diatom* sp. and *Synedra* sp. The maximum pupolation of bacillariophyceae was recorded in January. Cyanophyceae was also more among the phytoplankton assemblage of the reservoir. Their population was maximum in May. It was mainly represented by *Oscillatoria sp.* and *Anabaena sp.* Euglenophyceae was represented by 7 species. The maximum population of euglenophyceae was recorded in May.

Table 11.23. Phytoplankton diversity in Parola Reservoir

Chlorophyceae	*Ankistrodesmus falcatus, Chlamydomonas conferta, Chlorella conglamerata, C.valgoris, Cladophora, Closterium limeticum, Cosmarium contractum, Helimada species, Hydrodictyon, Micrasterias species, Oedogonium patulum, Pediastrum duplex, P.simplex, Scendesmus armatus, S.carinatus, Spirogyra, Ulothrix zonata, Zygnema* sp.
Bacillariophyceae	*Bacillaria paradoxa, Diatom sp., Diatom vuloare, Fragillaria capurina, Navicula gracilis, N.rediosa, N.viridula, Nitzschia subtilis, Synedra affinis, Synedra ulna.*
Cyanophyceae	*Anabaena constricta, Anacysitis sp., Aphanothee nidulanus, Merismopedia punctata, Microcystis aerugenose, Nostoc, Oscillatoria chlorine, O.limosa, O.tenuis, Phormidium muciola.*
Euglenophyceae	*Euglena pisciformes, E.acus, E.stellata, E.viridis, E.anabaena ver.mima.*

Jayabhaye *et al.* (2006) reported 11 species of fishes belonging to 5 orders (Table 11.24).

Table 11.24. Fishes recorded in Parola Reservoir

Order : Cypriniformes	*Catla buchanani, Labeo rohita, Cirrhina mrigala, Barilius bendelisis, Chela untrahi*
Order : Ophiocephaliformes	*Ophiocephalus gachua, Ophiocephalus micropeltes*
Order : Clupeiformes	*Notopterus kapirat*
Order : Mastacembeliformes	*Mastacembelus guentheri*
Order : Siluriformes	*Wallgao attu, Macrones leucophasis*

KALAMNURI RESERVOIR

Waghmare amd Mali (2007) studied phytoplankton diversity in Kalamnuri reservoir and reported 26 species of phytoplankton belonging to four groups (Table 11.26). The diversity of chlorophyceae with 10 sepcies was the highest to be followed by bacillariophyceae with 7 species, cyanophyceae with 5 and euglenophyceae with 4 species.The phytoplankton population varied from 90 to 685/ml at station I, 180 to 877/ml at station II and 172 to 766/ml at sampling station III.

Table 11.25. Morphometry of Kalamnuri Reservoir

1. Name of Project	Kalamnuri Reservoir
2. Location	Longitude 77°15′ Latitude 19°40′
3. Basin	Godavari
4. Sub-Basin	Kayadhu
5. Name of River/ Nala	Local nala
6. Catchment Area (sq.km.)	56.28
7. Type	Minor
8. Year of completion	1963
9. Type of dam	Earthen dam
10. Gross storage (mm^3)	4.823
11. Live storage (mm^3)	4.208
12. Dead storage (mm^3)	0.6157
13. FRL (m)	444.09
14. Spillway Crest Level (m)	444.09
15. MDDL (m)	441.05
16. Maximum dam height (m)	11,89
17. Sill level of head regulator/s (m)	441.05

Table11.26. Checklist of phytoplankton species recorded in Kalamnuri Reservoir

Bacillariophyceae: *Bacillaria paradoxa, Diatom* sp., *Diatom vuloare, Fragillaria capurina, Navicula radiosa, Navicula raiosa, Navicula viriduia.*
Chlorophyceae: *Cosmarium contractum, Chlorella valgoris, Chorella conglamerata, Ankistrodesmus falcatus, Pediastrum duplex, Odedogonium patulum, Spirogyra, Chlmydomonas conferata, Hydrodiction, Cladophora.*
Cyanophyceae: *Anabaena constricta, Anacystis* species, *Nostoc, Oscillatoria chlorina, Oscillatoria tenuis.*
Euglenophyceae: *Euglena viridis, Euglena stellata, Euglena acus, Euglena pisciformes.*

VISHNUPURI RESERVOIR

The idea of this project was put forward and pursued by former Chief Minister of Maharashtra late Shri Shankarrao Chavan. Hence in his remembrance the Govt. of Maharashtra has named the water reservoir as the Shankar Sagar Jalashaya. It was constructed across the river Godavari near Asarjan village, at about 8 k.m. from Nanded city. This is one of the largest lift irrigation projects in Asia. The project was completed in the year 1988. The back water covers 40 k.m. length of the river Godavari. Culturable Command Area of project is 23222 ha and irrigable command area is 19514 ha. Up till now irrigation Potential of 15856 ha is created. The command area of this project is distributed in Nanded, Kandhar and Loha taluka of Nanded district. Live storage of project is 80.79 Million cubic meters Out of which 43.95 million cubic meters storage is reserved for drinking purpose for Nanded city and 10.26 million cubic meters storage is reserved for Industrial purpose. The barrage has 18 vertical gates. With the huge water reservoir, presence of Kaleshwar temple, Landscaping, the Ghat and the Ratneshwari hills in the surrounding the place has been attracting pilgrims and tourists. Every year on the day of Mahashivratri lacks of pilgrims assemble for the Yatra. This is one of largest lift irrigation project in Asia sub-continent located on Godavari river.

Fig. 11.2. A panoramic view of Vishnupuri Reservoir of Nanded District

Fig.11.3. Coracle used in Vishnupuri Reservoir of Nanded

The nets are operated by using the coarcale (Fig. 11.3). Cast net, gill net and hook and lines are commonly used for fishing. The fish fauna comprises *Notopterus* spp, *Channa spp, Tillapia mossambica, Wallago attu, Rohtee ogilbii, Chela bacila, Puntius* spp, *Catla catla, Labeo rohita, Cirrhinus mrigala, Cyprinus carpio, Hyopophthalmichthys molitix, Mystus seenghala, Mystus cavassius, Clarius batrachus, Macrognathus pancalus,Salmostoma clupeoides* and *Rasbora daniconius*.

The zooplanktons recorded in the Vishnupuri water are *Indialona ganapati, Moina micrura, Alona rectangular, Ceriodaphnia cornulu, Diaphanosoma excisum, Diaptomus marshianus, Mesocyclops leukarti, Mesocyclops hyalinus, Stenocypris* sp, *Cypris obensa, Stenocypris* sp, *Cyclocypria osborni, Brachionus flacatus, Brachionus calyciflorus, Brachionus diversicornis, Filia longiseta, Keratellla tropica, Keratella quadrata, Trichocera longiseta, Vorticella* spp., *Arcella* spp. and *Paramecium* sp.

SUDHA RESERVOIR

The Sudha Dam is constructed across river Sudha about 4 km from Bhoker in Nanded district. It is the hilly area. The Sudha dam was constructed earlier and it is on the Kinwat road and at Bhoker towards eastern. The dam is situated 19015′ latitude and 73043′ longitude. The catchments area of the dam is 106 sq.km. Sudha river is emerging from Sitakhandi near Bhoker. The flow of water is from west to east in the direction.The area covered by this project is about 103 ha.It supplies water to Bhoker and several near by villages.Reservoir water is the below the barium, fluoride and pH permissible limits (Sayyed and Bhosale, 2012). Water temperature ranged from 22.75°C to 25.25°C. Fluoride levels showed only narrow variation range i.e. 0.21 mg/L to 1.26 mg/l with an average of 0.533 mg/L. Barium ranged from 0.11 mg/l to 0.26 mg/l and pH from 7.27 to 7.47.

A preliminary survery on fish and fisheries of Sudha reservoir revelaed that the fishery is supported by Indian major carps and local varieties like *Puntius sophore, Puntius ticto, Chela phulo, Channa marulius, Channa striatus, Channa gachua, Mastacembelus armatus, Rohtee cotio, Mystus seenghala, Mystus cavassius, Mystus aor, Ambassis ranga* and *Notopterus notopterus*.

TALWADE RESERVOIR

The Talwade reservoir is situated near village Ravalgaon in Malegaon taluka of Nashik district. It is situated at 74°20′9″ latitude and 20°28′ longitude. The length of reservoir is 1418 meters, maximum width is 86 meters and maximum height is 13.85 meters. The water storage capacity of this reservoir is 13.85 mcft. The reservoir plays an important role in supplying water to Malegaon city and in also maintaining water table. The Water of this reservoir in also used by local people for washing clothes, irrigation and for cattle washing. Shastri and Bhogaonkar (2006) studied water quality of Talwade reservoir. The values of physico-chemical parameters are given in Table 11.27.

The temperature of the water varied from 18°C to 26°C. The minimum temperature 18°C was recorded in December at all the sampling stations. The highest 26°C was recorded in May at the three stations. The gradual increased in temperature was observed from January to onwards.

Minimum pH value of 6.5 was recorded in the month of June, while maximum pH of 9.16 in the month of July.

Maximum dissolved oxygen (36.24 mg/l) was recorded in July, while minimum of 5.23 mg/l in February. A strong correlation between pH and dissolved oxygen was observed (Shastri and Bhogaonkar, 2006).

Table 11.27. Water Quality of Talwade reservoir

Parameters	Range
Temperature	18°C – 26°C
pH	6.5 – 9.16
Dissolved oxygen	5.23 – 36.24
Free CO_2	00 – 132
Calcium	7.21 – 40.08
Total Hardness	112 – 176
Total Alkalinity	PA = 00 – 2.5 TA = 20 – 35
Phosphate	00 – 0.52
Nitrate	00 – 0.019

(All values in mg/l, PA = Phenolphthalein Alkalinity, TA = Total Alkalinity)

Free carbon dioxide was totally absent from January to June. The maximum value of free CO_2 level *i.e.,* 132 mg/l was recorded in the month of November. Water showed free CO_2 in morning samples due to biotic community respiration. During day time water was generally CO_2 free due to photosynthesis (Shastri and Bhogaonkar, 2006).

Calcium is very important element influencing the flora of ecosystem which plays potential role in the metabolism and growth. The maximum calcium concentration 40.08 mg/l was recorded in November, minimum content of calcium i.e. 7.21 mg/l was recorded in the month of February. Total hardness varied from 12mg/l to 176mg/l.

Phenolphthaline alkalinity was absent from July to December. Methyl orange alkalinity however, exhibited a range of wide fluctuation. The minimum (22.5 mg/l) and maximum (35 mg/l) values were recorded in February and October respectively.

Overall the phosphate concentration was low. The maximum of 0.52 mg/l was recorded in February. It was absent in April. Zero values of phosphate may be due to the low productivity of reservoir.

Water samples showed lower concentration of nitrate. Generally water bodies polluted by organic matter exhibit higher values clearly indicate oligotrophic nature of low nutrient levels.

From the above study of above parameters, it can be concluded that the reservoir water is hard and calcium rich. The low values of nitrate and phosphate suggest that reservoir is oligotrophic in nature.

DAHIKHUTA RESERVOIR

Shastri and Pendse (2001) studied hydrobiology of Dahikhuta reservoir of Nashik district. The reservoir is located near Malegaon.The water temperature varied from 18°C to 29°C, showing higher temperature value (29°C) in September and (18°C) in July 1998. Higher temperature in September is may be due to clear atmosphere and greater solar radiation. Minimum water temperature can be explained on the basis of frequent clouds, high humidity, high current velocity and high water levels. pH of the reservoir water varied between 7.38 to 7.91. Maximum value of pH in the month of September and minimum value in the month of October was observed. Low pH in the month of October may be due to the temperature conditions. Minimum 94.83 mg/l of dissolved oxygen was recorded in July, while maximum of 12.08 mg/l was recorded in the month of November. Free CO_2 was totally absent from August to November. The maxmum value of free CO_2 level *i.e.,* 26.4 mg/l was recorded in the month of December. Water showed free CO_2 in morning samples due to the biotic community respiration. During day time due to photosynthesis water was generally CO_2 free. Phenolphthalein alkalinity showed a range from 00 to 40 mg/l. Zero Phenolphthalein alkalinity was recorded in the month of July, October, November and December. Total Alkalinity showed a variation between 10 to 90 mg/l. Minimum value (10 mg/l) was recorded in the month of July, while maximum of (90 mg/l) was in August. The low alkalinity in the month of July was due to dilution effect.The range of calcium content varied from 19.23 to 40.08 mg/l. The maximum calcium concentration 40.08 mg/l was recorded in August, minimum content of calcium *i.e.,* 19.23 mg/l was recorded in the month of October. The total hardness values in reservoir water varied within a range of 82 to 134 mg/ l. The highest value of hardness (134 mg/l) was observed in the month of October. The lowest value of 80 mg/l was observed in the month of September. The maximum amount of chlorides was (63.9 mg/l) in the month of October and minimum (31.24 mg/l) in the month of July. The maximum amount of chloride in October conferred by the influx of highly contaminated domestic sewage. The phosphate concentration was low. The minimum of 0.12 mg/l was recorded in the month of July and minimum of 00 mg/l in the month of December. Zero value of phosphates is may be due to low productivity of reservoir.Reservoir water showed lower concentration of nitrites *i.e.,* nil to 0.09 mg/l. TDS value ranged from 0.146 to 0.207 ppm. The minimum values were recorded in the month of July and maximum value in the month of September.

Shastri and Pendse (2001) also reported algal diversity from the same reservoir. They listed three algal groups *i.e.,* cyanophyceae, chlorophyceae and bacillariophyceae. Altogether 19 algal taxa were reported in the reservoir (Table 11.28).

Shastri and Pendse (2001) finally concluded that the reservoir water is moderately hard to hard. The presence of species like *Oscillatoria, Clostorium, Fragilaria, Synedra, Phormidium* and *Gomphonema* cleanly indicate that the reservoir water is polluted.

Table 11.28. Algal Diversity In Dahikhuta Reservoir

Cyanophyceae: *Oscillatoria homogenea, Fremy, Phormidium* spp., *Merismopedia Punctata meyen, Merismopedia glavca* (Ehr.) Nag.
Chlorophycesae: *Ulothrix* spp., *Oedogonium* spp., *Closterium tumidum, Chlamydomonas chrenbergii, Scenedesmus obliqus and Paze.*
Bacillanophyceae: *Nitzschia apiculata, Fragilaria capusina, Synedra tubulata, Achnanthes nagpurensis, Achnanthes inflata, Stauroneis phoenicenteron, Cymbella hungarica, Gomphonema subtile, G. parvlum, G. olivaceum.*

The information on fisheries of seed stocking and fish production from 14 reservoirs of Nashik district in depicted in Table 11.29.

Table 11.29. Details of few Reservoirs of Nashik District

Reservoir	Taluka	W.S. Area (ha)	Seed stocked (Lakh)	Fish Production (MT)	Fish Production (kg/ha/yr)
1. Darna	Igatpuri	3475	2.77	117	33.66
2. Vaitarna	Igatpuri	2702	12.00	60	22.20
3. Mukne	Igatpuri	1067	6.50	12	11.24
4. Gangapur	Nashik	1156	15	15	12.97
5. Valdevi	Nashik	207	5	22	106
6. Kashap	Nashik	205	2	17	82.92
7. Alandi	Dindori	252	4	8	31.74
8. Palkhed	Dindori	340	5	10	29.41
9. Karanjvan	Dindori	698	10	18	25.78
10. Panegaon	Dindori	211	5	40	189.57
11. Ozarkhed	Dindori	465	6	15	32.25
12. Chankapur	Kalvan	664	5.50	105	158.13
13. Haranbari	Satna	297	1.50	3.50	11.78
14. Girna	Malegaon	3290	10	30.50	9.36

SONVAD RESERVOIR

The sonvad reservoir is located near the village Dongaon in Shindkheda taluka of Dhule district. It is constructed on river Bhat.The reservoir is located at Latitude 21°4′ North and Longitude 74°50′ East. Nandan and Jain (2005) reported algal communities, which are used as indicators of organic pollution (Table 11.30). Altogether 16 pollution tolerant species were observed (Nandan and Jain, 2005).

Table 11.30. Pollution tolerant species of algae from Sonvad reservoir

Scenedesmus quadricuda, Oscillatoria tenuis, Syedra ulna, Ankistrodesmus falcatus, Chlorella valgaris, Cyclotella meneghiniana, Navicula cryptocaphala, Pediastrum boryanum, Phacus pleronectus, Navicula viridula, Phormidium autumnate, Cocconeis placentula, Scenedesmus diamorphus, Navicula cuspidate, Closterium leibleini, Tetraedron muticum.

DEVBHNAE RESERVOIR

The Devbhane reservoir is situated near village Devbhane of Dhule taluka. It lies at 21°2′ North Latitude and 74°48′ East Longitude. Nandan and Jain (2005) reported 12 pollution tolerant species of algae from Devbhane reservoir.

Table 11.31. Pollution tolerant species of Algae from Devbhane Reservoir

Euglena viridis, Euglena proxima, Scenedesmus quadricuda, Oscillatoria tenuis, Synedra ulna, Ankistrodesmus falcatus, Chlorella vulgaris, Navicula cryptocaphala, Oscillatoria spledida, Scenedesmus dimorphus, Closterium leibleii, Oscillatoria brevis.

Information on few reservoirs in Jalgaon district is given below. Jalgaon District is located in the North-West region of the state of Maharashtra. It is bounded by Satpuda mountain ranges in the North, Ajanta mountain ranges in the South, Dhule District in the West and Buldhana district in the East. Jalgaon is rich in volcanic soil which is well suited for cotton production. Jalgaon district receives an average rainfall of about 750 mm and the temperature varies from 10°C to 48°C in peak summer with an area of about 11,700 sq km, Jalgaon district has a population of about 4 million.

BORI RESERVOIR: The reservoir was constructed in 1977 across river Bori in Parola taluka of Jalgaon district. The water spread area of the reservoir is 630 hectares with gross storage capacity of 40960 10^3m^3. The length of the reservoir is 3365 m and the height of dam is 20 (Table 11.32).

Table 11.32. Morphometry of Bori reservoir of Jalgaon district

1. Year of construction	1977
2. River	Bori
3. Type	Earthfill
4. Height of dam (m)	20
5. Lenght of dam (m)	3365
6. Gross storage capacity (10^3m^3)	40960
7. Reservoir area (10^3m^2)	8460
8. Designed spillway capacity (m^3/sec)	4206
9. Volume content (10^3m^3)	5534
10. Effective storage capacity (10^3m^3)	25020
11. Water spread area (ha)	630
12. Purpose	Irrigation

Seed of Indian major carps is regularly stocked and the fish production ranged from 1 kg to 18 kg/ha/yr (Table 11.33). Catch structure is biased in the favour of predatory catfishes, mainly *Wallago attu, Mastacembelus armatus, Mystus seenghala* and *Mystus cavasius*. Their higher percentage along with the other predators like *Channa striatus, Channa marulius* and *Channa gachua* lowers the fish production by diverting the energy through a longer food chain. Predatory fishes along with the uneconomic species always outnumber the Indian major carps and the stocking done so far had only a limited impact on the catch structure.

Table 11.33. Stocking and fish production in Bori reservoir of Jalgaon district

Year	Stocking	Fish production (kg)	Fish production (kg/ha/year)
2007-08	8,00,000	3491	5.54
2008-09	6,80,000	2577	4.090
2009-10	5,75,000	1156	1.83
2010-11	13,60,000	716	1.136
2011-12	12,50,000	11373	18.05

Manyad Reservoir: It was constructed in year 1973 by damming the river Manyad in Chalisgaon taluka of Jalgaon district. The average water spread area is 360 ha.The salient morphometric features are presented in Table 11.34.

Table 11.34. Morphomteric features of Manyad Reservoir of Jalgaon district

1. Year of construction	1973
2. River	Manyad
3. Type	Earthfill
4. Height of dam (m)	45
5. Lenght of dam (m)	1677
6. Gross storage capacity (10^3m^3)	53980
7. Reservoir area (10^3m^2)	8710
8. Designed spillway capacity (m^3/sec)	3755
9. Volume content (10^3m^3)	896.5
10. Effective storage capacity (10^3m^3)	40257
11. Water spread area (ha)	360
12. Purpose	Irrigation

The details of fish seed stocking in the reservoir for the period of 2007-08 to 2011-12 are furnished in Table 11.35. Stocking was highest at 1528 nos./ha in 2009-10 while lowest seed was stocked in year 2011-12 (792 nos./ha). The average seed stocking was calculated at 1149 nos./ha.

Table 11.35. Details of stocking and fish production in Manyad reservoir

Year	Stocking	Stocking (nos/ha)	Fish Production (kg)	Fish production (Kg/ha/yr)
2007-08	4,00,000	1111	20,000	55.55
2008-09	3,75,000	1042	21000	58.33
2009-10	5,50,000	1528	37,800	105.00

The fish production for five years (2007-08 to 2011-12) ranged from 34.41 kg/ha/yr (year 2011-12) to 105 (2009-10) kg/ha/yr (Table 11.25) with an average of 62.57 kg/ha/yr, which is comparatively high as compared to other reservoir of Jalgaon district. High catches were recorded during July-October (the period of coinciding with receeding water level) followed by a sharp decline thereafter. The common species occur in the commercial catches are *Catla catla, Labeo rohita, Cirrhinus mrigala, Cyprinus carpio, Channa gachua, Channa marulius, Channa striatus* and *Puntius* spp. High catches were recorded during July-October (the period of coinciding with receeding water level) followed by a sharp decline thereafter.

BAHULLA RESERVOIR

It was consturucted in Pacchora taluka of Jalgaon district. The average water spread area is 239 ha. The fish production varied from 10.69 kg/ha/yr to 35.57 kg/ha/yr (Table 11.36). The fishery is supported by Indian major craps, common carp and catfishes.

Table 11.36. Bahulla reservoir in Pacchora taluka

Year	Stocking	Fish Production (kg)	Fish production (Kg/ha/yr)
2007-08	2,50,000	4414	11.80
2008-09	2,00,000	6660	17.80
2009-10	1,41,000	4000	10.69
2010-11	5,00,000	13304	35.57
2011-12	3,41000	6059	16.20

Table 11.37. Dhaigaon Dam Taluka Pacchora

Year	Stocking	Fish Production	Fish production (Kg/ha/yr)
2007-08	Nil	960	4.016
2008-09	Nil	940	3.93
2009-10	Nil	1000	4.184
2010-11	Nil	1152	4.820
2011-12	Nil	1200	5.020

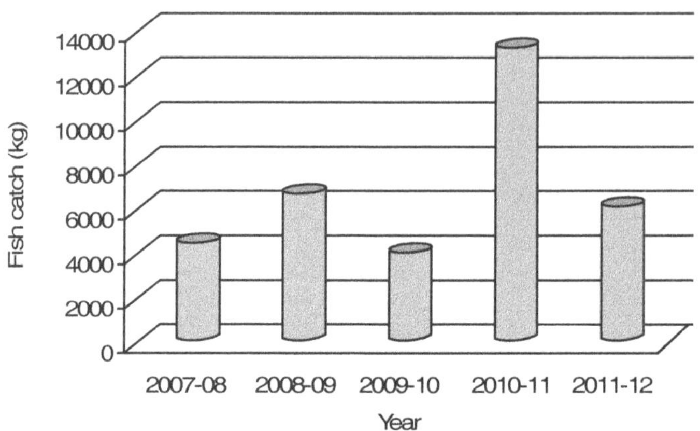

Fig. 11.4. Changes in the total catch

Table 11.38. Hivra Taluka Pachora: 576 hectare

Year	Stocking	Fish production (kg)	Fish production (kg/ha/year)
2007-08	10,00,000	1791	3.10
2008-09	10,00,000	1791	3.10
2009-10	14,00,000	3066	5.32
2010-11	24,00,000	5030	8.73
2011-12	17,00,000	20322	35.28

WAGHUR RESERVOIR

Waghur dam is earthfill dam on local river near Nandgaon. Waghur river from its source near Ajanta flows through the Khandesh region. The main purpose of the dam is to supply water to Jalgaon city and also for irrigation purposes. The work of right and left bank canal is in progress. This project turned to be a boon for the citizens of Jalgaon city. Due to construction of this dam Kandari village was submerged in the backwaters. The dam is 18 km from Jalgaon.

Table 11.39. Salient features of Waghur reservoir of Jalgaon district

1. Year of construction	1978
2. River	Waghur
3. Location	Nandgaon
3. Type	Earthfill
4. Height of dam (m)	13.6
5. Lenght of dam (m)	690
6. Capacity (km^3)	1060

Contd...

Table 11.39: Contd...

7. Surface area (km^2)	467
8. Gross storage capacity (km^3)	1220
9. Volume (km^3)	70
11. Water spread area (ha)	2372 ha
12. Purpose	Irrigation

Table 11.40. Salient Features of Hathnur reservoir of Jalgoan district

1. Year of construction	1982
2. River	Tapi
3. Location	Bhusawal
3. Type	Earthfill
4. Height of dam (m)	25.5
5. Lenght of dam (m)	2580
6. Gross storage capacity (10^{3}m^3)	388000
7. Reservoir area (10^{3}m^2)	48160
8. Designed spillway capacity (m^3/sec)	26415
9. Volume content (10^{3}m^3)	3850
10. Effective storage capacity (10^{3}m^3)	255000
11. Water spread area (ha)	5100
12. Purpose	Irrigation

RANGAVALI DAM

Rangavali Dam is known as Rangavali river project in government documents. In the year 1972, it is built over Rangavali river near Nagziri village of Nandurbar district. The average annual rain fall in the surrounding area is 1227 mm, the catchment area of dam is 99.20 sq. km. The geographic location of the dam is 73° 52′ 0″ longitudinal and 21°0″ latitude. The gross capacity of the dam is about 15.02 Mcum and capacity of the dead storage 2.13 mcum Dam is earthen of rolled filled maximum height of the dam in the river is 25.63 meter and the length is 1878 meter. The earthwork of the dam is 1.042 mcum, concrete 113.00 Mcum, masonry 7770 mcum and excavation 327134 mcum. The entire area is declared as a tribal area by the central government of India. This dam has not received much attention by limnologist and this prompted us to sample the fishes throughout the year to assess the diversity of fishes.

Jaiswal and Ahirrao (2012) reported 28 fish species from Rangavali reservoir (Table 11.41).

Table 11.41. Fish fauna of Rangavali Reservoir,Maharashtra

Order Clupeiformes : *Notopterus notopterus , Notopterus Chitla.*
Order Cypriniformes : *Nemacheilus moreh, Catla catla, Cirrhinus mrigala, Labeo rohita, Cyprinus carpio, Rasbora daniconias, Puntius sophore, Thynicthys sandkhol, Salmostona novacula and Garra mullya.*
Order Siluriformes : *Mystus aor, Mystus seenghala, Wallago attu, Ompak bimaculatus* and *Rita rita.*
Order Channiformes : *Clarius batrachus, Channa gachua, Channa marulius* and *Channa punctatus.*
Order Synnbranchiforms : *Mastacembelus armatus.*
Order Perciformes: *Oreochromis mossambicus, Chanda nama, Glossogobius giuris giuris, Xentodon cancilla* and *Parambassis ranga.*
Order Mugiliformis : *Rhinomugly carsula.*

TULSHI RESERVOIR

The reservoir is located at radhanagari in Kolhapur district with 3 TMC capacity and has brought 6000 hectares of land under irrigation, six surrounding villages have been affected due to this project. The reservoir covers 348 hecatares area.

Table 11.42. Morphometry of Tulshi reservoir in Kolhapur district

1. Year of construction	1978
2. River	Tulshi
3. Location	Radhanagri
4. Type	Earthfill-gravity
5. Height (m)	48.68
6. Length (m)	1512
7. Volume content (10^3M^3)	25
8. Gross storage capacity (10^3M^3)	98290
9. Rseservoir area (10^3M^2)	533
10. Effective storage capacity (10^3M^3)	89910
11. Designed spillway capacity (M^3/Sec)	640
12. Purpose	Irrigation & Hydel power generation

Koli and Muley (2012) studied zooplankton diversity and seasonal variation with special reference to physico-chemical parameters in tulshi reservoir. A total of 39 species of zooplanktons have been found, of which 15 species of rotifer, 12 species of copepod, 10 species of cladocera and 25 species of ostracoda have been found. Zooplankton population showed positive significant correlation with physico-chemical parameters like temperature, alkalinity, phosphate, hardness and BOD, whereas negatively correlated with rainfall and salinity. The seed stocking for three years (2009-10 to 2011-12) varied from 5,40,000 to 6,00,000. In the same period fish production was in the range of 30 to 41 kg/ha/yr (Table 11.43).

Table 11.43. Fish seed stocking and fish production in Tulshi reservoir

Year	Seed stocking	Stocking (ha/yr)	Fish production (kg)	Fish production (kg/ha/yr)
2009-10	5,40,000	1552	14424	41
2010-11	5,40,000	1552	10434	30
2011-12	6,00,000	1724	10684	31

RANKALA LAKE

Rankala lake is a popular evening spot and recreation centre in Kolhapur city. In recent years it has been observed that the lake water was covered by *Eichhornia crassipes* (Mart.) Solms and by an aquatic pteridophyte *Salvinia Seguir* sp. Alongwith these plants, the mass of certain pollution indicating algae has been observed in water body indicating increased water pollution.It was found that the amount of two parameters *i.e.,* total solids and total dissolved solids was increased in monsoon season, while amount of other parameters was found to be less in monsoon season than the summer one.The variability in physico-chemical properties of water collected in two different seasons is tabulated in Table 11.44.

Table 11.44. Physico-chemical properties of Rankala Lake

Sr.No.	Parameter	Summer season	Monsoon season	WHO limits
1.	pH	7.34	7.1	6.5-8.5
2.	Temperature (°C)	27.7	26.2	20-32
3.	Dissolved oxygen (mg/l)	6.7	6.1	5.0
4.	Chlorides (mg/l)	44	39	200-300
5.	Magnesium (mg/l)	20	13	30
6.	Sulphate (mg/l)	40	36	200
7.	Nitrate (mg/l)	0.004	0.003	45
8.	Calcium (mg/l)	40	35	100
9.	Total solids (mg/l)	878	989	500
10.	Total dissolved solids (mg/l)	654	732	500

Urbanization and development in most of the fields are unending and necessary processes, but it causes major problems of water pollution, air pollution, sound pollution etc. which creates hazardous effects on all living organisms. In above case the domestic waste and sewage water should have to be treated before it mixes with water body for reducing the pollution level.

VADGAON RESERVOIR

Vadgaon reservoir is situated 15 km away from Kolhapur city. It lies between 16046′ 26.56″ N latitude and 74018′ 21.54″E longitude. The area of the reservoir is 120 ha.

Manjare *et al.* (2010) studied the water quality of Vadagon reservoir with special reference to zooplankton. The zooplanktons were represented by four groups viz rotifera, cladocera, copepoda and ostracoda. Rotifera comprised 7 species, cladocera 5 species, copepoda 1 species and ostracoda 1 species. The rotifera population recorded peak in the month of October (57.89%) with its maximum composition in July (37.60%) respectively. Among these *Brachionus angularis, B. caudatus, B. fulcatus, B. calyciflorus and B. vulgaris* were most common forms. The *B. rubens and Keratella tropica* was observed with minimum population density. *B.angularis, B. caudatus, B. fulcatus,* comprises maximum population density as compare to other species of rotifera, whereas their absence was noted in the month of June. The copepod population was recorded at distinct peak in the month of September (20.96%) with their maximum density in January (20.90%) while minimum in the month of June (8.75%). The copepod species includes *Cyclopoid copepod, Calanoid copepod, Eucyclop* spp, *Mesocyclops,Neodiaptomus, Paracyclops, Diaptomus copepod etc.* Among these copepod most dominant species were *Calanoid copepod, Mesocyclops, Rhinediaptomus* spp., *Paracyclops, Diaptomus copepod, Cyclopoid copepod* were observed in the winter and summer seasons, whereas their absence was noted in the month of September. The maximum percentage of ostracoda was recorded in June (45.77%) and minimum in November (11.31%) with their absence from July to September. The Ostracoda group was represented by *Hemicypris fossulate*. Cladocera was observed maximum in the month of July (98.64%) and minimum in October (12.95%) the copepods were represented as *Diaphanosoma sarsi, Diaphanosoma excisum, Monia brachiata, Daphnia pulex, Macrothrix laticornis.*

ANJANI RESERVOIR

Khabade and Mule (2009) studied monthly study of water quality and insect fauna of Anjani reservoir. Monthly collection of water sample was done from the period July 2004 to June 2005, by using samplers i.e. plastic containers of 5 litre size. It has been found that the species diversity was higher during monsoon when rainfall was maximum. Table 11.45 shows the rainfall data during July 2004 to June 2005. The results on water quality assessment in terms of a number of physico-chemical parameters are summarized in the Table 11.46. During rainy season water temperature was also optimum which affect the species diversity and it becomes higher. During summer, in the month April 2005 when water temperature was high about 31°C, the insect species diversity was minimum in this month.It is also observed that the species diversity was minimum during summer. In the situation, only pollution tolerant aquatic insects will be present and pollution intolerant species decline. The pollution tolerant species can grow more rapidly without competition for space, nutrients, predation and other extrinsic and intrinsic factors too. This results in heavy dominance of these species leading to the decline in the values of species diversity and also in the evenness of species.

The rainfall was recorded during July 2004 to October 2004 and also during June 2005. The higher rainfall about 94.2 mm was reported during September 2004. In the remaining months no rainfall was recorded in the Savlaj Block. The air temperature and humidity reported during rainy season was optimum. The maximum air temperature of about 38°C was reported during month May 2005 and minimum air temperature of about 26°C was reported during months December 2004 and January 2005. The maximum humidity 79.00% was reported during February 2005 while minimum humidity 31.00% was reported during January 2005. The optimum range of air temperature and humidity was also responsible for maximum insect species diversity during rainy season.

Light transparency of water depends on the total solids, total dissolved solids and total suspended solids of the water. In present study it is found that these physical parameters not cause any adverse effect on the distribution and abundance of insect diversity, in Anjani water reservoir during study.

The maximum water temperature about 31°C was recorded during April 2005 and minimum water temperature about 24°C was recorded during December 2004 and January 2005. The maximum light transparency about 136.0 cm was recorded during October 2004 while minimum light transparency about 16.0 cm was recorded during July 2004. The maximum total solids reported was 5215 mg/L, during May 2005 while minimum total solids reported was 700 mg/L, during October 2004. The maximum total dissolved solids reported was 3200 mg/L during June 2005 while minimum total dissolved solids reported was 336 mg/L during November 2004. The maximum total suspended solids reported was 3210 mg/L during July 2004 while minimum total suspended solids reported was 100 mg/L during January 2005. The analysis of historical records reveals a number of inter-relationship among key water quality parameters used in the assessment of cumulative impacts. The recommended concentration of TSS is 80 mg/L. It is also found that the chemical parameters such as pH of water, electrical conductivity, total alkalinity, hardness, magnesium, calcium, chloride, dissolved oxygen, acidity, residual chlorine, free CO_2, hydrogen sulphide, sodium, potassium, nitrate, phosphate may not cause any adverse effect on the distribution and abundance of insect fauna.

Table 11.45. Monthwise record of rainfall at Anjani Reservoir

Sr. No.	Month	Rainfall (mm)	Name of the block	Water Reservoirs coming under Savlaj Block
1	July 2004	59.6	Savlaj Block	Siddhewadi
2	August 2004	75.00		Anjani
3	September 2004	94.20		Balgawade
4	October 2004	14.00		Bastawade
5	November 2004	—		
6	December 2004	—		
7	January 2005	—		
8	February 2005	—		
9	March 2005	—		
10	April 2005	—		
11	May 2005	—		
12	June 2005	14.00		

No Rainfall.

Table 11.46. Solubility of oxygen in pure water exposed to water-saturated air at mean sea level pressure of 760 mm Hg

Temperature (°C)	Dissolved oxygen (mg/litre)	Temperature (°C) (mg/litre)	Dissolved oxygen (mg/litre)
0	14.16	18	9.18
1	13.77	19	9.01
2	13.40	20	8.84
3	13.05	21	8.68
4	12.70	22	8.53
5	12.37	23	8.38
6	12.06	24	8.25
7	11.76	25	8.11
8	11.47	26	7.99
9	11.19	27	7.86
10	10.92	28	7.75
11	10.67	29	7.64

Contd...

Table 11.46: Contd...

12	10.43	30	7.53
13	10.20	31	7.42
14	9.98	32	7.32
15	9.76	33	7.22
16	9.56	34	7.13
17	9.37	35	7.04

Source: Hand book of common methods in Limnology by Lind O. T. 1974

GHODPETH RESERVOIR

Ghodpeth reservoir is a beautiful perennial freshwater reservoir located in Ghodpeth villege of Bhadrawati taluka of Chandrapur district on Nagpur – Chandrapur state highway. This is a rain fed reservoir having a water spread area of about 20 acres approximately with a depth of about 15 feet in rainy season. The Ghodpeth reservoir is 6 kms from Bhadrawati and its catchment area contains a variety of flora and fauna.

This reservoir is biodiversity rich ecosystem and harbours a variety of local as well as migratory birds due to abundant food available throughout the year in the form of crustaceans, insects, worms, molluscs as well as aquatic weeds. The reservoir harbours different kinds of aquatic as well as terrestrial weeds in its catchment area which supports a large number of fauna on which the birds thrive very well.

Harney *et al.* (2013) prepared a check list of birds from Ghodpeth reservoir. The studies show that this reservoir is having a rich avi-fauna having 44 different kinds of bird species recorded in the catchment area.

This reservoir harbors a number of aquatic weeds in the submerged as well as floating state on which thrive a large number of organisms. Due to abundant food available throughout the year in this reservoir in the form of aquatic crustaceans, insects, molluscs etc. The reservoir always attracts a large number of birds throughout the year. The banks of this reservoir are infested with weed *Ipomea aquatica* which provide suitable resting place for the birds. Apart from this the lake periphery is covered with bushes and trees which provide suitable habitat for many migratory as well as resident birds.

Altogether 39 different bird species including aquatic and non aquatic birds were recorded from the Ghodpeth Reservoir (Table 11.49). It was observed that the maximum bird species were recorded during spring, early monsoon and late winter season, while comparatively less number of bird species was observed during late summer, late rainy season and early winter. The lake harbours a large number of fauna which attracts the migratory as well as non migratory birds which shows that the entire lake basin is highly productive and conducive to all kinds of birds.

In the Present investigation the recorded birds belong to 11 different orders *viz.*, podicipediformes, anseriformes, ciconiformes, falconiformes, galliformes,

Table 11.47. Meterological Parameters of Anjani Water Reservoir from July 2004 to June 2005

Month & Year	July 2004		August 2004		Sept. 2004		Oct. 2004		Nov. 2004		Dec. 2004		Jan. 2005		Feb. 2005		Mar 2005		April 2005		May 2005		June 2005	
Parameters Sites	S_1	S_2	S_1	S_2	S_1	S_2	S_1	S_2	S_1	S_2	S_1	S_2	S_1	S_2	S_1	S_2	S_1	S_2	S_1	S_2	S_1	S_2	S_1	S_2
Air Temperature (°C)	31	31	32	33	33	34	31	32	29	28	27	26	26	26	30	30	32	32	37	36	35	38	32	31
Humidity (%)	72	67	72	65	73	62	52	52	52	45	52	52	50	31	73	79	73	46	36	41	67	65	55	55

All values are mean of four readings.

Table 11.48. Meterological Parameters of Anjani Water Reservoir from July 2004 to June 2005

Month and Year	July 2004		August 2004		Sept. 2004		Oct. 2004		Nov. 2004		Dec. 2004		Jan. 2005		Feb. 2005		Mar 2005		April 2005		May 2005		June 2005	
Parameters Sites	S_1	S_2	S_1	S_2	S_1	S_2	S_1	S_2	S_1	S_2	S_1	S_2	S_1	S_2	S_1	S_2	S_1	S_2	S_1	S_2	S_1	S_2	S_1	S_2
Air Temperature (°C)	31 ± 0.88	31 ± 0.88	32 ±	33	33	34	31	32	29	28	27	26	26	26	30	30	32	32	37	36	35	38	32	31
Humidity (%)	72	67	72	65	73	62	52	52	52	45	52	52	50	31	73	79	73	46	36	41	67	65	55	55

All values are mean of four readings.

pelecaniformes, charadriformes, coumbiformes, psittaciformes, passeriformes, and coraciformes.

The little grebe, common sand piper, little cormorant and black winged stilt were observed in October. Common sandpiper, rufosubacked shrike, brahminy starling and purple moorhen in November, asian open bill stork, black winged kite in February, shikra was recorded in month of March. The brahminy starling was recorded in winter. Black necked stork a threatened species was also recorded during winter months in the reservoir basin. The overall check list prepared show that 39 different kinds of birds have visited the reservoir water for feeding and breeding activities during the year, as abundant food available in the reservoir water.

Table 11.49. Check List of Birds of Ghodpeth Reservoir

Common Name/Order/Family	Scientific Name	Habit
Order – Podicipediformes **Family Podicipedidae**		
Little grebe	*Tachybaptus ruficollius*	R
Order – Ciconiformes **Family – Ardeidae**		
Grey Heron	*Ardea cinerea*	RM
Indian Pond Heron	*Ardeola grayii*	R
Cattle Egret	*Bubulcus ibis*	RM
Large Egret	*Casmerodius albus*	RM
Family – Ciconidae		
Asian Open Bill Stork	*Anastomus Osciatans*	R
Black Necked Stork	*Ciconia episcopus*	M
Family – Threskiornithidae		
Black Ibis	*Pseudibis papillosa*	R
Order – Anseriformes **Family – Anatidae**		
Spot Bill Duck	*Anas poecilorhyncha*	RM
Order – Falconiformes **Family – Accipitridae**		
Black Winged Kite	*Elanus caeruleus*	R
Black Kite	*Milvus migrans*	R
Order – Galliformes **Family - Phasinidae**		
Grey Francolin	*Francolinus pondicerianus*	R

Contd...

Table 11.49: Contd...

Order – Gruiformes **Family – Gruidae**		
White Breasted Water Hen	*Amaurormi Phoenicurus*	R
Purple Moorhen	*Porphyrio porphyrio*	R
Common Coot	*Fulica atra*	RM
Order – Pelecaniformes **Family – Phalcrocoracidae**		
Little Cormorant	*Phalacrocorax niger*	RM
Order – Charadriformes **Family – Recurvirostridae**		
Black Winged Stilt	*Himantopus himantopus*	*R*
Family – Charadridae		
Red Wattled Lapwing	*Vanellus indicus*	*R*
Family – Scolopacidae		
Common Sandpiper	*Actitis hyoleucos*	M
Order – Columbiformes **Family – Columbidae**		
Little Brown Dove	*Streptopelia senegalensis*	R
Order – Psittaciformes **Family – Psittacidae**		
Rose Ringed Parakeet	*Psittacula krameri*	R
Family – Cuculidae		
Asian Koel	*Eudynamis scolopacea*	R
Greater Coucal	*Centropus sinensis*	R
Order – Coracifores **Family – Alcedinidae**		
Small Blue Kingfisher	*Alcedo athis*	RM
White Breasted Kingfisher	*Halcyon smyrnesis*	R
Family – Meropidae		
Small Green Bee Eater	*Merops orientalis*	R
Family – Coraciidae		
Indian Roller	*Coracias benghalensis*	RM
Family – Upupidae		
Common Hoopoe	*Eupopa epops*	RM

Contd...

Table 11.49: Contd...

Order – Passeriformes **Family – Lanidae**		
Rufousbacked shrike	*Lanius schach*	R
Family – Dicrudidae		
Black drongo	*Dicrurus macrocercusq*	R
Family – Sturnidac		
Common myna	*Acridotheres tristis*	R
Brahminy starling	*Sturnus pagodarum*	R
Family – Pycnonotidae		
Red Vented bulbul	*Pycnonotus cafer*	R
Family – Muscicapidae		
Jungle babbler	*Turdoides striatus*	R
Indian Robin	*Saxicolodies fulicata*	R
Family – Necatarinidae		
Purple Sunbird	*Nectarinia asiatica*	R
Family – Passeridae		
House sparrow	*Passer Domesticus*	R
Pheasant Tailed Jackana	*Hydrophasianus chirurgus*	R
Order – Passeriformes **Family – Hirudinidae**		
Common Swallow	*Hirundo rustica*	RM

R – Resident, RM – Resident Migratory, M – Migratory.

FUTALA LAKE

Meshram ad Nasare (2011) documented the phytoplankton community of Futala Lake of Nagpur. Altogether 44 genera of phytoplankton belonging to cyanophyceae, chlorophyceae and bacillariophyceae were recorded. Members of cyanophyceae *viz. Gloeocapsa, Microcystis, Nostoc, Spirulina, Oscillatoria, Anacystis, Gleotrichia, Anabaena, Rivularia, Scytonema, Stigonema, Cylindrospermum, Tolypothrix, Oscillatoria* were observed throughout the investigation period. Amongst them *Nostoc, Anabaena, Oscillatoria, Anacystist, Microsystis* were found to be dominant.

Member of chlorophyceae *viz., Chalmydomonas, Pandorina, Eudorina, Scenedesmus, Draparnaldia, Fritschiella, Oedogonium, Zygnema, Cosmarium, Hydrodicyton, Spirogyra, Voucheria, Chara, Nitella, Volvox, Pediastrum, Mougeotia, Pithophora, Cladophora,Protococcus, Stigeoclonium, Coleachaete, Chaetophora, Ulothrix, Chlorella,* were observed throughout the study period Amongs them *Vaucheria, Cosmarium, Spirogyra, Volvox, Chara and Oedogonium* were found to be dominant.

Five members of bacillariophyceae viz *Diatom, Cyclotella, Navicula, Nitzschia, Rhopalodia* have been recorded. Amongst bacillariophyceae *Diatoms* was found to be dominant.

Table 11.50. Phytoplankton diversity of Futala Lake of Nagpur, Maharashtra

Sr.No.	Genera/ Species	Months (2006-2007)					
A	CYANOPHYCEAE	Sept	Oct	Nov	Dec	Jan	Feb
1.	*Gloeocapsa* sp.	2	7	13	18	15	17
2.	*Microcystis* sp.	12	17	19	23	12	16
3.	*Nostoc* sp.	13	17	22	38	52	37
4.	*Spirulina* sp.	0	2	8	10	13	17
5.	*Oscillatoria* sp.	17	22	25	32	51	42
6.	*Anacystis* sp.	17	21	27	13	35	12
7.	*Gleotrichia* sp.	3	7	9	2	7	14
8.	*Anabaena* sp.	10	13	17	22	45	38
9.	*Rivularia* sp.	5	7	13	2	13	17
11.	*Scytonema* sp.	2	1	7	2	5	7
12.	*Stigonema* sp.	0	1	3	7	2	5
13.	*Cylindrospermum* sp.	3	7	9	13	12	9
14.	*Tolypothrix* sp.	3	9	7	5	11	7
B	CHLOROPHYCEAE	Sept	Oct	Nov	Dec	Jan	Feb
1	*Chlamydomonas* sp.	7	2	11	5	4	9
2	*Pandorina* sp.	2	7	3	1	0	2
3	*Eudorina* sp.	1	5	7	3	9	2
4	*Scenedesmus* sp.	11	9	10	7	11	5
5	*Draparnaldia* sp.	7	11	13	17	22	7
6	*Fritschiella* sp.	8	2	5	7	6	11
7	*Oedogonium* sp.	13	14	9	22	17	15
8	*Zygnema* sp.	2	9	7	11	5	4
9	*Cosmarium* sp.	12	17	22	32	29	33
10	*Hydrodictyon* sp.	2	13	16	9	11	5
11	*Spirogyra* sp.	19	17	22	27	13	29
12	*Vaucheria* sp.	13	19	32	37	33	29
13	*Chara* sp.	15	12	17	19	23	19
14	*Nitella* sp.	2	7	1	0	11	9

Contd...

Table 11.50: Contd...

15	*Volvox* sp.	11	13	17	29	31	27
16	*Pediastrum* sp.	7	9	12	17	15	13
17	*Mougeotia* sp.	2	1	0	1	0	0
18	*Pithophora* sp.	2	7	0	0	1	0
19	*Cladophora* sp.	7	9	2	1	7	9
20.	*Protococcus* sp. (*Pleurococcus* sp.)	3	9	11	15	2	7
21.	*Stigeoclonium* sp.	3	11	13	17	2	9
22.	*Coleochaete* sp.	2	7	9	11	15	7
23.	*Chaetophora* sp.	3	1	0	5	7	0
24.	*Ulothrix* sp.	2	3	7	19	2	17
25.	*Chlorella* sp.	3	7	2	12	9	11
C	**BACILLARIOPHYCEAE**	**Sept**	**Oct**	**Nov**	**Dec**	**Jan**	**Feb**
1.	*Diatom* sp.	16	18	29	33	37	42
2.	*Cyclotella* sp.	2	13	9	5	6	2
3.	*Navicula* sp.	7	0	2	0	0	1
4.	*Nitzschia* sp.	7	2	5	3	0	2
5.	*Rhopalodia* sp.	5	11	2	1	3	0

*The numbers in table indicates no. of organisms recorded per ml.

EKRUK RESERVOIR

Ekruk reservoir near is very close Solapur city in Maharashtra. The reservoir was constructed in year 1871 across the river Aghili. It is the oldest reservoir in district with water spread area of 1,842 ha. The reservoir has an opportunity to irrigate 2610 hectares of agricultural land. The detailed morphomteric measurements of the reservoir are depicted in Table 11.51.

The results of physico-chemical parameters are depicted in Table 11.51. The water temperature fluctuated from 19 to 35°C, values were high during summer. Summer temperature increase was due to greater heating of water and insolent heat from the sun. The secchi disc transparency fluctuated from 89 to 170 cm. The pH was alkaline with a highest value of 8.2 during April 2004. The lower value of pH during September 2004 may be attributed to heavy rainfall during this month. Optimal growth and survival of aquatic plants and animals require that their environment pH should be confined with a very short range below or above while they will be subjected to various kinds of stresses and diurnal fluctuations of pH of water body should remain in the range of 6.4 to 8.5 in order to support optimum fish growth.The average pH value (7-9) of Ekruk reservoir is suitable for optimum growth of fish. In case of Ekruk reservoir the average pH value during the study period was observed to be above 7.9.

Table 11.51. Salient features of the Ekruk reservoir, Maharashtra

River	Aghilia
Year of construction	1871
Water spread area (ha)	1842
Length of dam (m)	2134
Depth of dam (m)	21.45
Irrigation potential (ha)	2610
Rainfall at catchment area (mm)	723.90
Water temperature (°C)	19-35
Transparency (cm)	89-170
pH	7.9 – 8.2
Dissolved oxygen (mg/l)	5.9-12.4
Free carbon dioxide (mg/l)	2.3-4.2
Total hardness (mg/l)	83.2-120
Total alkalinity (mg/l)	153.2-289
Total Dissolved Solids (mg/l)	217-307.5

The dissolved oxygen was found to be ranged between 5.9 to 12.4 mg/l, indicating the good water quality for fish survival.

The carbon dioxide content of water depends upon the temperature of water, depth of water, rate of respiration, decomposition of organic matter, chemical nature of the bottom and the geographical and physiological feature of the terrain surrounding the water. The free carbon dioxide in Ekruk reservoir was observed in the range of 2.3 to 4.2 mg/l.

Alkalinity of freshwater ecosystems is generally caused by carbonates and bicarbonates or hydroxides of calcium, magnesium, sodium, potassium, ammonium and iron. Natural water bodies usually show a wide range of fluctuations in total alkalinity depending upon the location, plankton population, nature of bottom deposits etc. Waters of hilly streams, sandy, rocky, flooded rivers in rainy season and waters infested with submerged weeds usually have a low total alkalinity. On the other hand, stagnant waters in low rainfall areas during summer season are likely to have high total alkalinity values. In Ekruk reservoir total alkalinity was observed in the range of 153.2 to 289 mg/l. During present investigation the total alkalinity was more than 100 mg/l, indicating that the reservoir water is highly productive.

The hardness of water is mainly due to the presence of calcium and magnesium. Maximum values were found during summer and lowest values were found during winter months. Productive waters should have hardness value above 20 mg/l, calcium content above 5 mg/l and magnesium content above 2 mg/l. Very hard water (> 300 mg/l) also becomes uncogenial for fish production because of higher pH. Optimum hardness for fish culture has been observed to be around 75 to 150 mg/l. In Ekruk reservoir, the values of total hardness ranged between 83.2 to 120 mg/l. Hence the water is suitable for fisheries.

During the present study the total dissolved solids were high in summer followed by winter and monsoon. It was due to the dilution of water.

Plankton Diversity: The phytoplankton constitute baseline of food webs in an aquatic ecosystem. In the present study phytoplankton belonging to three groups i.e., myxophyceae, chlorophyceae, and bacillariophyceae and zooplanktons belonging to rotifera, cladocera, copepoda and ostracoda were identified. The plankton population of the reservoir showed summer peak coinciding with the mean depth, influx and outflow of water. Rotifera, copepoda, ostracoda, myxophyceae and chlorophyceae contributed to the summer peak of planktonic population.

Table 11.52. Plankton diversity of Ekruk reservoir

A. PHYTOPLANKTON
1. **Myxophyceae:** *Anabaena, Microcystis, Merismopedia*
2. **Chlorophyceae:** *Nostoc, Periastrum, Closteridium, Microspora*
3. **Bacillariophyceae:** *Fragillaria, Navicula, Nitzschia, Tabellaria*
B. ZOOPLANKTON
1. **Rotifera:** *Brachionus* sp, *Euchlanis dilata, Filinia ingiseta, Keratella tropica, Trichocera Porcellus*
2. **Ladocera:** *Ceriodaphnia cornuta, Moina micrura, Biapertura karna, Indialona ganapali*
3. **Copepoda:** *Cyclops, Mesocyclops, Neodiaptomus lindbergi, Nauplius*
4. **Ostracoda:** *Cypris* sp., *Cyclocypris globosa, Stenocypris* sp.

Fish Fauna: The fish fauna is an important aspect of fishery potential of a waterbody. Fish fauna of Indian reservoirs has been studied by several workers and it has been found that the distribution of fish species is quite variable because of geographical and geological conditions of the area. During the present investigation altogether 34 species of fishes belonging to 24 genera falling under 7 orders have been identified (Table 11.53). Of the 7 orders, order cypriniformes dominated with 17 species falling under 12 genera of which genus *Puntius* is abundant with 4 species of them dominating the catch. Next in abundance are the fishes coming under the order siluriformes in which *Mystus* is dominant with 2 species, while the order anguilliformes has only one dominant species.

Table 11.53. Fish fauna of Ekruk reservoir

Order : Osteoglossiformes
Suborder : Notopteroidei
Family : Notopteridae
1. *Notopterus notopters* (Pallas)
2. *Notopterus chitala* (Ham.)

Contd...

Table 11.53: Contd...

Order : Anguilliformes **Suborder : Anguilliformes** **Family : Anguillidae** 3. *Anguilla bengalensis* (Gray)
Order : Cypriniformes **Suborder : Cyprinedei** **Family : Cyprinidae** 4. *Catla catla* (Ham.) 5. *Cirrhinus mrigala* (Ham.) 6. *Cirrhinus Cirrhosus* (Ham.) 7. *Labeo rohita* (Ham.) 8. *Labeo calbasu* (Ham.) 9. *Ctenopharyngodon idella* (Val.) 10. *Hypophthalmichthys molitrix* (Val.) 11. *Cyprinus carpio* (Linn.) 12. *Osteobrama cotio* (Ham.) 13. *Rohtee ogilbii* (Sykes) 14. *Puntius kolus* (Sykes) 15. *Puntius sophore* (Ham.) 16. *Puntius ticto ticto* (Ham.) 17. *Puntius sarana sarana* (Ham.) 18. *Salmostoma clupeoides* (Day) 19. *Rasbora daniconius* (Ham.) 20. *Amblypharyngodon mola* (Ham.)
Order : Perciformes **Suborder : Percoidae** **Family : Centropomidae** 21. *Chanda nama* (Ham.) 22. *Chanda ranga* (Ham.)
Order : Siluriformes **Family : Siluridae** 23. *Wallago attu* (Schneider) 24. *Ompak bimaculatus* (Bloch)
Family : Bagaridae 25. *Rita rita* 26. *Mystus seenghala* (Sykes) 27. *Mystus cavassius* (Ham.)
Family : Clariidae 28. *Clarias batrachus* (Linn.) 29. *Heteropneustes fossilis* (Bl.)
Order : Synbranchiformes **Family : Mastacembelidae** 30. *Mastacembelus armatus* (Lac.)

Contd...

Table 11.53: Contd...

Family : Gobidae 31. *Glossogobius giuris giuris* (Ham.)
Order : Channiformes **Family : Channidae** 32. *Channa marulius* (Ham.) 33. *Channa gachua* (Ham.) 34. *Channa punctatus* (Ham.)

HOTAGI LAKE

Pawar and Mushan (2012) studied the zooplankton diversity of lakes in Solapur district and recorded rotifers, cladocerans, copepods, ostracods, protozoans and euglenoids from Hotagi lake near Solapur (Table 11.54). They also recorded temperature, pH, dissolved oxygen,total hardness values at 15°C, 6.5, 5.8 mg/l and 330 mg/l respectively. Hotagi lake is used for fish culture. Acidic pH of lake water is due to mixing of effluents from nearby sugar factory.

Table 11.54. Zooplankton diversity in Hotagi Lake near Solapur

Rotifers: *Keratella* sp., *Euchalnis* sp., *Brachionus* spp.
Cladocerans: *Bosmina* sp., *Cerodaphnia* sp., *Daphnia* sp., *Moina* sp.
Copepods: *Cyclops* sp., *Eucyclops* sp., *Mesocyclops* sp., *Microcyclops* sp., *Macrocyclops* sp., *Diaptomus* sp., *Copepod nauplii.*
Ostracods: *Cypris* sp., *Cyclocypris* sp., *Stenocypris* sp.
Protozoa: *Amoeba* sp., *Vorticella* sp.
Eugelnazoa: *Euglena* sp.
Others: *Dragonfly larvae, Water bettle, Nematodes, Gastropod Larvae, Mayfly Larvae, Mosquito Larvae.*

UJANI RESERVOIR

The Ujani reservoir was formed when a dam constructed on the Bhima river near the village Ujani in Madha taluka of Solapur district. The Bhima river is a major tributary of the Krishna river in Maharashtra. The reservoir site is located at Lat. 1804′24″N and 7507′15″S.

The temperature of reservoir varies between 35°C to 44°C during summer and 12°C to 30°C during the winter. The summer highs are reached between 15th April and 15th May and winter lows fall generally between late December and early January. The large difference between summer and winter temperature coupled with low humidity makes for a day tropical climate throughout the year. The average annual rainfall recorded at site of reservoir is 500-600 mm. The climate over the entire catchment area varies from moist tropical in the source region of the rivers to dry tropical in the immediate vicinity of dam.

To face the water scarcity, the dam was constructed o the river Bhima, a major tributary of river Krishna. The construction was started in year 1964 and completed in June 1980. The river Bhima rises on the crest of the Sahyadri ear the famous temple of Joytirling Bhimashankar, about 80 kms southeast to Mumbai. 51 villages were submerged due to the construction of the reservoir. It includes 23 villages from Solapur district in Madha taluka, 25 villages from the Pune district and 3 villages from Ahemadnagar district in Karjat taluka. The agricultural lad of 82 villages was submerged due to the construction of Ujani reservoir. It includes 30 villages from Solapur district, 38 villages from Pune and 14 from Ahemadnagar district. The main purpose of reservoir is to provide water for irrigation in Solapur district. The reservoir water is also used for drinking purpose to Solapur, Pandharpur, Sangloa, Mangalwedha ad Akkalkot. Many villages on the bank of the river Bhima are also getting drinking water from Ujani reservoir.

Fig. 11.5. A panoramic view of Ujani reservoir

Sarwade and Khillare (2010) reported the occurrence of 60 fish species belonging to 6 orders, 15 families and 36 genera. The order cypriniformes was most dominant constituting 66.66% followed by order perciformes constituting 11.66%, siluriformes constituting 10%, beloniformes constituting 8.33%, osteoglossiformes constituting 3.33% and synbranchiformes constituting 1.66% of the total fish species. The members of order cypriniformes were dominated by

Table 11.55. Morphometric features of Ujani reservoir

1. River	Bhima
2. Length of dam (m)	2228
3. Masonary (m)	913
4. Earthen (m)	1626
5. Height of dam (m)	56.40
6. Annual rainfall (mm)	500-600
7. Water storage capacity (TMC)	117.25
8. Live storage capacity (TMC)	53.71
9. Dead storage capacity (TMC)	63.54
10. Free catchment area (km^2)	14850
11. Water spread area (ha)	39000

with 40 species followed by perciformes with 7 species, siluriformes with 6 species, beloniformes with 5 species, osteoglossiformes with 2 species and synbranchiformes with 1 species. Order cypriniformes was dominant group with 40 species in the assemblage composition in which *Catla caltla, Cirrhinus mrigala, Cyprinus carpio,* and *Labeo rohita* were found most abundant. *Cirrhinus reba, Ctenopharyngodon idella, Labeo boggut, Labeo calbasu, Puntius conchonius, Puntius sarana, Puntius sophore, Puntius ticto, Schismatorhynchus nukta, Garra mullya* and *Nemacheilus botia* were found abundant form. *Cirrhinus fulungee, Hypselobarbus curmuca, Amblypharyngodon mola, Labeo fimriaus, Labeo kawrus, Labeo potail, Osteobrama bakeri, Osteobrama bhimensis, Osteobrama cotio cunma, Osteobrama vigorsii, Osteobrama niilli, Tor khudree, Chela cahius, Salmostoma untrahi, Barillius bakeri, Barillius bendelisit, Barillius evezardi, Danio aequipinnatus, Parluciosoma daniconius, Nemacheilus denisonii* and *Lepidocephalus guntea were found less abundant. Salmopharia novacula, Salmostoma bacaila* and *Salmostoma boopis* were found rare abundant. Followed by order perciformes was 7 species in the assemblage composition in which *Oreochromis mossambicus* was found most abundant form. *Channa marulius* and *Channa orientalis* were found abundant form. *Rhinomugil corsula, Glossogobius giuris, Chanda nama* and *Pseudoambassis ranga* were found less abundant. Followed by order siluriformes was 7 species in the assemblage composition in which *Mystus bleekeri* and *Wallago attu* were found abundant form. *Aorichthys aor, Aorichthys seenghala, Mystus malabaricus* and *Ompok bimaculatus* were found less abundant. Followed by order beloniformes was 5 species in the assemblage composition in which *Xenentodon cancila, Aplocheilus lineatus, Hemirampus georgii, Hyporhamphus limbatus* and *Gambusia affinis* were found less abundant. Followed by order osteoglossiformes was 2 species in the assemblage composition in which *Notopterus notopterus* were found less abundant. *Notopterus chitala* were found rare abundant. Followed by order synbranchiformes was 1 species in the assemblage composition in which *Mastacembelus armatus* was found less abundant (Table 11.56).

Table 11.56. Fish diversity in Ujani Reservoir in Solapur district

1. *Notopterus notopterus* (Pallas)
2. *Notopterus chitala* (Pallas)
3. *Catla catla* (Ham-Buch)
4. *Cirrhinus fulungee* (Sykes)
5. *Cirrhinus mrigala* (Ham-Buch)
6. *Cirrhinus reba* (Ham-Buch)
7. *Cyprinus carpio* (Linn.)
8. *Ctenopharyngodon idella* (Val)
9. *Hypselobarbus curmuca* (Ham-Buch)
10. *Amblypharyngodon mola* (Hamilton)
11. *Labeo Boggut* (Sykes)
12. *Labeo calbasu* (Ham-Buch)
13. *Labeo fimriaus* (Bloch)
14. *Labeo kawrus* (Sykes)
15. *Labeo potail* (Sykes)
16. *Labeo rohita* (Ham-Buch)
17. *Osteobrama bakeri* (Day)
18. *Osteobrama bhimensis* (Singh and Yazdani)
19. *Osteobrama cotio cunma* (Day)
20. *Osteobrama vigorsii* (Sykes)
21. *Osteobrama niilli* (Day)
22. *Puntius conchonius* (Ham-Buch)
23. *Puntius sarana* (Ham-Buch)
24. *Puntius sophore* (Ham-Buch)
25. *Puntius ticto* (Ham-Buch)
26. *Salmopharia novacula* (Valenciennes)
27. *Schismatorhynchus nukta* (Sykes)
28. *Tor khudree* (Sykes)
29. *Chela cahius* (Ham–Buch)
30. *Salmostoma bacaila* (Ham-Buch)
31. *Salmostoma boopis* (Day)
32. *Salmostoma untrahi* (Day)
33. *Barillius bakeri* (Day)
34. *Barillius bendelisit* (Ham–Buch)
35. *Barilius evezardi* (Day)
36. *Danio aequipinnatus* (Mc Clelland)
37. *Parluciosoma daniconius* (Ham-Buch)
38. *Garra mullya* (Sykes)
39. *Nemacheilus botia* (Ham–Buch)
40. *Nemacheilus denisonii* (Day)
41. *Lepidocephalus guntea* (Ham-Buch)
42. *Aorichthys aor* (Ham-Buch)
43. *Aorichthys seenghala* (Sykes)
44. *Mystus bleekeri* (Day)
45. *Mystus malabaricus* (Jerdon)

Contd...

Table 11.56: Contd...

46.	*Ompok bimaculatus* (Bloch)
47.	*Wallago attu* (Schneider)
48.	*Xenetodon cancila* (Ham-Buch)
49.	*Aplocheilus lineatus* (Val.)
50.	*Hemirampus georgii*
51.	*Hyporhamphus limbatus* (Valenciennes)
52.	*Gambusia affinis* (Baird and Girard)
53.	*Mastacembelus armatus* (Lacepede)
54.	*Oreochromis mossambicus* (Peters)
55.	*Rhinomugil corsula* (Ham-Buch)
56.	*Glossogobius giuris* (Ham-Buch)
57.	*Ambassidae Chanda nama* (Ham-Buch)
58.	*Channa marulius* (Ham-Buch)
59.	*Channa orientalis* (Bloch and Schneider)
60.	*Pseudoambassis ranga* (Hamilton)

Zooplankton diversity: Kumbhar (2008) recorded 27 zooplankton species, out of which 12 belonged to rotifera, 5 to copepoda, 4 to cladocera, 3 to ostracoda and 3 to protozoa (Table 11.57). The average zooplankton was found maximum during summer and minimum during winter.

The rotifera was most dominant group contributed 47.51% and 48.69% of the total zooplankton population during year the first and second year of the study period, 2003 to 2005. The rotifera was represented by 12 genera. The most dominant being *Branchionus*, represented by 4 species, *viz. Brachionus calciflorus, B. diversicornis, B. folculus* and *Branchionus* spp. The species *Brachionus calciflorus* was found to be numerically most dominant. The genus *Keratella* was represented by one species *i.e., Keratella tropica.* The other species such as *Euchlanata dilatata* and *Lacune bulla.* were also noticed in moderate number. The *Trichocera spp, Filina longiseta, Notholca acumita, Eilina* spp. and *Rotaria* spp. were occurred however they were found to be seasonal. Maximum rotifera population was reported during summer season and minimum during winter.

The copepods were another important group represented by 5 genera. These are *Diaptamus* spp., *Mesocyclops* spp., *Paracyclops* spp., *Cyclops viridis* and *Nauplii.* The copepoda contributed 24.06% and 19.55% of the total zooplankton population during first and second year of investigation respectively. Numerically the *Mesocyclops* spp. and *Nauplii* were found most dominant over other members, followed by *Diaptamus* spp. The highest population density of copepoda was observed during summer and lowest during monsoon.

The cladocera were represented by 4 genera. *viz. Moina* spp., *Cereodaphnia* spp., *Alona rectangula* and *Indialona ganapati.* Numerically the cladocera was found to be the insignificant, contributed only 13.10% and 11.28% of the total zooplankton during first and second years respectively. The cladocera did not show marked seasonal trend in their spacial distribution. Among the cladocera the *Cereodaphnia,* and *Moina* spp. were prolifically represented while *Alona rectangula* and *Indialona ganapati* occurred in less numbers.

The ostracoda were represented by 2 species such as *Cypris subglobasa, Stenocypris* spp. and *Candocypria* spp. Ostracoda contributed only 66.1% and 14.85% of zooplankton population during first and second year respectively. The maximum population of ostracoda was recorded during summer and minimum during winter.

The protozoa represented by 3 species *viz. Paramecium caudatum, Arcella* spp. and *Difflugia* spp. Numerically the *Paramecium caudatum* was found most dominant over other species. They contributed about 8.51% and 5.59% of total zooplankton during first and second year respectively. The maximum population density of protozoa was observed during summer and minimum during winter.

Figs. 11.6 & 11.7. Ecological studies in Ujani Reservoir

Table 11.57. Zooplankton diveristy in Ujani Reservoir

ROTIFERA
1. *Brachionus calciforus*
2. *Brachionus diversicornis*
3. *Brachionus falculus*
4. *Brachionus* spp.
5. *Keratella tropica*
6. *Lacune bulla*
7. *Euchlanis dilatata*
8. *Notholca acuminate*
9. *Rotaria* spp.
10. *Filina* spp.
11. *Trichocera* spp.
12. *Filina longiseta*
COPEPODA
1. *Nauplii*
2. *Cyclops viridis*
3. *Paracyclops* spp.
4. *Mesocyclops* spp.
5. *Diaptamus* spp.
CLADOCERA
1. *Moina* spp.
2. *Ceirodaphnia* spp.
3. *Alona rectangular*
4. *Indialona ganapati*
OSTRACODA
1. *Stenocypris* spp.
2. *Cypris* spp.
3. *Candocypris* spp.
PROTOZOA
1. *Paramoecium caudatum*
2. *Arcella* spp.
3. *Diffluga* spp.

Table 11.58. Average values of physico-chemical parameters of Ujani reservoir during 2004-05

Month	Air Temp. (°C)	Water Temp. (°C)	Trans-parency (cm)	TDS (mg/l)	pH	Magnes-ium (mg/l)	Total Alkalinity (mg/l)	Dissolved Oxygen (mg/l)	Free CO_2 (mg/l)	Hardness (mg/l)	Chlorides (mg/l)	Nitrate (mg/l)	Calcium (mg/l)
July 04	31.4	31.2	40	321	8.6	18.9	154	8.1	3.8	134	42.1	0.04	20
Aug	32.4	30.2	47	300	8.6	13.5	160	8.2	5.2	124.1	40.1	0.09	18.5
Sept	30.2	28.6	38	231.6	8.5	15.9	172	8	4.2	117.1	34.2	0.04	14.5
Oct	30.1	29.8	35	205	8.5	13.2	178	7.9	3.2	109.2	32.9	0.09	15.3
Nov	30.1	29.2	49	201	8.2	13.9	176	8.4	3	102.6	27.1	0.09	19.3
Dec	26.4	24	47	252.6	8.4	14	181	8.2	2.9	116.2	24.9	0.05	22
Jan 05	23	21	50.2	268.7	8.6	16.1	199	7.8	2.6	119.6	28.7	0.02	22.1
Feb	28	24.4	73	270.1	8.8	20.1	192	7.9	1.2	128.2	40.1	0.04	19.8
March	37.4	34.2	73	301.2	8.8	18.5	201	8.3	1.0	129.6	39.5	0.01	17.5
April	40.3	35.3	79.5	328.1	8.8	18.7	198	8.6	0.9	127.1	44.5	0.09	22.2
May	44.1	35.7	68.2	348.2	8.9	17.1	210	8	1.2	126.8	48.5	0.04	26.3
June	37.2	34.3	53.4	353.6	8.9	15.3	191	8	2.9	128.1	42.1	0.03	24.3

Water quality: Kumbhar (2008) studied water quality of Ujani reservoir from different four locations. The average values of water quality parameters are depicted in Table 11.58. The air and water temperature raged from 23 to 44.1 and 21 to 35.7°C respectively. Transparency and pH values ranged between 35 to 79.5 cm and 8.2 to 8.9 repectively. The parameters like magnesium, total alkalinity, dissolved oxygen, carbon dioxide, hardness, chlorides, nitrate and calcium were in the range of 13.2 to 18.9, 154 to 210, 7.8 to 8.6, 0.9 to 5.2, 102.6 to 134, 24.9 to 48.5, 0.02 to 0.1, 14.5 to 26.3 mg/l respectively.

BHATGHAR RESERVOIR

Reservoir is located within the coordinates of Latitude 18°10′N and Longitude 27°52′E across the river Yelwandi near Bhor village of Pune district. The length, height and width of the dam are 1625 m, 43 m and 38 m respectively. The salient morphometric features of the reservoir are given in Table 11.59.

Table 11.59. Morphometric features of Bhatghar reservoir

1.	Year of construction completion	1928
2.	Purpose of dam	Irrigation and power generation
3.	Length of dam	1625 meters
4.	Maximum height from river bed	43 m
5.	Full sill level	623.28 m
6.	Top of dam evel	626.63 m
7.	Gross storage capacity	672.6 million cubic meter
8.	Discharge capacity	1600 cuses
9.	Length of spillway	325.5 m
10.	Number of irrigation sluices (south)	6
11.	Number of irrigation sluices (north)	6
12.	Head of water above sill level	26 m
13.	Catchment area	336 sq.km.
14.	Maximum length of reservoir	45 km at FRL
15.	Mean depth	24.02 m

Physico-chemical characteristics of soil and water: CIFRI (1997) under the All India Coordinated Projects on Ecology and Fisheries of fresh water reservoirs carried out ecological investigations on Bhathghar reservoir. The texture of bottom soil is sandy, reddish in colour ad poor in quality. The soil is alkaline in nature and there is very little variation in pH range. Organic carbon, available phosphorus and available nitrogen were relatively high during post monsoon months. The overall percentage of sand is always high. The percentage of clay increased during post-monsoon months due to the heavy influx of rain water from the catchment area.

In the lentic sector, the minimum temperature was recorded in winter (December) while maximum in the moth of April. In the intermediate sector it was minimum in December and maximum in April. In the lotic sector it was minimum in January and maximum in May.

In the lentic sector, the minimum values of pH were recorded in March, while the maximum was in January. In the intermediate and lotic sectors the minimum and maximum values were as in the lentic sector.

Low values of transparency in the lentic sector were recorded during July and August while the high values in October/January. In intermediate and lotic sectors, the low values were recorded during the months of July and August. It may be due to the heavy rains and winds of high velocity. The high values were recorded from January to May and again from October to December probably due to low and moderate velocity of winds.

The maximum and minimum values of dissolved oxygen in lentic sector were recorded in May and October. In the intermediate sector it was maximum in April and minimum in October. In the lotic sector it was maximum and minimum as in the lentic sector.

In the lentic sector, the maximum value of free carbon dioxide was recorded in January while the minimum was in the month of September. In the intermediate and lotic sectors it was maximum and minimum as in the lentic sector.

In the lentic sector of reservoir, the maximum values of total alkalinity were recorded in July while the minimum was in December. In the intermediate sector it was maximum in May and minimum in December. In the lotic zone it was maximum in May and minimum in January. The value of total alkalinity was always very poor in all the three sectors of the reservoir.

The maximum values of phosphates in lentic sector were recorded in July/ August. Similar situation prevailed in the other sectors also.

In the lentic sector, the maximum value of chlorides was recorded in July, while the maximum was in April/May. In the intermediate sector it was maximum in August and minimum in March. In the lotic zone it was maximum in June and minimum in March.

In the lentic sector silicates were maximum in March and minimum in July.

Total hardness was maximum in July and minimum in December in the lentic sector. While in the intermediate sector it was maximum in July and minimum in June. In the loti sector its maximum value was recorded in January and minimum in June.

In lentic sector maximum specific conductivity was in March and minimum in October. In the intermediate sector it was maximum in January and minimum in June. In the lotic sector it was maximum in December and minimum in July.

In lentic zone the maximum values of total dissolved solids were recorded in March and minimum in October. In the lotic sector it was maximum in December and minimum in July. In the intermediate sector it was maximum in January and minimum in June.

Table 11.60. Physcio-chemical characteristics of Bhatghar reservoir

Sr.No.	Parameters	Lentic		Intermediate		Lotic	
		Range	Average	Range	Average	Range	Average
1.	Air temperature (°C)	15-36.4	23.95	22-36.5	28.14	24-37.5	30.37
2.	Water temperature (°C)	17-30.8	24.78	22.5-31	26.42	21.8-31	26.14
3.	pH	6.2-8.2	7.2	6.8-8	7.4	6.7-8	7.3
4.	Transparency (cm)	5-198.5	49.23	20-266	109.75	16-249	132.5
5.	Dissolved oxygen (ppm)	3.2-12.2	7.08	2.4-16	7	2-16	6.75
6.	Free carbon dioxide (ppm)	1.6-16	3.18	2-6	3.07	2-6	3.14
7.	Total alkalinity (ppm)	10-36	23.15	18-28	22	10-26	22.28
8.	Phosphate (ppm)	Traces-0.48	0.36	Traces	—	Traces	—
9.	Chlorides (ppm)	4-14	8.80	6-15	9.36	6-17.5	10.78
10.	Silicates (ppm)	2.5-35	11.09	—	—	—	—
11.	Hardness (ppm)	60-300	130.42	44-240	101.45	52-140	97
12.	Specific conductivity (µ mhos)	50-84.25	57.78	45-67	56.12	38.5-79.1	55.14
13.	Total Dissolved Solids (ppm)	23.90-42.10	27.64	22.2-33.3	30.64	19.2-39.9	27.50

(Source : CIFRI,1997)

Plankton Diversity: Group chlorophyceae formed 29.75% of the total planktoic polulation and was represented by *Gonatozygon* sp., *Hormidium* sp., *Eudorina* sp., *Closteridium* sp., and *Kirchneriella* sp. The maximum density of chlorophyceae was in November and minimum in September.

Myxophyceae formed 36.14% of the total plankton population. Three pulses, first in April, second in June and third in December in lesser magnitude were observed. The dominant forms were *Gomphosphaerium* sp., and *Anacystis* sp.

Group Desmidaeceae formed 2.98% of the total planktonic population ad was mainly represented by *Staurastrum* sp. This group was totally absent during June to September.

Dinophyceae group ranked 5th in order of abundance and was absent during June to October and December. On an average this group constituted 1.87% of the total plankton population. It was mainly represented by *Ceratium* sp.

Rotifers ranked second in order of abundance among zooplankters. On an average group rotifer formed 5.95% of the total plankton.Its peak wasobserved in March (63.41%), followed by October (44.44%) and July (27.27%). Rotifers were absent during August-September. The dominant forms were *Brachionus* sp., *Keratella* sp., *Cephalodella* sp., and *Asplanchna* sp.

Group cladocera ranked third in order of abundance among zooplankters and formed 1.54% of the total plankton. It was most abundant in August (50%) followed by May (18.75%) and December (11.76%). The group was absent during July and September to November. It was mainly represented by *Chydorus* sp., *Bosmina* sp., *Bosminopsis* sp. and *Diaphanosoma* sp.

Group copepod was the most dominant group among the zooplankters and formed 7.99% of the total plankton. It was most abundant in January (69.53%) followed by April (39.13%) and July (72.73%).The group was totally absent during August to September. The group was mainly represented by *Diaptomus sp., Cyclops sp.,* and their larvae (*Nauplii*).

Protozoa formed 0.83% of the total plankton. It was abundant during September (100%) followed y March (9.76%) and May (12.50%). Protozoans were absent during January, July and October-November. The group was represented by *Arcella* sp., *Actinophyrys* sp., and *Heterophrys* sp.

Fisheries: Altogether 48 species belonging 12 families were recorded under investigations by CIFRI. The list of fishes recorded in the reservoir is depicted in Table 11.61.

Table 11.61. Fish fauna of Bhatghar reservoir

Order : Clupeiformes **Suborder : Notopteridae** **Family : Notopteridae** 1. *Notopterus notopterus* (Pallas) 2. *Notopterus chitala* (Ham.)
Order : Perciformes **Suborder : Percoidei** **Family : Centropomidae** 3. *Chada ranga* (Ham.) 4. *Chanda nama* (Ham.)
Order : Cypriiformes **Suborder : Cyprineidei** **Family : Cyprinidae** 5. *Salmostoma boepia* (Day) 6. *Salmostoma untrahi* (Day) 7. *Rasbora elanga* (Ham.) 8. *Rasbora daniconius* (Ham.) 9. *Danio equipinnatus* (Mc.clelland) 10. *Danio devario* (Ham.) 11. *Catla catla* 12. *Cirrhinus mrigala* (Ham.) 13. *Cirrhinus fulungee* (Sykes) 14. *Cyprinus carpio* (Lin.) 15. *Garra lamta* (Ham.) 16. *Labeo rohita* (Ham.) 17. *Labeo calbasu* (Ham.)

18. *Labeo sindensis* (Day)
19. *Labeo porcellus* (Heckel)
20. *Hypophthalmichthys molitrix* (Val.)
21. *Osteobrama cotio cotio* (Ham.)
22. *Osteobrama vigorsii* (Sykes)
23. *Osteobrama neilli* (Sykes)
24. *Rohtee ogilbii* (Sykes)
25. *Tor khudree* (Sykes)
26. *Tor mussulah* (Sykes)
27. *Puntius kolus* (Sykes)
28. *Puntius sophore*
29. *Puntius ticto ticto* (Ham.)
30. *Puntius sarana sarana* (Ham.)
31. *Puntius dobsonii* (Day)
32. *Puntius subnastutus* (Val.)
33. *Puntius amphibious* (Val.)

Family : Cobitidae

34. *Nimachilus evazardi* (Van-Hornell)
35. *Nimachilus denisonii* (Day)
36. *Nimachilichthys rueprel* (Day)

Order : Siluriformes

Family : Siluridae

37. *Ompak bimaculatus* (Bloch)
38. *Wallago attu* (Schneider)
39. *Pseudotropius etherinoides*
40. *Sylondia sykesii* (Sykes)

Family : Bagridae

41. *Mystus cavasius* (Ham.)
42. *Mystus seenghala* (Sykes)

Family : Sisoridae

43. *Gogata itchkea* (Day)

Order : Beloniformes

Family : Belonidae

44. *Xenetodon cancilla* (Ham.)

Order : Mugiliformes

Family : Mugilidae

45. *Rhinomugil corsula* (Ham.)

Order : Channiformes

Family : Channidae

46. *Channa gachua* (Ham.)
47. *Chana marulius* (Ham.)

Order : Symbranchiformes

Family : Gobidae

48. *Glossogobius giuris giuris*

(Source: CIFRI, 1997)

The State Fisheries Department has leased the reservoir to Gajanan Fisheries Co-operative Society; Bhor with rights to fish in the reservoir with membership of 120, the society has been doing fishing since 1986. The society is not in a position to stock the reservoir because of the monetary reasons. Even though the State Fisheries Department is stocking the reservoir since 1979-80, there is no record of fish landings available with the department.

Crafts ad Geras used in Fishing: Fishing is carried by nylon gill-nets. The net is without foot rope ad has floats made up of thermocol. The net has a mesh bar range of 20 mm-150 mm. The fishing unit consists of two fishermen and ten to fifteen gill nets of different mesh bars. Boats are not used in fishing in the reservoir excepting a single fisherman of the lentic sector who is in possession of a mechanized boat.

Fish Production: The total fish production from the reservoir was estimated to be 24506.75 kg (1987). The production decreased onwards 12721.5 kg (1988), 9278.88 (1989) and 8999.05 kg (1990). It showed an increase in 1991 (12099.5 kg). The average production was 13521.14 from the reservoir during the entire period of CIFRI's investigation. The present fish yield from the reservoir is about 4.8 kg/ha/yr.

Catch Per Unit Effort (CPUE): Commercial gill-nets with a mesh bar range of 20 mm to 150 mm were sampled for CPUE. Table 11.62 gives the details of CPUE. The average catch per unit effort was estimated to e 1.092 kg/day.

Table 11.62. Catch Per Unit Effort in Bhatghar Reservoir

Year	No.of units	Total catch (kg)	Catch/unit/effort (kg/day)
1988	112	116.75	1.048
1989	259	263.92	0.913
1990	146	357.65	2.449
1991	88	156.77	1.238

Escape of Fish Through the Gates: Loss of fish stock can be prevented if the gates are provided with some kind of mesh filters to retain the fish.

Closed Season: In Bhatghar reservoir, fishing is carried out throughout the year. No 'closed season' has been declared by the fisheriesdepartment or the society. The fishing should be banned during July and August and this period should be considered as the closed season for fishing.

PAUNA RESERVOIR

In Pune district, Pauna is an important tributary of river Mula-Mutha. Thus, it forms the party of Bhima basin, a part of larger Krishna basin. The average water spread area is 1594 ha. The morphometry of reservoir is given in Table 11.63.

Table 11.63. Morphometric features of Pauna Reservoir

1. Commencement of work	1966
2. Completion of work	1972
3. Location	Pauna Nagar
4. Rainfall (mm)	2030
5. Maximum height (m)	42
6. Catchment area (sq.km)	113
7. Water spread area (ha)	2365
8. Storage capacity (TMC)	10.76
9. Usable capacity (TMC)	9.67
10. Irrigation potenatial (ha)	5946

Muley and Patil (2006) studied physico-chemical parameters of Pauna reservoir (Fig. 11.8 to Fig. 11.17). The reservoir water temperature varies in the range of 23.5 to 29°C. The maximum water temperature was recorded during February and lowest in October. The pH values indicates slight alkaline nature of water. Dissolved oxygen varies between 4.87 to 9.34 mg/l. The free carbon dioxide was observed in the range of 6.6 to 28.6 mg/l. In reservoir water, the chlorides fluctuated between 8.52 to 24.14 mg/l. Nitrate range was from 0.7 to 6.83 mg/l. Phosphate concentration in reservoir water varies from 0.024 to 0.18 mg/l. Maximum values of BOD and COD were recorded during February and May respectively.

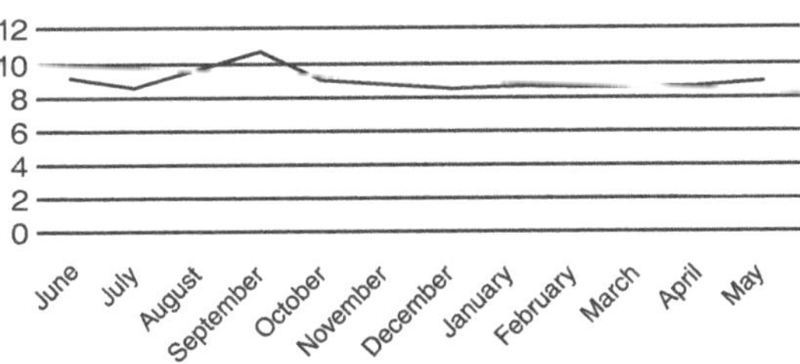

Fig. 11.8. Monthly variations in pH of Pauna water from June 2002 to May 2003

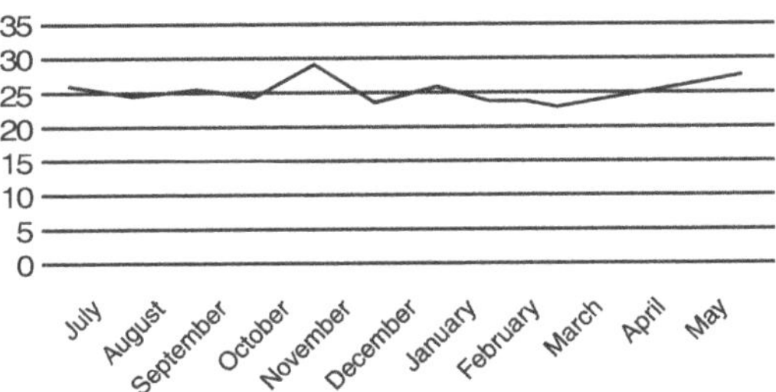

Fig. 11.9. Monthly variations in water temperature of Pauna water from June 2002 May 2003

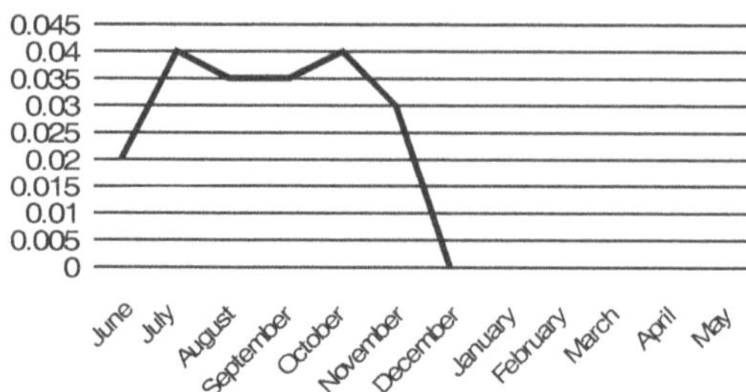

Fig. 11.10. TDS values of Pauna water from June 2002 to May 2003

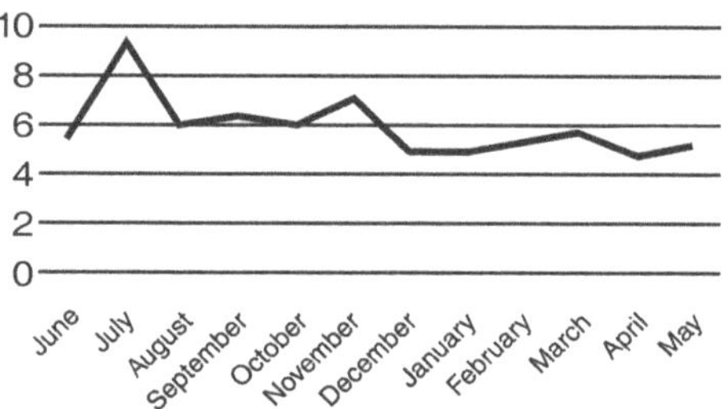

Fig. 11.11. Dissolved oxygen in Pauna water from June 2002 to May 2003

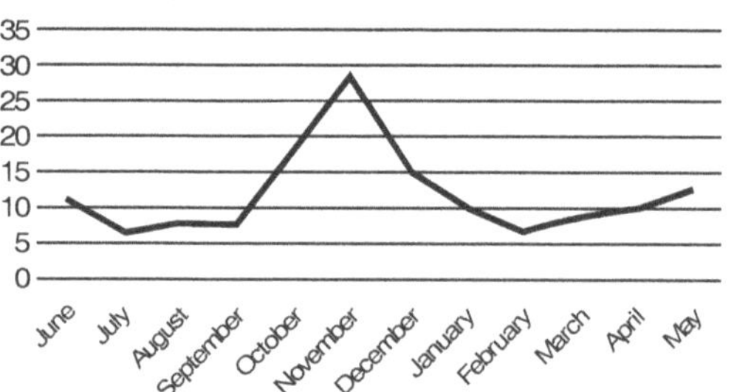

Fig. 11.12. Monthly variations in carbon dioxode from Pauna water from June 2002 to May 2003

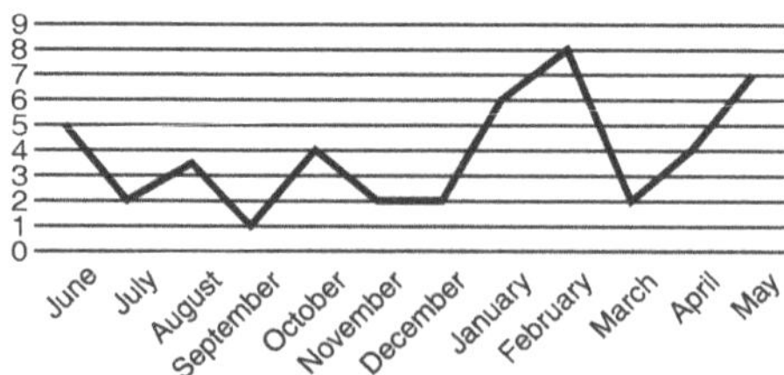

Fig. 11.13. Monthly variations in BOD in Pauna water from June 2002 to May 2003

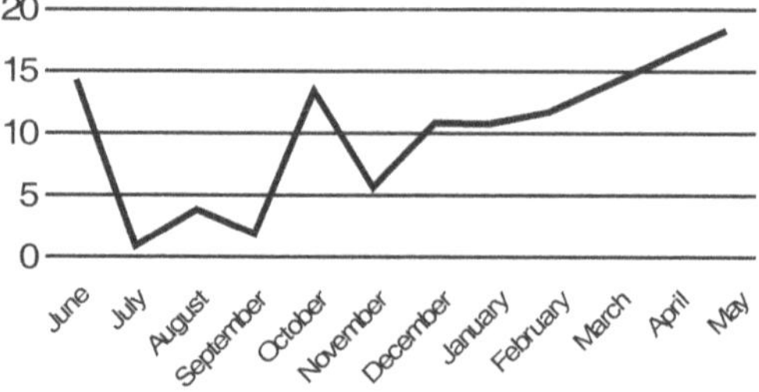

Fig. 11.14. Monthly variations in COD of Pauna water from June 2002 to May 2003

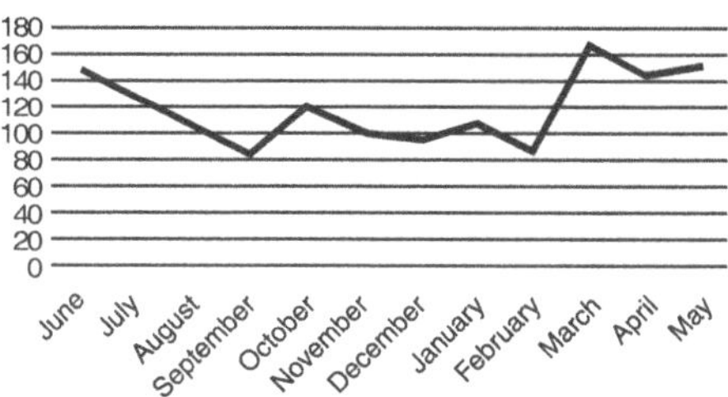

Fig. 11.15. Monthly variations in alkalinity of Pauna water from June 2002 to May 2003

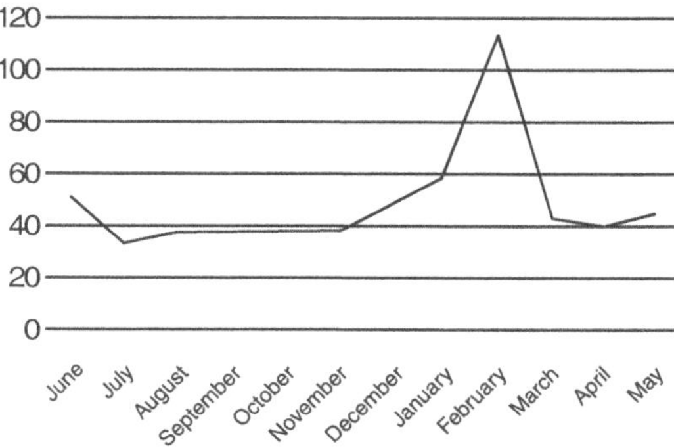

Fig. 11.16. Monthly variations in hardness of Pauns water from June 2002 to May 2003

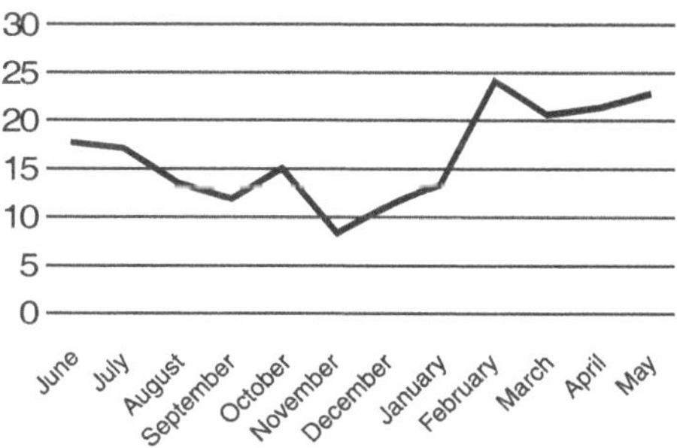

Fig. 11.17. Monthly variations in Chlorides of Pauna water from June 2002 to May 2003

Muley and Patil (2006) confirmed the occurrence of 25 fish species belonging to 3 orders at Ambegaon village, an important fish landing centre on the bank of reservoir (Table 11.64). The species such as *Notopterus notopterus, Clarias batrachus, Oreochromis mossambicus, Channa marulius, Channa punctatus, Glossogobius giuris* and *Xenentedon cancila* were not recorded in the reservoir. These species are common in Pauna river. The absence of these species in reservoir may be due to low organic material, deforestation, followed by sedimentation and construction of dam.

Table 11.64. Fish fauna of Pauna Reservoir of Pune district

ORDER : CYPRINIFORMES
1. *Labeo calbasu* (Hamilton)
2. *Labeo bogo* (Hamilton)
3. *Cirrhinus reba* (Hamilton)
4. *Puntius kolus* (Sykes)
5. *Puntius thomassi* (Day)
6. *Puntius sarana* (Hamilton)
7. *Puntius ticto* (Hamilton)
8. *Barbus duvancelli* (Cuvier-Blecker)
9. *Barbus punctatus* (Malabar)
10. *Tor mossal* (Sykes)
11. *Rohtee ogilbii* (Sykes)
12. *Salmostoma bacaila* (Hamilton)
13. *Rasbora elanga* (Hamilton)
14. *Rasbora buchanani* (Jerdon)
15. *Osteobrama vigorsi* (Sykes)
16. *Garra mullya* (Sykes)
17. *Nemachilus denisonni* (Day)
ORDER : SILURIFORMES
18. *Mystus bleekeri* (Day)
19. *Mystus vittatus* (Bloch)
20. *Mystus lamarii* (Cuvier)
21. *Wallago attu* (Schneider)
22. *Ompak bimaculatus* (Bloch)
ORDER : PERCIFORMES
23. *Aplocheilus lineatus* (Valenciennes)
24. *Pseudoambassis ranga* (Hamilton)
25. *Mastacembelus armatus* (Lacepede)

DIMBHE DAM

The Kukadi project is an integrated project comprising of five dams (*viz.* Yedgaon, Manikdoh, Dimbhe, Wadaj, and Pimpalgaon Joge Dam) on tributaries of river Bhima in Krishna river basin along with canal system. The project on completion will provide irrigation to an area of 1,56,278 ha of land in Pune, Ahemadnagar and Solapur districts of Maharashtra. In addition there are two powerhouses, one at Dimbhe dam and other at Manikdoh having installed capacity of 5 MW and 6 MW respectively.

Dimbhe is a masonry dam across river Ghod near village Dimbhe to impound a gross storage of 382.22 M.cum (13.50 TMC). The dam is a masonry dam. Maximum height of the dam is 72.10 meters. It has spillway in portion with 5 Nos. of gates of size 12 × 8.50 M each with a designed discharge capacity of 2122 cumecs. Left bank canal takes off directly from the storage and the right bank canal takes off from left bank canal through an aqueduct across Ghod river. Both the canals are proposed to irrigate areas in the Ghod and Mina valleys, the balance water would be supplied through Dimbhe left bank canal and utilized by Yedgaon through Kukadi left bank canal. The hydro electric power house with the generation of 5 mw is completed at the foot of the dam. Out of total 116 Km. length of Dimbhe Right Bank Canal 86 km. length is almost complete. Remaining work of main canal, lining and distribution system are in progress.

Total number of submerged villages due to construction of dam is 24 and area under submergence is 2202 ha. Construction of dam was started in 1977 and it was completed in 2000. The reservoir has two canals-one is Dimbhe left bank canal having a length of 55 km and other is Dimbhe right bank canal with a length of 116 km.

Table 11.65. Morphometry of Dimbhe Reservoir

River	Ghod
Year of completion of dam	1995
Type of dam	Masonry
Catchment area (Sq.km.)	298
Gross Storage (TMC)	13.500
Dead Storage(TMC)	1.00
Live Storage(TMC)	12.500
Full Reservoir level (m)	719.145
Maximum height of dam from river bed (m)	67.65
Total length of dam (m)	852
Average rainfall at reservoir site (mm)	382.42
Total Irrigation Potential (ha)	22016

Nineteen dam-displaced villages, populated almost completely by tribals, are situated on the fringes of Dimbhe reservoir which has an average fishing area of 1,278 hectares. Shashwat, an NGO based in Ambegaon in Pune district started organizing the fisherfolk in 2003. In 2006, Kusum Karnik, co-founder of Shashwat,helped the community to get the fishing lease. The then-Divisional Commissioner of Pune district saw the commitment of the locals and Shashwat, and decided to make a development plan for the region. He roped in the State tribal and fisheries departments, as well as CIFE, Mumbai. Government, NGO and community put together a convergence of schemes, under which the families living by the dam waters were supported for fish seed stocking, more boats and nets. Swiss Aid India has continuously supported and stood by us and the Rotary Club of Pune, Tilak Road, also helped with funds. The boats were built by the people, who chipped in with wood and their labour.Today, fishing at Dimbhe dam is community-run and scientific. 203 tribal families (of which 89 families are of the Katkari primitive tribe) run the Dimbhe Tribal Co-operative Fishing Society that harvests up to 27 tonnes of fish (Rohu, Catla, Mrigala and Chela) per year. In 2010, the total catch has been 14721 kg, with 4381 kg Chela and 10340 kg Indian Major Carps (IMC), in 147 fishing days.

The community has built 138 flat-bottomed boats, owns over 2000 kg nets, and has the use of a fibre-glass motorboat to ply the catch across the reservoir.Guided by the CIFE, the tribals now use state-of-the art cage culture, where fish fingerlings 35 mm long are grown in floating cages until they are 100-175 mm long. They are then released into the open waters of the reservoir, increasing their chances of survival. Since 2009 they have also successfully implemented pen culture to rear advanced fingerlings.

Aggressive stocking of IMC (Catla, Rohu and Mrigala) is done as advised by the CIFE. The tribal development dept. has introduced the Dimbhe type-boat to other tribal areas in the state. The fisheries department has supported us with boats and nets, fish seed, etc. Net size regulation and no fishing in the closed season are strictly maintained by the society. Back in 2003, at a meeting in Digad, all these fishers had resolved that they would not use poison or explosives for fishing.The non-fishing monsoon months fall hard on the villagers, especially the landless Katkari and Thakar. So we have introduced rearing of ornamental fish – gorgeous orange and silver creatures. Sixteen more cages have just been introduced for this, with NFDB support. 20 women have been trained in the upkeep of these fish, and a new and exciting phase to the Dimbhe fishery project is to begin.There is a long way to go. The total catch this year has been rather low. The fishers say the mesh size of their nets is too small now as the fish have grown bigger. We hope to help the fisher-folk of the 6 dams in the adjacent blocks, Khed and Junnar, to increase their fish catch, to break the endless cycle of poverty and to bring a smile on the faces of the fisherwomen and children.

Shashwat has been chosen for the 'Equator Prize 2012', of the United Nations Development Program, for its work on local development solutions for people, nature and resilient communities. The NGO will receive the US$ 5,000 award to be presented in the 2012 United Nations Conference on Sustainable Development in Brazil, to be held on June 20 to 22. The Equator Prize is awarded to recognise outstanding community efforts to reduce poverty through conservation of biodiversity.

Kusum Karnik and Anand Kapoor from Shashwat have been working with tribals in the catchment area of the Dimbhe dam, which submerged eleven villages and destroyed cropland in another 13 since 1981. The NGO was registered in 1996.

The organisation helps the local community to develop small-scale fishing activities in the dam reservoir and the locals are supported to obtain fishing leases, boats and nets. The locals have banned dynamite fishing to ensure fish stocks are replenished. Dynamite fishing is a practice of using explosives to kill fish for easy collection. The banning of this practice has helped increase the fish size and number, thereby enhancing their income too.

Sehgal *et al.* (2013) studied zooplankton diversity of Dimbhe dam. The diversity study revealed four groups of zooplankton *viz.* Rotifera, Cladocera, Copepoda, Ostracoda. In Dimbhe, seven species of zooplankton were recorded during winter as well as summer season, of which only one species belongs to rotifera, one species of cladocera, four species of copepoda and only one species of ostracoda was present. The total standing crop of zooplankton showed peak population during summer.

In Dimbhe, seven species of zooplankton were recorded during winter as well as summer season, of which only one species (*Keratella tropica*) belonged to rotifera, one species (*Bosmina* sp.) to cladocera, four to copepoda and only one species (*Cypris* sp.) to ostracoda. Among copepoda *Phyllodiaptomus* sp was dominant in both the season. In Dimbhe, zooplankton density followed the order of copepoda > ostracoda > rotifera > cladocera during winter as well as summer season.

The maximum zooplankton density recorded in Dimbhe was during summer season with 5123 no./100 litre, while in winter it was 1314 no./100 litre. A total of seven species of zooplankton were recorded each during both the seasons with copepoda being most commonly observed order and *Cypris* sp. the most dominant.

Air temperature was always higher than the water temperature and showed direct effect on water temperature.During summer season, maximum air and water temperature was 36.3°C and 33.3°C in Dimbhe reservoir. The pH value during winter as well summer season was 7.3. Dissolved oxygen content ranges from 5.46 to 6.33 mg / l during winter and summer seasons (Sehgal *et al.* 2013).

YEDGAON RESERVOIR

The construction weir cum storage across Kukadi river at Yedgaon in Pune district with a gross capacity of 93.00 M.cum (3.30 TMC) has been completed in 1977. It is an earthen dam with waste weir in right side saddle which consisted 11 gates of size 12 × 5.00 meter each with a designed discharge capacity of 3844 cumecs, maximum height of the dam is 23.60 meter and length 4470 meter. The main Kukadi left Bank canal off takes from this weir. The works of Kukadi left bank canal are atmost completed.

Table 11.66. Salient Features of Yedgaon Reservoir

River	Kukadi
Year of completion of dam	1977
Type of dam	Earthen
Gross Storage (TMC)	3.3
Dead Storage (TMC)	0.500
Live Storage (TMC)	2.800
Full Reservoir level (m)	641.00
Maximum height of dam from river bed (m)	23.6
Total length of dam (m)	4470
Total Irrigation Potential (ha)	93396

The fishes which commonly occurred in reservoir water are *Labeo rohita, Catla catla, Cirrhinus mrigala, cyprinus carpio, Oreochromis mossambicus* and *Ambassis* spp.

WADAJ RESERVOIR

It is composite small dam across river Meena near village Wadaj in Pune district to impound a gross of 48.13 M.cum (1.70 TMC). Construction work of this dam is completed by June 1981 including erection of gates. It is an earth dam. Maximum height is 26.42 M and length 1830 M. Its spillway has 5 Nos. of gates of size 12 × 5 m each with design discharge of 1426 cumecs.

Table 11.67. Salient Features of Wadaj Reservoir

River	Meena river tributary of Ghod river.
Year of completion of dam	1983
Type of dam	Earth
Gross Storage (TMC)	1.271
Dead Storage (TMC)	0.100
Live Storage (TMC)	1.171
Full Reservoir level (m)	717.50
Maximum height of dam from river bed (m)	26.42
Total length of dam (m)	1830
Total Irrigation Potential (ha)	20265

KASAR SAI RESERVOIR

The Kasar Sai dam is situated near Hinjewadi, Pune. This medium sized dam is mainly constructed for irrigational purposes. It is about 20 km away from Pune city and is surrounded by hills, agricultural fields and village Kasar Sai. It is located in 180 37.30′ 13″ N Latitude and 730 39.22′ 55″ E Longitude. This dam is built on a natural source of water. A seasonal water stream surrounded by hills was developed into a fresh water resource by constructing a dam. It has a spread area of approximately 336 hectare, mainly constructed for irrigational purpose.

Bhalerao (2012) studied the seasonal fluctuations in water quality and fish fauna of Kasar Sai reservoir and suggested some measures for the improvement of the fish production to uplift the economic condition of the fishermen. The water temperature ranged from 20°C to 26°C. The maximum temperature (26°C) was recorded in the month of June (summer) and minimum (20°C), in the month of December (winter). It showed higher temperature in summer and relatively lower temperature in winter. The values of DO fluctuated from 7.50 mg/l to 11.82 mg/l. The maximum value (11.82 mg/l) was recorded in the month of June (summer) and minimum value (7.50 mg/l) was recorded in the month of

December (winter). The pH values were ranging from 8.35 to 9.63. The maximum pH value (9.63) was recorded in the month of December (winter) and minimum (8.35) in the month of April. Total alkalinity ranges from 48.69 mg/l to 65.93 mg/l. The maximum value (65.93 mg/l) was recorded in the month of June (summer) and minimum value (48.69 mg/l) in the month of December (winter). The maximum alkalinity was observed in June (summer) due to increase in bicarbonates in the water. The values of chloride contents and suspended solids were negligible. This may be due to absence of any polluting industry in nearby areas. The sewage content was also negligible as the area is sparsely populated.The fishes found in the reservoir were major carps, exotic carps, local variety and very limited variety of crabs. The fish production is approximately 250 to 350 kg per month from December to February and raises up to 1000 kg per month from March to May (Bhalerao, 2012).

HINGNI (PANGAON) RESERVOIR

The Hingni (Pangaon) reservoir, one of the artificial water bodies in Barshi tahsil in Maharashtra is used for irrigation as well as fisheries. The reservoir was constructed during year 1976 in between villages Hingni and Pangaon on river Bhogawati.The average rainfall at reservoir area was recorded at 582 mm. The reservoir starts filling by the late July. The water level reaches its peak generally in the middle of August. During September and October it becomes somewhat stable. From the month of January, the level declines steadily. The reservoir water is used for irrigation and it has an opportunity to provide drinking water to about 15 villages of Solapur district. The reservoir water is also used for Bhogawati sugar factory at Vairag. The detail morphometric features of Hingni (Pangaon) reservoir are depicted in Table 11.65. The reservoir is about 40 km away from Tuljapur town, on Tuljsapur-Barshi state highway.

Table 11.68. Salient features of Hingni (Pangaon) reservoir

1. Name of river	Bhogawati
2. Year of construction	1976
3. Average rainfall (mm)	582
4. Length of dam (m)	1987.80
5. Mean depth (m)	21.87
6. Water spread area (ha)	563
7. Irrigation potential (ha)	5625
8. Dead storage capacity (million m^3)	13.54
9. Usable storage capacity (million m^3)	31.97
10. Gross storage capacity (million m^3)	45.91
11. Catchment area (km^2)	401.45
12. Maximum flood (m)	504

Sakhare (2005) studied water quality of Hingni (Pangon) reservoir and its significance to fisheries.Water temperature in the reservoir ranged from 19 to 26°C with an average of 23.5. Water temperature influences aquatic life and concentration of dissolved gases like CO_2, O_2 and chemical solutes. The minimum temperature was recorded in November while maximum in the month of April.

Table 11.69. Water quality of Hingni (Pangaon) reservoir

Water parameters	Range
1. Water temperature (°C)	19 – 26
2. pH	8.1 – 8.6
3. Transparency (cm)	15 – 43
4. Dissolved oxygen (mg/l)	3.7 – 9.4
5. Free carbon-dioxide (mg/l)	Nil – 4.1
6. Total alkalinity (mg/l)	49 – 69.8
7. Total hardness (mg/l)	48 – 53.2
8. Chlorides (mg/l)	14.3 – 59.2
9. Total dissolved solids (mg/l)	39 – 43

Transparency ranged from 15 to 43 cm (av. 33 cm). Low values were recorded during September, while the high values in May. The low values in month of September might be due to low and moderate velocity of winds.

The pH of the reservoir water varied from 8.1 to 8.6 (av. 8.2) indicating alkaline range throughout the study period. The pH in a water body normally depends on the prevailing biological activity. A high pH value of waters is usually associated with higher photosynthetic activity, which results in utilization of carbon dioxide from bicarbonates and formation of carbonates. According to Jhingran (1990) the pH range of Indian reservoirs is from 6.5 to 9.2 and the pH in between 6.0 to 8.5 indicates the medium productive reservoir. The reservoir under present investigation falls under medium productive category. It is due to its average pH value recorded at 8.2.

Total alkalinity is the sum of phenolphthalein and methyl orange alkalinity. During present investigation total alkalinity value ranged from 49 to 69.8 mg/l with an average value of 52.5 mg/l. The minimum value was recorded in March, while the maximum was during November.

Dissolved oxygen in the water is greatly affected by the biological activity, such as photosynthesis or from air due to turbulence of the water surface by wind action, which is dissolved into the water at its surface. The dissolved oxygen content of warm water fish habitat should not be less than 5 mg/l during at least 16 hours of any 24 hour period. It may be less than 5 mg/l for a period of not exceeding 8 hours within any 24 hour period and no time shall the dissolved oxygen content be less than 3 mg/l. It has been found that fish can survive at dissolved oxygen in the range of 1 to 5 mg/l, but its growth and reproduction are retarded.Dissolved oxygen in Hingni (Pangaon) reservoir varied from 3.7 to 9.4 mg/l with maxima in October and minima in March.

In an aquatic ecosystem sources of CO_2 are community respiration and decomposition, while it is consumed in the photosynthesis. The values of free carbon dioxide varied between Nil to 4.1 mg/l with an average value of 2.8 mg/l.The maximum concentration (4.1 mg/l) was recorded during August and it was totally absent during months of November.

Chlorides occur naturally in all types of waters. High concentration of chlorides is considered to be the indicator of pollution due to high organic wastes of animal or industrial origin. The present study found a range of 14.3 to 59.2 mg/l of chloride. The maximum concentrations of chlorides were recorded during April and minimum during summer.

Total hardness value was found maximum (53.2 mg/l) in April and minimum (48 mg/l) in October. Epizootic Ulcerative Syndrome outbreak has been observed to be more frequent in waters with low hardness. The productive water bodies should have hardness above 20 mg/l. Optimal hardness for fish culture has been observed tobe around 75 to 150 mg/l. The total hardness values recorded in Hingni (Pangaon) reservoir indicate that the reservoir is productive.

Total dissolved solids ranged from 39 to 43 mg/l with an average value of 41.9. The reservoir water showed maximum value in the month of May, while its minimum was during August.

MANJARA RESERVOIR

The Manjara reservoir, a purely irrigation project was constructed on river Manjara, an important tributary of river Godavari in year 1981.The reservoir is bounded by latitudes 18°25′ to 18°55′N and 75°15′ to 76°15′E. The water of reservoir is used for irrigation and as drinking water to towns like Kallam, Kaij, Ambajogai, Kille Dharur, Latur and another 20 small villages. The fishing rights of the reservoir are given on lease to the fishermen's co-operative society. The reservoir is having catchment area of 2371.59 km. The live storage and dead storage capacity of reservoir is 173.32 mm^3 and 77.38 mm^3 respectively. The salient morphometric features of reservoir are depicted in Table 11.70.

Table 11.70. Salient feature of Manjara reservoir, Maharashtra

River	Manjara
Construction started	1978
Year completion of project	1981
Location	Dhanegaon
Altitude	641.87 m above sea level.
Latitude	18°25″ to 18°55″ North
Longitude	75°15″ to 76°15″ East
Surface area	4260 mt^2

Contd...

Table 11.70: Contd...

Maximum depth	217m
Gross storage capacity	250.70mm^3
Live storage capacity	173.32 mm^3
Dead storage capacity	77.38 mm^3
Total catchment area	2371.59 km
Irrigation potential	18222 hectares

Rainfall: The normal annual rainfall over the reservoir area for the period of October 1999 to September 2000 varies from nil to 404 mm. The area has experienced moderate, severe and acute drought conditions for last 5 years. Hence the entire area comes under the category of drought area.

Physico-Chemical Environment: The physico-chemical parameters of the reservoir during October 1999 to September 2000 are given in Table 11.71. The water temperature ranged from 3.12 to 28.1°C. The minimum temperature was recorded in November, while the maximum temperature was come across the month of March. pH was in the range of 7.1 to 8.4. The minimum and maximum values of pH were recorded in the months of December and July respectively. Transparency ranged from 25 to 154 cm, indicating low values during July and high values in the month of May. Total Dissolved Solids (TDS) ranged from 114 to 234.98 mg/l, and its maxima and minima were recorded in February and September respectively. The range of dissolved oxygen content ranged from 5.2 to 9.15 mg/l and that of free carbon dioxide from 1.9 to 2.75 mg/l. The total

Table 11.71. Physico-chemical profile of Manjara reservoir

Sr.No.	Parameters	Range
1.	Air temperature (°C)	3.25 to 35
2.	Water temperature (°C)	3.12 to 28.1
3.	Rain fall (mm)	0 to 404
4.	Water volume (mm^3)	102 to 251.99
5.	pH	7.1 to 8.4
6.	Transparency (cm)	25 to 154
7.	Total Dissolved Solids (mg/l)	114 to 234.98
8.	Dissolved oxygen (mg/ l)	5.2 to 9.15
9.	Free carbon-dioxide (mg/l)	1.9 to 2.75
10.	Total Alkalinity (mg/l)	130.6 to 254
11.	Total Hardness mMg/l)	50 to 104.5
12.	Chlorides (mg/l)	17.55 to 32.6
13.	Phosphates (mg/l)	0.1 to 0.34
14.	Nitrates (mg/l)	0.1 to 0.47

alkalinity was in the range of 130.6 to 254 mg/l. The minimum and maximum values of total alkalinity were recorded in the months of December and April respectively. The range of total hardness observed in between 50 and 104.5 mg/l and those chlorides from 17.55 to 32.6 mg/l. Phosphates and nitrates were in the range from 0.1 to 0.43 mg/l and 0.1 to 0.47 mg/l respectively.

Zooplankton Diversity: Among the zooplankton, the rotifera was the dominating group accounting for about 55.59% of the total population of the zooplankton. It was followed by copepoda accounting for about 21.60% population, where as cladocera accounts about 15.20% population. The ostracoda was least in abundance regarding diversity and density and accounting only about 7.60% population.

Information of Ichthyofauna: Sakhare (2005) reported 28 species of fishes belonging to 4 orders (Table 11.72). Of the 4 orders, order cypriniformes dominated with 12 species falling under 9 genera of which genus *Puntius* is abundant with 4 species. Next in abundance is the order Perciformes in which genus *Channa* is dominant with 3 species. In terms of species abundance genus Puntis tops the list. Out of 4 species occurring in the reservoir, *P. kolus,* and *P. sarana sarana* are predominant and occurred in all the areas. *Heteropneustes fossilis* and *Clarias batrachus* also occurred in all localities but in very less numbers. The fishes, which commonly occurred in reservoir water, are *Labeo rohita, Catla catla, Mystus* spp and *Channa* spp.

Table 11.72. List of fishes recorded from catches of Manjara reservoir

Order : Osetoglossiformes
Family : Notopteridae
1. *Notopterus notopterus* (Pallas)
2. *Notopterus chitala* (Hamilton-Buchanan)
Order : Cypriniformes
Family : Cyprinidae
3. *Catla catla* (Hamilton-Buchanan)
4. *Cirrhinus mrigala* (Hamilton-Buchanan)
5. *Labeo rohita* (Hamilton-Buchanan)
6. *Cyprinus carpio communis* (Linnaeus)
7. *Hypophthalmichthys molitrix* (Valenciennes)
8. *Puntius sarana sarana* (Hamilton-Buchanan)
9. *Puntius sophore* (Hamilton-Buchanan)
10. *Puntius ticto* (Hamilton-Buchanan)
11. *Puntius kolus* (Sykes)
12. *Chela bacila* (Hamilton-Buchanan)
13. *Rohtee ogilbii* (Sykes)
14. *Chela bacila* (Hamilton-Buchanan)

Order : Siluriformes
Family : Siluridae
15. *Ompak bimaculatus* (Bloch)
16. *Wallago attu* (Schneider)
Family : Bagridae
17. *Mystus cavassius* (Hamilton-Buchanan)
18. *Mystus seenghala* (Sykes)
Family Clariidae
19. *Clarias batrachus* (Linnaeus)
Family : Heteropneustidae
20. *Heteropneustes fossilis* (Bloch)

Order : Perciformes
Family : Ambassidae
21. *Chanda nama* (Hamilton-Buchanan)
22. *Chanda ranga* (Hamilton-Buchanan)
Family Gobiidae
23. *Glossogobius giuris* (Hamilton-Buchanan)
Family : Channidae
24. *Channa striata* (Bloch)
25. *Channa marulius* (Hamilton-Buchanan)
26. *Channa gachua* (Hamilton-Buchanan)
Family : Mastacembelidae
27. *Macrognathus pancalus* (Hamilton-Buchanan)
28. *Mastacembelus armatus* (Lacepede)

Crafts and gears used in fishing: The crafts used in Manjara reservoir is of primitive type, and no modern crafts have been introduced for fishing. This craft is nothing but a platform of 6 × 3 feet size with a depth of 4-10 inches and is locally known as 'Nav'. This is prepared from the thermocoel and covered by a plastic sheathing. At present, almost cent percent crafts are of this type. The cost of one craft varies from Rs. 500-800. Previously the wooden country boats were also used. The vehicle tyre tubes are common as crafts along the bank of reservoir. There are so many drawbacks of such type of crafts such as (1) They are not suitable to carry large nets, having heavy weight, (2) when fishermen get large quantities of fish catches, it become difficult to carry them to the landing centers.There are no fishermen regularly operating in this reservoir and part-time fishermen who employ gill nets, cast-nets and rod and line for raising extra income carry out the fishing. Rod and line is mostly used by the fisher children who help their parents in improving their income. Rod and lines are operated near the shore of the reservoir. Cockroaches, insects or other small organisms are used as bait for the capture of carnivorous fishes. A gill net measures 10 × 2 m with a mesh size varying from 10 to 30 mm.

Fish Catch: Most of the fishermen are the local farmers and farm labourers who changed their profession in favour of fishing. Average catch per individual fishermen on a normal day is 3 to 6 kg, while during the peak fishing season the catch per individual fishermen is about 10-12 kg. The catch is sold to middleman who carts the fish everyday by road to market. The womenfolk in the fishermen families often hawk fish in local markets and residential areas.

Fishing Days and Closed Season: In Manjara reservoir, fishing is carried out throughout the year. No 'closed season' has been declared by the State Fisheries Department or the society. However, on the suggestions of state fisheries department, the society has framed some rules for the conservation of fish. For example, catching of fish belonging to Indian Major Carps less than 1.5 kg in weight has been prohibited. Such types of fish if caught, it should be immediately released in the reservoir. This rule is not strictly followed and a large number of Indian Major Carps weighing less than 1.5 kg are caught.

Poaching and Fisheries Conservation: The strength of staff for checking of the poaching of fish from the reservoir is inadequate. As there are a large number of landing centers, the present strength of fisheries staff is not sufficient to book the poachers and other fishery offenders under the Indian Penal Act for fish theft and/or under the Indian Fisheries Act, 1897, for violating fishery rules and regulations. It is worthwhile to state that during the present study, a good number of landing centers were visited and presence of many Indian major carps weighing less than 1.5 kg was observed in catches of the fishermen.

Recommendations for Improvement of Fisheries:

1. The state government should provide sufficient funds to the society so that it could purchase good quality nets and boats for members.
2. The government may also depute adequate technical as well as supervisory so as to guide the society for formulating, implementing and monitoring of the development programmes and also to check poaching and violation of fishery rules and regulations.
3. The society should undertake the transport and marketing of fish itself and eliminate the intermediaries.
4. A committee comprising active members of the society should be constituted for ensuring the stocking of the reservoir with adequate number of advanced sized fingerlings of desired fish species in a definite ratio.
5. The society should offer more incentives to its members for catching more and more large catfishes and minnows from the reservoir so that their populations could be reduced considerably in the reservoir.
6. Poachers can be punished under Indian Penal Code for the theft or for offences committed in violation of rules made under the fisheries act.
7. A well-equipped small field laboratory may be established at Manjara for monitoring the productivity, hydrobiological parameters and other ecological conditions of the reservoir. This laboratory should also maintain the data on the stocking of fish seed, fish production and socio-economic conditions of fishermen.
8. The most unfavourable feature of the reservoir from a fisheries point of view is the complete withdrawal of reservoir water for drinking and

irrigation, which results in fish kill. Hence a definite quantity of water must be reserved for the survival of fish.

MAJALGAON RESERVOIR

The Majalgao dam is part of the Jayakwadi Scheme (2400 km^2), one of the largest irrigation areas in Maharashtra. The first phase of this scheme was financed by world bank and began in 1970s. Limited water resources have reduced the first stage irrigation area to 580 km^2, although it is planned to ultimately irrigate 1190 km^2.

It is located at 19°8′59″ North and 76°13′48″ East in Majalgaon region of Beed district on river Sindphana, which is the tributary of the river Godavari. The river Godavari is one of the most important rivers in India. It originates at Trimbakeshwar (hill) in Nashik district of Maharashtra and travels some 1500 km to join the Bay of Bengal. It has a catchment area of 312.812 sq. km of which 63.3 % lies in Maharashtra. Its main tributaries are Pravara, Purna, Penganga, Wainganga, Manjara, and Sindphana. There are many pilgrims center on the bank of Godavari out of which Trimbakeshwar, Nashik, Paithan and Nanded are major one where millions of peoples gather on festivals. Many cities and villages are situated on the bank of this river which utilises the river water. Hence Godavari river in considered as the Ganga of Deccan.

Fig. 11.18. Sluice gates of Majalgon reservoir

Fig. 11.19. Hydelpower generation unit at Majalgaon Reservoir

Fig. 11.20. A view of Majalgaon Reservoir

River Sindphana is the main tributary of the Godavari River. On the right bank, its origin is in the Balaghat range 40 to 50 km away from Majalgaon city. It merges in Godavari River near Manjarath village. On the river Sindphana this well known dam has been built near 2 km from Majalgaon city. It is a irrigation and power production project. The reservoir has an opportunity to provide water for sugarcane, wheat, cotton, bajra, jowar, groundnut etc.

The climate of this region is dry except in south west monsoon season (June to September). There are wide seasonal variations in temperature of the region. December is the coldest month of the year. Sometimes the region is affected by cold waves, which are associated with the westward passage of the western disturbances across north India. In the period from March to May there is continuous rise in both day and night temperatures. May is the hottest month of the year. With the one set of monsoon by the second week of June, the temperature goes down and the weather is very pleasant throughout the southwest monsoon. The relative humidity is high during the southwest monsoon season. The summer season is the driest period of the year.

Table 11.73. Salient Features of Majalgaon Reservoir

Attribute	Value	Attribute	Value
Name of Dam	Majalgaon	Minimum Draw Down Level (MDDL Minimum Draw Down Level)	426.11
Nearest City	Majalgaon	Gross Storage Capacity (MCm)	453.64
District	Bid	Live storage capacity (MCM)	311.3
State	Maharashtra	Design Flood (cumec)	15500
River Name	Sindphana	Type of Spillway	Ogee
Basin	Godavari	Length of Spillway (meter)	239

Contd...

Table 11.73: Contd...

Type of Dam	Earthen	Type of Spillway Gates	Radial
Purpose of Dam	Irrigation, Hydroelectric	Number of Spillway Gates	16
Year of Completion	1987	Size of Spillway Gates (m × m)	12 × 8
Catchment Area (Th ha)	384	Land affected-Total (Th ha)	7.458
Length of Dam (meter)	6488	Land affected-Culturable (Th ha)	07.458
Maximum Height above foundation (meter)	31.19	Land Affected-Forest (Th ha)	—
Maximum Water Level in (meter)	434.8	Land Affected- Others (Th ha)	—
Full Reservoir Level in (meter)	431.8	Number of Villages Affected	20

The society generally purchased fish seed from the private fish seed farm, about 10 kms from the reservoir site. During 2009-10 the society purchased 22,50,000 fish seed, while during 2010-11 society could purchase 27,00,000. During 2011-12 society stocked 36,00,000 fish seed. Recently during 2012-13 reservoir was not stocked. It was due to insufficient water in reservoir.The seed of catla, rohu and mrigal is stocked.

Table 11.74. Quantity of seed stocked in Majalgaon Reservoir

Year	Seed stocked		Total stocking
	Fish seed	Prawn seed	
2009-10	22,50,000	13,00,000	35,50,000
2010-11	27,00,000	15,00,000	42,00,000
2011-12	36,00,000	7,00,000	43,00,000
2012-13		Not stocked	

The fishery is exploited by fishermen who have been organized from different villages into co-operative society. The details of the fish production from the reservoir for the period 2009-10 to 2010-11 is shown in Table 11.75. The study of the data reveals that the maximum fish production (from 2009-10 to 2010-11) was recorded at 59066.75 kg during 2010-11 *i.e.,* 7.68 kg/ha/yr.The minimum production was during 2009-10 at 51288.25kg, giving an average fish production of 6.67 kg/ha/yr. Local fishes showed the higher catches than those of Indian major carps (Fig. 11.21 and Fig. 11.22).

Table 11.75. Fish catch from Majalgaon Reservoir

Year	Local fish	Carps	Total catch	Fish catch (kg/ha/yr)
2009-10	33326.25	17962	51288.25	6.67
2010-11	39695.25	19371.5	59066.75	7.68

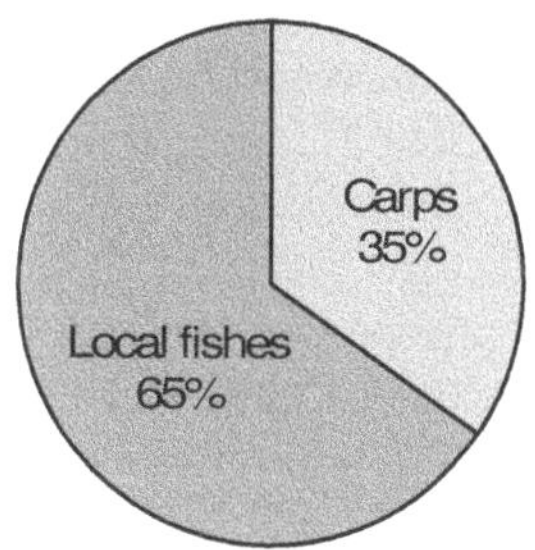

Fig. 11.21. Fish catch composition in Majalgaon reservior during 2009-10

Fig. 11.22. Fish catch composition in Majalgaon reservior during 2010-11

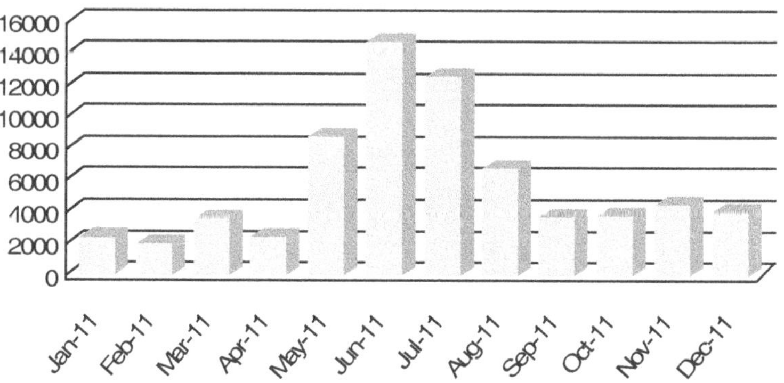

Fig. 11.23. Fish production (kg) in Majalgaon Reservoir (year 2011)

Table 11.76. Water quality of Majalgaon Reservoir

Water parameters	Range
1. Air Temperature (°C)	26 to 38
2. Water Temperature (°C)	21 – 29
3. pH	7.6 – 8.25
4. Conductivity (ppm)	350 – 410
5. Total Dissolved Solids (mg/l)	190 – 240
6. Salinity (mg/l)	43.6 – 56.5
7. Hardness (mg/l)	160 – 198
8. Calcium (mg/l)	27.5 – 35.54
9. Magnesium (mg/l)	24.8 – 27.52
10. Chlorides (mg/l)	24.12 – 31.24

BORNA RESERVOIR

Borna reservoir was constructed in year 1983 across river Borna with gross storage capacity of 10908 10^3m^3. He height and length of dam is 22.3 m and 866 m respectively.The reservoir is located in Parli taluka of Beed district. Soon after the formation of reservoir, the Maharashtra state fisheries department leased the reservoir for fisheries to Mandwa Fishermen's Co-operative Society. The society has 65 members, out of them 5 members belong to SC, 15 to ST, 8 muslims and 37 belongs to other castes. Fry of catla,rohu,mrigal and common carp is stocked .Prawn seed was also stocked during 2010-11 and 2011-12. But prawn catch was very low. It may be due to presence of catfishes which feedson seed of prawn and carps. Stocking done so far has little impact on the fisheries. The small size at stocking time made them an easy prey to the carnivores in the ecosystem. The policy of stocking never had an ecology base or a consideration to the production potential of the reservoir. Neither had it taken care of growth rate nor the feeding habits of the different species in the ecosystem.

Fig. 11.24. A haul of tilapias in from Borna reservoir

It was revelaed that tilapia was the most dominant species in Borna reservoir and which may be reducing the number of other indigenous fish species. It may be due to its capacity to adjust with any aquatic conditions and rapid productivity. Tilapia destroys the eggs of other fishes and can steadily replace them. It may be one of the reasons in reducing the carp fishery of reservoir. The tremendous disappearance of carps from the ecosystem may be attributed to the invasion of fish, Tilapia mossambica. Performance of tilapia in Borna reservoir is discouraging due to its early maturity,continuous breeding, over polulation and dwarfing. It is found that the growth rate of *Catla catla* has adversely affected by Tilapia.

Major component (40%) of the landings is of tilapia. Catfishes form 35% of catch, followed by carps (10%) and other fishes (15%).

WAN RESERVOIR

The Wan reservoir (Fig. 11.25) is a medium sized reservoir near Parli (Vaijyanath) city in Marathwada region. The reservoir was constructed in year 1963 across the river Wan. The reservoir is bounded by latitude 18°53′N and Longitude 76°27′. It is the oldest reservoir in district with water spread area of 347 ha. The reservoir has an opportunity to irrigate 7567 ha of agricultural land. Besides supply of water for irrigation,the reservoir also supplies drinking water to Parli Vaijnath town and Vaijnath Co-operative Sugar Factory. The area around reservoir comprises forest-covered hills. The reservoir has two small canals through which water supply goes to agricultural lands of the villages like Nagapur P angri, Deshmukhtakli, Talegaon, Limbota, Kowthali, Sangam, Gopalpur, Mandekhel, Bhilegaon, Tandoli, Selu, Parchondi, Sirsala, Kanadi and Pimpalgaon. The salient features of the reservoir are depicted in Table 11.74.

Fig. 11.25. Map of Wan Reservoir near Parli

Table 11.77. Salient features of Wan reservoir, Maharashtra

River	Wan
Year of construction	1963
Water spread area (ha)	347
Height of dam (m)	19.81
Length of dam (m)	2188
Gross storage capacity (mcm)	25.181
Catchment area (km^2)	379.92
Full tank level	454.87
High flood level (m)	456.40
Irrigation potential (ha)	7567
Length of canal • Right canal (m) • Left canal (m)	12.039.65

Rainfall at catchment area (mm)	535
Water temperature (°C)	19-35
Transparency (cm)	89-170
pH	7.9 – 8.2
Dissolved oxygen (mg/l)	5.9-12.4
Free carbon-dioxide (mg/l)	2.3-4.2
Total hardness (mg/l)	83.2-120
Total alkalinity (mg/l)	153.2-289
Total Dissolved Solids (mg/l)	217-307.5

Physico-chemical Environment: The results of physico-chemical parameters are depicted in Table 11.54. The water temperature fluctuated from 19 to 35°C, values were high during summer. Summer temperature increase was due to greater heating of water and insolent heat from the sun. The secchi disc transparency fluctuated from 89 to 170 cm. The pH was alkaline with a highest value of 8.2 during April 2004. The lower value of pH during September 2004 may be attributed to heavy rainfall during this month. The average pH value (7-9) of Wan reservoir is suitable for optimum growth of fish. The dissolved oxygen was found to be ranged between 5.9 to 12.4 mg/l. The carbon dioxide content of water depends upon the temperature of water, depth of water, rate of respiration, decomposition of organic matter, chemical nature of the bottom and the geographical and physiological feature of the terrain surrounding the water. The free carbon-dioxide in Wan reservoir was observed in the range of 2.3 to 4.2 mg/l. Alkalinity of freshwater ecosystems is generally caused by carbonates and bicarbonates or hydroxides of calcium, magnesium, sodium, potassium, ammonium and iron. Natural water bodies usually show a wide range of fluctuations in total alkalinity depending upon the location, plankton population, nature of bottom deposits etc. Waters of hilly streams, sandy, rocky, flooded rivers in rainy season and waters infested with submerged weeds usually have a low total alkalinity. On the other hand, stagnant waters in low rainfall areas during summer season are likely to have high total alkalinity values. In Wan reservoir total alkalinity was observed in the range of 153.2 to 289 mg/l. The hardness of water is mainly due to the presence of calcium and magnesium. Maximum values were found during summer and lowest values were found during winter months.Values of total hardness ranged between 83.2 to 120 mg/l. The total dissolved solids were high in summer followed by winter and monsoon. It was due to the dilution of water.

Plankton Diversity: The phytoplankton constitute baseline of food webs in an aquatic ecosystem. In the present study phytoplankton belonging to three groups *i.e.*, myxophyceae, chlorophyceae, and bacillariophyceae and zooplanktons belonging to rotifera, cladocera, copepoda and ostracoda were identified. The plankton population of the reservoir showed summer peak coinciding with the mean depth, influx and outflow of water. Rotifera, copepoda, ostracoda, myxophyceae and chlorophyceae contributed to the summer peak of planktonic population (Table 11.78).

Table 11.78. Plankton diversity of Wan reservoir

A. PHYTOPLANKTON
1. **Myxophyceae:** *Anabaena, Microcystis, Merismopedia*
2. **Chlorophyceae:** *Nostoc, Periastrum, Closteridium, Microspora*
3. **Bacillariophyceae:** *Fragillaria, Navicula, Nitzschia, Tabellaria*
B. ZOOPLANKTON
1. **Rotifera:** *Brachionus* sp., *Euchlanis dilata, Filinia ingiseta, Keratella tropica, Trichocera Porcellus*
2. **Cladocera:** *Ceriodaphnia cornuta, Moina micrura, Biapertura karna, Indialona ganapati*
3. **Copepoda:** *Cyclops, Mesocyclops, Neodiaptomus lindbergi, Nauplius*
4. **Ostracoda:** *Cypris* sp., *Cyclocypris globosa, Stenocypris* sp.

Fish Fauna: The fish fauna is an important aspect of fishery potential of a waterbody. Fish fauna of Indian reservoirs has been studied by several workers and it has been found that the distribution of fish species is quite variable because of geographical and geological conditions of the area. During the present investigation altogether 34 species of fishes belonging to 24 genera falling under 7 orders have been identified (Table 11.79). Of the 7 orders, order cypriniformes dominated with 17 species falling under 12 genera of which genus *Puntius* is abundant with 4 species of them dominating the catch. Next in abundance are the fishes coming under the order siluriformes in which Mystus is dominant with 2 species, while the order anguilliformes has only one dominant species.

Table 11.79. Fish fauna of Wan reservoir

Order : Osteoglossiformes Suborder : Notopteroidei Family : Notopteridae 1. *Notopterus notopters* (Pallas) 2. *Notopterus chitala* (Ham.)
Order : Anguilliformes **Suborder : Anguilliformes** **Family : Anguillidae** 3. *Anguilla bengalensis* (Gray)
Order : Cyprniformes **Suborder : Cyprinedei** **Family : Cyprinidae** 4. *Catla catla* (Ham.) 5. *Cirrhinus mrigala* (Ham.) 6. *Cirrhinus Cirrhosus* (Ham.)

7. *Labeo rohita* (Ham.)
8. *Labeo calbasu* (Ham.)
9. *Ctenopharyngodon idella* (Val.)
10. *Hypophthalmichthys molitrix* (Val.)
11. *Cyprinus carpio* (Linn.)
12. *Osteobrama cotio* (Ham.)
13. *Rohtee ogilbii* (Sykes)
14. *Puntius kolus* (Sykes)
15. *Puntius sophore* (Ham.)
16. *Puntius ticto ticto* (Ham.)
17. *Puntius sarana sarana* (Ham.)
18. *Salmostoma clupeoides* (Day)
19. *Rasbora daniconius* (Ham.)
20. *Amblypharyngodon mola* (Ham.)

Order : Perciformes
Suborder : Percoidae
Family : Centropomidae

21. *Chanda nama* (Ham.)
22. *Chanda ranga* (Ham.)

Order : Siluriformes
Family : Siluridae

23. *Wallago attu* (Schneider)
24. *Ompak bimaculatus* (Bloch)

Family : Bagaridae

25. *Rita rita*
26. *Mystus seenghala* (Sykes)
27. *Mystus cavassius* (Ham.)

Family : Clariidae

28. *Clarias batrachus* (Linn.)
29. *Heteropnestes fossilis* (Bl.)

Order : Synbranchiformes
Family : Mastacembelidae

30. *Mastacembelus armatus* (Lac.)

Family : Gobidae

31. *Glossogobius giuris giuris* (Ham.)

Order : Channiformes
Family : Channidae

32. *Channa marulius* (Ham.)
33. *Channa gachua* (Ham.)
34. *Channa punctatus* (Ham.)

Fig. 11.26. Fish catch at Wan reservoir

Fig. 11.27. A fisherman with *Wallago attu*

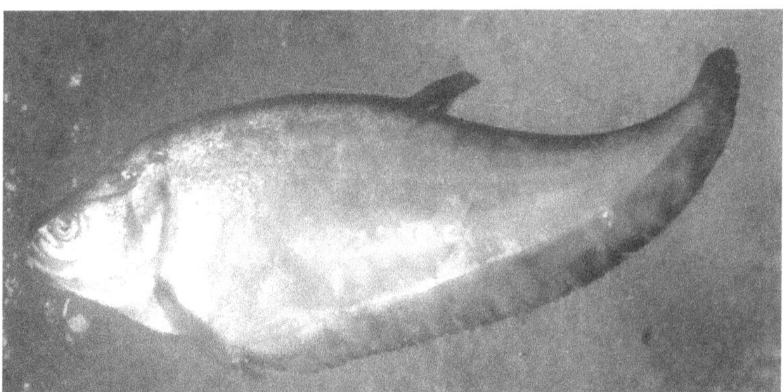

Fig. 11.28. *Notopterus notopterus*

Fig. 11.29. *Catla* spp.

VANJARWADI RESERVOIR

Patil *et al.* (2010) studied physico-chemical parameters of Vanjarwadi reservoir (70° 38′N L and 18°58′E L) near Beed in Martahwada region.Air temperature ranged 22°C to 35.5°C. Water temperature ranged 18°C to 32°C. pH ranged between 8.0 to 8.6. Values of pH may vary in high temperature due to discharge of CO_2. Increase in pH may be due to the absence of pollutants, low CO_2 content, very high photosynthesis and rich algal diversity. Dissolved oxygen ranged from 5.2 mg/L to 8.8 mg/L. It fluctuated in every month. Carbon dioxide ranged between 1.20 mg/L to 3.50 mg/L. The biological oxidation demand ranged between 1.62 mg/L to 7.02 mg/L.

Table 11.80. Physico-chemical profile of Vajarwadi Reservoir in Beed district

Air Temp (°C)	Water Temp (°C)	Dissolved Oxygen (mg/l)	Free-carbon Dioxide (mg/l)	pH	BOD (mg/l)	Month
27	26.5	5.3	1.80	8.5	3.15	June 05
26.5	25.0	5.84	1.50	8.4	2.02	July 05
26.0	24.0	6.50	1.53	8.1	3.01	Aug 05
25.0	24.5	7.1	1.20	8.0	1.62	Sept 05
23.5	23.0	8.3	1.36	8.1	3.01	Oct 05
23.0	22.0	8.4	1.50	8.2	4.85	Nov 05
22.0	18.0	8.6	3.50	8.2	5.25	Dec 05
24.0	23.0	8.8	2.90	8.3	7.02	Jan 06
25.0	23.8	7.5	1.50	8.1	3.01	Feb 06
30.5	26.0	7.2	1.60	8.4	3.51	Mar 06
34	30	6.2	1.70	8.5	3.01	Apr 06
35.5	32	5.1	1.50	8.6	2.01	May 06
32.22	24.75	7.07	1.79	8.28	3.45	Mean
±6.80	±1.62	±1.22	±0.64	±0.22	±1.48	SD
21.10%	6.5%	17.25%	35.75%	2.65%	12.3%	CV

SD = Standard Deviation; CV = Coefficient of variation

GANGAPUR RESERVOIR

The Gangapur dam is located near Gangawadi village and it is 10 km from the Nashik city. It is the earthen dam constructed in year 1954. Total length of dam is 3810 meters and maximum height is 36.56 meters. The reservoir is located within the geographical coordinates of 74°07′ Longitude E and 20°00′ Latitude N and elevation above sea level 1992 feet. The reservoir came into existence with the impounding of river Godavari. The detail morphometry of Gangapur reservoir presented in Table 11.81.

Table 11.81. Morphometry of Gangapur Reservoir in Nashik district

1. Year of construction	1965
2. River	Godavari
3. Type	Earthfill
4. Height of dam (m)	36.59
5. Lenght of dam (m)	3902
6. Gross storage capacity ($10^3 m^3$)	215880

7. Reservoir area ($10^3 m^2$)	22860
8. Designed spillway capacity (m^3/sec)	2293
9. Volume content ($10^3 m^3$)	4612
10. Effective storage capacity ($10^3 m^3$)	203880
11. Purpose	Irrigation

Rathod and Khedkar (2011) reported 26 fish species belonging to 5 orders.Order cypriniformes was dominant with 14 species in which *Puntius ticto* and *Salmostoma navacula* were found most abundant. *Catla catla, Cyprinus carpio* and *Garra mullya* were found abundant form.*Puntius fraseri, Rasbora daniconius, Thynicthys sandkhol, Osteobrama vigorsii, Namachilus moreh*, and *Nemachilus botia* were found less abundant. *Chela labuca,Puntius dorsalis* and *Puntius filamentosus* were found rare abundant. Followed by order perciformes were six species in the assemblage composition in which *Chanda nama* was found most abundant form.*Glossogoius giuris* were found abundant form. *Channa marulius, Channa punctatus, Channa orientalis* and *Parambassis ranga* were found less abundant. Followed by order siluriformes were four species in the assemblage composition in which *Aorichthys aor* were found abundant form. *Mystus cavasius* were found less abundant. *Wallago attu* and *Heteropneustes fossilis* were found rare abundant. Followed by order synbranchiformes were one species in the assemblage composition in which *Macrognathus pancalus* were found abundant. Followed by order osteoglossiformes were only one species in the assemblage composition in which *Notopterus notopterus* were abundant (Table 11.82).

Table 11.82. Fish fauna of Gangapur Reservoir in Marathwada Region of Maharashtra

I. Family : Notopteridae
1. *Notopterus notopterus*
II. Family : Cyprinidae
2. *Puntius ticto*
3. *Puntius fraseri*
4. *Puntius filaentosus*
5. *Puntius dorsalis*
6. *Salmostoma navacula*
7. *Chela labuca*
8. *Rasbora daniconius*
9. *Thynicthys sandkhol*
10. *Osteobrama vigorsii*
11. *Catla catla*
12. *Cyprinus carpio*
13. *Garra mullya*
III. Family : Balitoridae
14. *Nemachilus morech*
15. *Nemachilus botia*

IV. Family : chanidiae
16. *Chanda nama*
17. *Parambassis ranga*
V. Family : Gobiidae
18. *Glossogobius giuris giuris*
VI. Channidae
19. *Channa marulius*
20. *Channa punctatus*
21. *Channa orientalis*
VII. Family : Bagiridae
22. *Aoirichthys aor*
23. *Aoirichthys cavasius*
VIII. Family : Siluridae
24. *Wallago attu*
IX. Heteropneustidae
25. *Heteropneustes fossilis*
X. Mastacembelidae
26. *Macrognathus pancalus*

In Gangapur dam order cypriniformes was most dominat constituting 55% followed by order perciformes at 22%, siluriformes at 12%, osteoglossiformes at 8% and synbranchiformes at 3% of the total fish species (Rathod and Khedkar, 2011).

Bhadane (2012 a) reported monthly variations in water quality parameters of Gangapur reservoir. The dissolved oxygen values ranged from 3.0 – 7.8 mg/ lit. Maximum dissolved oxygen was noted in winter and minimum was recorded and in summer months. Range of biological oxygen demand was 3.5 - 24.7 mg/ lit. The biological oxygen demand values were found more during monsoon and low in winter may be due to lesser quality of solids and microbial population. Values of chemical oxygen demand values ranged from 4.4 – 56.7 mg/ lit.Maximum values of COD were noted in Monsoon may be due to inflow of organic dead matter and minimum values are found in winter may be due to settlement and dilution effect.

Bhadane (2012b) studied zooplankton dynamics of Gangapur reservoir. The major groups of zooplankton include copepods, ceriodaphnia, ostracodes and rotifers. The maximum number of copepods are found in May and minimum in the Month of July, at the beginning of rainy season. *Ceriodaphnia* were found maximum during summer, where as they are very less or absent in the month of July at all the stations. Ostracod number found less as compared to copepods and ceriodaphnia at all the three stations.The maximum number of rotifers were recorded during the month of April and May, while their population is quite less during the period of monsoon.

BHANDARDARA RESERVOIR

The Bhandardara reservoir (19°31′45″N and 73°45′5″E) located in the western part of Ahemadnagar district in the middle of Sahyadri mountain hills. The reservoir is also known as 'wilson dam'. It is the oldest dam which was built in year 1910 on the river Pravara about 500 feet above the mean sea level. The detail moephometry of the reservoir is presented in Table 11.83.

Fig. 11.30. Bhandardara reservoir

Table 11.83. Morphometry of Bhandardara Reservoir in Ahemednagar district

1. River	Pravara
2. Year of completion	1926
3. Length of dam (m)	2717
4. Maximum height above foundation (m)	82.35
5. Maximum water level (m)	746.04
6. Full reservoir level (m)	744.725
7. Gross storage capacity (mcm)	312.4
8. Live storage capacity (mcm)	307.31
9. Length of spillway (m)	214.95
10. Number of spillway gates	2

Information on fisheries of few reservoirs in Ahmednagardistrict in depicted in Table 11.84. The entire Ahmednagar district is covered by basaltic lava flows of the Deccan Traps belonging to the Upper Cretaceous to Eocene age. The Deccan Trap basalts are overlain by alluvial deposits along the river Godavari and occupy the northern part of the district. The stratigraphical sequence of the different lithological units occurring in the district is give below. It receives an annual average rainfall of about 556 mm. The rainfall varies between 360 mm in Pathardi tahsil to 1034 mm in Akole tahsil. However, the rainfall in Akole and Rahata tahsils is relatively more than that of the other tahsils. The cold season in the district commences from December and ends in the month of February. The period from March to the 1st Week of June is the hot season which is followed by southwest monsoon season which lasts till the end of September. October and November constitute the post monsoon season.

Table 11.84. Fisheries of small tanks in Ahmednagar District of Maharashtra

Sr.No.	Reservoir	Taluka	Area (ha)	Seed stocked	Fish production (MT)
1.	Baradari	Ahmednagar	30	50,000	25
2.	Cichondi Patil	Ahmednagar	48	60,000	27
3.	Kaudgaon	Ahmednagar	49	70,000	25
4.	Kamargaon	Ahmednagar	63	70,000	28
5.	Deulgaon siddhi	Ahmednagar	67	50,000	23
6.	Musalwadi	Rahuri	52	100,000	23
7.	Taklibhan	Shrirampur	92	125000	30
8.	Mutewadgaon	Shrirampur	31	80,000	16
9.	Shiral chichondi	Pathardi	30	75000	13
10.	Mohari (P)	Pathardi	40	50000	24
11.	Pampalgaon tappa	Pathardi	43	100000	20
12.	Kuttarwadi	Pathardi	38	70000	15
13.	Pargaon sudrik	Shrigonda	45	100000	25
14.	Autewadi	Shrigonda	60	70000	18
15.	Ghodegaon	Shrigonda	62	300000	22
16.	Limgewadi	Shrigonda	23	50000	20
17.	Kolgaon	Shrigonda	33	110000	20
18.	Rakshaswadi (K)	Shrigonda	54	200000	30
19.	Rui chatrapati	Parner	37	80000	20
20.	Bhalwani	Parner	36	80000	18
21.	Tikhol	Parner	35	50000	15
22.	Jadhavwadi	Parner	22	50000	25
23.	Thergaon	Karjat	27	80000	20
24.	Therwadi	Karjat	99	100000	40
25.	Gurav pimpri	Karjat	96	110000	38

Contd...

Table 11.84: Contd...

26.	Durgaon	Karjat	67	110000	30
27.	Bhairobawadi	Karjat	76	75000	25
28.	Takli (K)	Karjat	47	100000	25
29.	Rakshaswadi (B)	Karjat	43	100000	30
30.	Yesodi	Karjat	35	80000	22
31.	Naigaon	Jamkhed	57	100000	25
32.	Mohari (J)	Jamkhed	68	70000	22
33.	Bhutwada	Jamkhed	26	70000	19
34.	Pimpalgaon (A)	Jamkhed	48	100000	25
35.	Dhondpargaon	Jamkhed	37	80000	20
36.	Dhotri	Jamkhed	34	50000	18
37.	Kelewadi	Sangamner	15	50000	16
38.	Malunje	Sangamner	28	70000	18
39.	Sangwi	Akole	33	126000	10
40.	Balthan	Akole	27	50000	5

Ahmednagar is the largest district of the state covering an area of 17,412 sq.km. constituting 5.54% of the total area of the state. The main methods of irrigation in the district are large, medium and small dams, lift irrigation and wells.Mula and Bhandardara are the two major irrigation projects in the districts. Water for irrigation is also got via canals from Gangapur (Upper Godavari Canal), Nasik District, and Ghod and Kukdi projects in Pune District. The total irrigation capacity of these projects is about 208 thousand hectares.There are other medium scale projects at Visapur Adhla, Pargaon, Ghatshil, Manadohol, Bhojapur, Mahe Sangvi and Sina which can irrigate upto 71 thousand hectares, and about 90 small scale projects with a total irrigation capacity of 23 thousand hectares.

THE JAIKWADI DAM (NATH SAGAR RESERVOIR)

It is located at Latitude 19° 30.641′N Longitude 75°22.522′E°. The construction of dam started on 18th October 1965. The project was completed and officially dedicated to the nation by Late Prime Minister Mrs. Indira Gandhi on 24th February 1976. The river basin is very wide at the dam site. It is without even a single narrow stretch. The terrain is very flat. This has resulted in construction of an extremely long dam wall admeasuring 10.2 kms across the river Godavari. The dam impounds 2909 m.c.m. of water. The reservoir is with a submergence area of 339.80 sq.kms at its maximum capacity. Its catchment area is 21750 sq.kms. The construction of reservoir has caused displacement of 70,000 people and has submerged about 118 villages. The reservoir has an opportunity to supply water to towns like Jalna, Ambad, Gangapur, Shevgaon, Pathardi, Newasa and other 21 villages. The reservoir water is also used for industries in Aurangabad, Paithan,

Waluj and Chiklthana. About 25 to 28% reservoir area is under shallow habitat which is having the water depth less than one meter.These shallow regions are biologically productive. These regions are more productive since the nutrients from adjoining watersheds get accumulated in the form of compost, sediments and silts, which are rich in organic consistutents required for the growth of aquatic weeds, phytoplankton and zooplankton. Due to low water depths, the sun rays penetrates up to the bottom of this shallow habitat.The maximum water level in the reservoir is generally in months of August and October. It is due to the maximum inflow of water in these months. In last 35 years,the reservoir was completely filled only in 11 years. Jayakwadi dam have rocky bed with 39777 ha of water spread area. It is fed by rivers Godavari and Pravara 55 km Long and 27 km wide. Of this area, about 25-28% area is shallow, where the water depth is less than one meter. Most of the catchment area and the bed of the reservoir are having fertile black cotton soil and the bed is comparatively flat. The average depth is 7.3 m and reservoir is shallow.About 118 villages were submerged in the reservoir and the residents were resettled in 94 villges along the periphery. This shallow water spread, with recending water line is very attractive to a large number of waterfowl and waders. Taking into consideration its importance to waterfowl, the Government of Maharashtra declared this wetland as Jaikwadi Bird Sanctuary in 1986.

Fig. 11.31. Sluice gates of Jaikwadi dam of Paithan

Almost all important fish landing centres *viz.* Paithan, Dhakepal, Shewala, Saokheda, Dhaigaon are connected by the roads. The main fishery of the reservoir is supported by *Mystus seenghala, Mystus aor, Notopterus, Mastacembelus armatus, Channa* spp, *Puntius* spp. *etc.* The fish yield is about 7.6 kg/ha at full reservoir level (Desai, 1980). Fishermen are part time operators and the catches are normally at subsistence level.

The aquatic vegetation includes mainly the species of Chara, Spirogyra, Hydrilla, Potamogeton and Vallisneria. *Aregemone mexicana* and *Ipomoea fistulosa* are found in the surrounding area.

Yardi *et al.* (2008) recorded 64 residential and 24 nonresidential species of birds from Jaikwadi dam. Most of the bird species were observed in winter due to more food availability and favorable climatic conditions for nesting and roosting.

After careful observation and analysis, it has been found that the extinction of the migratory birds fauna from Jaikwadi dam is due to sudden climatic change, new diseases, reduced breeding potential, human interferences, habitat change, poaching, commercial and recreational exploitation of the dam, reclamation and encroachment of the dam for government planning, lack of proper legislation and administrative lapses, lack of proper environmental education and training of the people.

The dam bed is used as a sink for the domestic waste and unwanted materials, biodegradable and non-biodegradable. Polythene bags and floating waste materials enter the water during high tide and get entangled among the weeds in the dam. Due to these activities, the lake gets clogged. Conversion of the dam into terrestrial land will seriously affect the food availability for water birds. Efforts may be undertaken to remove the waste materials from the dam and to clean the area (Yardi *et al.* 2008).

Air pollution due to heavy traffic on highway is creating large amount of pollution due to the emission of toxic gases like sulphur dioxide, hydrocarbons and carbon monoxide, which is also a health issue to local residents. Encroachment of revenue land in the catchment area is a serious problem to the lake and its bird community. Land authorities should initiate required preventive measures against illegal encroachment.

MPCB in collaboration with CIFE studied the water quality of Jaikwadi dam for the period of May 2009 to February 2010.Atmospheric temperature and water temperature ranged from 15 to 21°C and 20.5 to 26°C.Water was alkaline (7.32 to 7.94).Transparency fluctuated from 90 to 95 cm.Values of dissolved oxygen and biochemical oxygen demand varied from 3.92 to 7.08 and 3.2 to 4.8 mg/l respectively.

Andhale *et al.* (2012) reported diversity of diatoms from Jaikwadi dam (Table 11.85). Genus *Gomphonema, Fragilaria, Lichmophora, Nitzschia* and *Surirella* occur dominantly in various locations. All these taxa are being reported for the first time from this area.

Table 11.85. Diatoms from Jaikwadi Bird Sanctuary

Gomphonema balatonis, G. gracile, G. intricatum, G. lanceolatum, G. lanceolatum, G. longiceps, G. magnifica, G. moniliforme, G. subventricosum, Fragilaria brevistriata, F. intermedia, Synedra ulna, Achnanthes inflate, A. inflata, Anomoeoneis lanceolata, A. sculpta, Stauroneis kirtikarii, S. phoenicenteron, Neidium affine, Mastogloia baltica, Lichmophora abbreviate, Nitzschia obtusa, N. obtusa, Surirella capronii, S. capronioides, S. ovate, S. robusta, S. tenuissima.

Table 11.86. Fish Fauna of Jaikwadi reservoir

1.	*Notopterus notopterus*
2.	*Amblypharyngodon mola*
3.	*Chanda nama*
4.	*Channa marulius*
5.	*Chela fasciata*
6.	*Glossogobius giuris*
7.	*Macrognathus aral*
8.	*Macrogranthus pancalus*
9.	*Mastacembelus armatus*
10.	*Mystus cavasius*
11.	*Mystus seenghala*
12.	*Mystus aor*
13.	*Nemacheilus botia*
14.	*Ompak malbaricus*
15.	*Osteobrama catio peninsularis*
16.	*Parambassis ranga*
17.	*Puntius jerdoni*
18.	*Puntius phutunio*
19.	*Punctius shalynis*
20.	*Punctius singhala*
21.	*Punctius terio*
22.	*Punctius ticto*
23.	*Salmonstoma bacaila*
24.	*Salmonstoma novacula*
25.	*Salmonstoma sardinella*
26.	*Strongylura leivra*
27.	*Strongylura strongylura*

Craft: Thermocole rafts are common in River Godavari in Maharashtra. It is made of 2 to 3 thermocole pieces tied together and useful only for laying and hauling gillnets, casting castnets and angling. Inflated tubes of motor vehicles are also used in Godavari in Maharashtra for fishing purposes. The boat, locally called *hodi,* is mainly owned by full-time fishermen. It is mostly used by those fishermen who do most of the fishing inside the river for large fishes. Besides, it is also used for the transport of nets and family belongings when the fishermen migrate to the other areas of the river. Now, in the Maharashtra stretch, it is rarely used because of the shallow nature of the river.

Gear: Gill nets, seines, cast nets, drag nets and several miscellaneous types of gear are employed in River Godavari. The type of gear operated is mainly determined by the target species and the conditions prevailing in the river. Set gillnet is usually the multifilament gillnet that is observed throughout the river course in Maharashtra. It is used all around the year, except the monsoon months. Mesh size varies from 12 to 50 mm. Large shore seine operated by 10 to 12 persons. In the earlier years, carps, catfishes and miscellaneous species formed the dominant catch, while in recent years, miscellaneous fishes and prawns account for the major portion of the catch. It was a popular gear, but its operation has come down significantly in recent years. In the present study, at Gangapur dam, we observed giant prawns, while at Raher, we could see large specimens of catla weighing 8 to 10 kg. Small seine is operated by 2 to 3 persons for exploiting prawns. Cast net is most commonly used by fishermen. Almost every fisherman owns a cast net to exploit prawns and small fishes. Few units with bigger mesh (15-20 cm) exploit large carps.

Plankton Diversity: The common planktons recorded in Jayakwadi water are Anabaena, *Asterionella, Brachionus, Chlorella, Closterium, Daphnia, Keratella, Micractinium, Microcystis, Moina, Nauplii, Navicula, Pediastrum, Planktosphaera, Rhizosolenia, Spirulina, Synedra, Thalassionema, Volvox, Zygnema, Synedra.*

Illeagal fishing in the Jaikwadi reservoir has been going on unabated for the past one deacde, despite a high court ban. The violation of the ban can be gauged from the fact that huge volumes of fish are being brought to the Aurangabad city for the sale from reservoir.

Being a part of the bird sanctuary, the local bench of Mumbai High Court had, in 1998, banned fishing activities in Jaikwadi reservoir. The total area of the sanctuary is over 340 square kilometers, out of which reservoir covers an area of 27500 hectares.

The Maharashtra Fisheries Corporation (MFC) grants licenses to the fishermen. Declaration of bird sanctuary in Paithan was made in October 1986 thereby making the Wildlife (Protection) Act (1972) applicable to the area. However, fishing activity continued till 1996.

In February 1996, the MFC had invited tenders and allotted contract of fishig to a private party for a period of five years. The Department of Forest swung into action ad objected to the MFC move saying that fishing was not allowed in Jaikwadi reservoir, as it has been declared a 'sanctuary'.

The private party (contractor) then challenged the Department of Forest's decision in the local high court bench in 1997 urging the court to either allow him fishing in the reservoir or repay his money with interest. Indict on November 7, 1998, the high court had directed the MFC to return the money along with an interest to the contractor. Since then, fishing has been banned.

Sources said that the concerned officials are well aware of the illegal fishing but they have turned a nelson's eye towards this problem for reasons best kown to them. This, sources said, indicates that some vested interests are misguiding over 5000 villagers or fishermen to continue illegal fishing in reservoir. With the quthorities adopting a soft approach, the problem of illegal fishing and biotic pressure is obce again acquiring grim proportions.

Under the provisional unstarred questions, the issue of fishing ban in Jaikwadi reservoir was raised in the Lok Sabha in March 2005. The government had admitted in the house that the ban on fishing was in place. The commissioner of fisheries had also informed the Loksabha that he had requested the ministry of agriculture to study the problem and urge the ministry of environment and forests to exclude the 'culturable' fish species from the term 'wildlife' and to lift blanket ban any fishing in the reservoir.

Prior to the ban,the MFC against the release of 1.37 lakh small fish used to collect a quantity of 1700 metric tones of fish of different species. It would ern them over Rs. 4 crore against the meager expenses of Rs. 70 lakh only. There are 14 reservoirs in Maharashtra where fishing activities have been banned by the forest department as they are declared as 'sanctuaries'.

YELDARI RESERVOIR

The Yeldari reservoir, a purely hydro-electric project was constructed in year 1962 in hilly area of Jintur tahsil (Fig. 11.32). The reservoir lies in between north Latitude 19°-43′-00″ and East Longitude 76°- 45′-00″. The water of reservoir is also used for irrigation through another reservoir and for supplying drinking water to Parbhani, Vasmat, Jintur and other towns. No canal for irrigation takes off directly from this reservoir. However, a few miles down stream, there is the Siddeshwar reservoir at Siddeshwar village, which is built on the same river. From the Siddeshwar reservoir, a system of irrigation canals originates. As it is mainly a hydro-elctric project, all rights of reservoir management are belonging to the irrigation Department and the State Electricity Board. The area around reservoir comprises forest-covered hills. The reservoir is having catchment area of 7,330.00 sq.km. The maximum level of reservoir is of 462.380 m. The reservoir consists of three outlets, water through which goes to the Siddeshwar reservoir after passing through the turbines and ten spillway gates.

The discharge of water from the Yeldari reservoir rotates the turbines for hydel-power generation. This project started functioning from the year 1965 and was managed by the Irrigation Department. But for the last 30 years, it is managed and governed by the Maharashtra State Electricity Board. This project was constructed using Swedish technology with capacity of 22.50 m.w. electricity through 3 electricity generating units or sets, each of 7.5 m.w. capacity. Whenever water from reservoir is released for irrigation, the hydroelectric units start functioning.

Fig. 11.32. Sluice gates of Yeldari Reservoir

Fig. 11.33. Fishing in progress

The maximum water level in the reservoir is generally in months of August and October. It is due to the maximum inflow of water during this period. From the water inflow data of last 20 years, an average inflow has been worked out to be 938.99 mm^3, but only in nine out of last 20 years the inflow has been more than 788.85 mm^3 *i.e.,* 75% dependable yield of the project.

The normal annual rainfall over the reservoir area varies from about 850 m to about 98-0 mm. Though the area is thought to belong to assured rainfall zone the area has experienced moderate, severe and acute drought conditions for more than 20 per cent of the last 92 years. Hence the entire area comes under the category of drought area.

The air is generally dry from February to May, the relative humidity during the afternoons being 20%. During the southeast monsoon season, the humidity is as high as 80% in the mornings and 60% in the afternoons.

The climate of the reservoir area is semiarid subtropical, characterized by a hot summer. Except for the monsoon period the weather is generally dry. The summer outsets in the area generally in March and continues up to May. This is followed by the southwest monsoon from June till September. However, there are periods of dry spells even within this monsoon period. The winter season stretches from December to the end of February.

In the study area, winds are generally moderate with appreciable increase in force during southwest monsoon season. During the monsoon season the wind is mainly from west and southwest directions. During the rest of the year, wind blows from north and east directions. The maximum-recorded wind speed during the month of May and August is 11.4 to 14.2 km/hour with annual average wind speed being of 8.6 km/hour.

The winter starts from about the middle of November and continues till the end of February. December is the coldest month of the year with perceptibly chilly nights. From March, the day temperature increases progressively and reaches a maximum in May, which is the hottest month. With onset of the southwest monsoon there is an appreciable drop in temperature.

Table 11.87. The salient features of Yeldari reservoir

1. Catchment area	733.00 km^2
2. Water spread area	106.84 km^2
3. Full reservoir level	461.772 m
4. Maximum water level	462.30 m
5. Top of the dam	465.90 m
6. Crest level	450.190 m
7. M.D.D.L. without carry over with carry over	 477.751 m 454.54 m
8. Length of masonry	350.215 m
9. Total length of earthen dam	4430.00 m
10. Length of spillway	149.65 m
11. Height of dam above the reservoir bed	51.35 m
12. Gross storage capacity	934.44 m.ha.m
13. Power generation	22500 kw
14. Design spillway At FRL At MWL	 9740.80 Cumecs 10874 Cumecs
15. Maximum Flood	10478 Cumecs
16. No. of irrigation outlets	3 Nos.(1.92 × 2.44 m)
17. No. of spillway gates	10 Nos.(Radial gate)
18. Size of radial gates	12.5 × 11.65 m
19. Power outlets service gates stop log gates	 3 Nos. 9 Nos.

Water Quality: Sakhare and Joshi (2010) reported water quality of Yeldari reservoir in relation to fisheries. Water temperature ranged from 21 to 34°C. Its minimum value was recorded in January, while its maximum (34°C) was recorded in April. pH in the reservoir varied from 7.7 to 8.3. The minimum pH was 7.7 was recorded in months of December and April, while the maximum recorded was 8.3 in februray. The observations indicate that the water was alkaline. By the secchi disc, the transparency of 65.5 was recorded in September and maximum 95 cm was observed during February. The less transparency might may be due to the silt brought into the reservoir during rainy season. Total dissolved solids irrespective of the seasons ranged from 96 to 262 mg/l. The maximum value of 262 mg/l in April and minimum of 96 mg/l in August. The observations on TDS clearly indicate that TDS values were high in summer followed by winter and monsoon. The dissolved oxygen ranged from 6.4 to 14 mg/l. Its minimum (6.4) was found in March and maximum of 14 mg/l was in August. It is seen that, the high dissolved oxygen was found during monsoon, while lower values were recorded in summer. Phenolphthalein alkalinity ranged between 20.5 mg/l in August to 48 mg/l in December. The minimum value of total alkalinity (189 mg/l) was noticed in August, while the maximum value (394) was in winter followed by summer and monsoon. Free carbon dioxide was totally absent in reservoir water. The minimum and maximum values of chloride were 14.18 mg/l and 56.22 mg/l in month of February and April respectively. Total hardness (226 mg/l) was found in May, its minimum (68) was recorded in January. The higher values of total hardness were recorded during summer and lower values were recorded in the winter season.

Zooplankton Diversity in Reservoir: Sakhare (2007) studied plankton diversity of Yeldari reservoir. During 1999-2000, copepoda dominated the zooplankton population (71.93 %) followed by cladocera (11.69%), rotifera (9.72%) and ostracoda (6.64%) are presented in Fig. 11.34. Total 25 different species of zooplankton were identified. Summer months exhibited higher population of zooplankton (Fig. 11.35). Rotifers, cladocerans, copepods as well as ostracods showed the summer maxima and minima in winter.

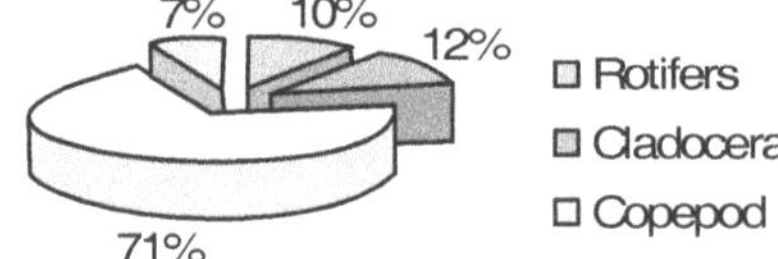

Fig. 11.34. The percentage of composition of zooplankton in Yeldari reservoir during 1999-2000

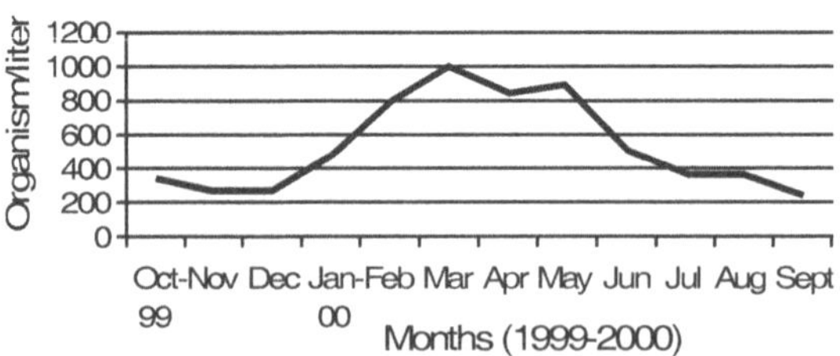

Fig. 11.35. Monthly fluctuations in zooplankton in Yeldari reservoir during year 1999-2000

Rotifers were represented by 8 species (Table 11.88). The species *Brachionus calcyflorus* accounted for 10 months, while Trichocera porellus was present only in month of July. The maximum rotifer population was recorded during March and April (80 Organisms/litre) and minimum population during August (10 organisms/litre)

Table 11.88. Species composition of rotifera (density-organisms/litre) during year 1999-2000

Species	Oct	Nov	Dec	Jan	Feb	Mar	Apr	May	Jun	Jul	Aug	Sept
1. *Brachionus Calyciflorus*	05	10	10	10	15	10	05	20	10	10	—	—
2. *B.diversicornis*	10	10	20	20	30	20	10	—	—	10	—	20
3. *B.falcatus*	-	—	10	20	10	20	20	30	—	10	—	—
4. *Euchlanis Dilatata*	10	—	—	—	—	10	10	20	10	05	—	10
5. *Filinia Longiseta*	—	05	—	—	—	10	10	—	—	10	—	10
6. *Keratella tropica*	—	—	10	10	20	10	20	—	—	—	10	—
7. *Lecane bulla*	—	—	—	10	—	—	—	10	10	—	—	—
8. *Trichocera porcelus*	—	—	—	—	—	—	—	—	—	10	—	—

Cladocerans were represented by 6 species (Table 11.89), and species *Ceriodaphnia cornuta* recorded for 11 months, while *Diphanosoma sarsi* was accounted for 4 months only. The peak period of cladocerans was observed during month of May (180 organisms/litre) and it was minimum during month of October (25 organisms/litre).

Table 11.89. Species composition of cladocera (density-organisms/litre) during year 1999-2000

Species	Oct	Nov	Dec	Jan	Feb	Mar	Apr	May	Jun	Jul	Aug	Sept
1. *Ceriodaphnia cornuta*	20	15	10	30	40	30	20	50	—	10	20	10
2. *Moina micrura*	—	05	10	30	10	10	10	20	15	20	—	—
3. *Diaphanosoma sarsi*	—	—	—	10	10	10	—	50	—	—	—	—
4. *Alona rectangular*	—	10	—	10	10	—	—	—	20	—	—	10
5. *Biapertura karna*	05	—	10	10	10	10	—	—	—	10	30	—
6. *Indialona ganapati*	10	—	10	20	60	—	20	—	10	—	—	—

Copepoda was represented by 7 species (Table 11.90). The maxima was 810 organism/litre during March and minima was 140-organisms/litre during December. *Diaptomus marshianus* was not seen in the major period of monsoon;

except in month of September. *Phyllodiaptomus annae* was absent during entire winter period and two months of monsoon (August and September). *Cyclops viridis* and *Mesocyclops leukarti* were present throughout the year.

Table 11.90. Species composition of copepoda (density-organisms/liter) during year 1999-2000

Species	Oct	Nov	Dec	Jan	Feb	Mar	Apr	May	Jun	Jul	Aug	Sept
1. *Cyclops viridis*	110	60	60	80	50	115	120	100	80	70	95	70
2. *Mesocyclops leukarti*	80	60	30	60	50	50	170	80	30	30	40	40
3. *M.hyalinus*	40	20	—	—	90	75	80	215	70	30	60	—
4. *Diaptomus marshianus*	10	15	20	40	110	120	110	40	—	—	—	30
5. *Phyllodiaptomus annae*	—	—	—	—	125	160	80	10	125	20	—	—
6. *Neodiaptomus lindbergi*	—	10	20	50	90	190	10	40	—	—	10	10
7. *Nauplius lava*	25	10	10	40	75	100	95	110	90	70	70	—

Ostracoda was represented by four species. The highest density was observed during June (50 organisms/litre), while lowest density was recorded during October, January and September (20 organisms/litre) (Table 11.91).

Table 11.91. Species composition of ostracoda (density-organisms/liter) during year 1999-2000

Species	Oct	Nov	Dec	Jan	Feb	Mar	Apr	May	Jun	Jul	Aug	Sept
1. *Cypris* sp.	10	05	20	05	20	20	10	10	10	15	15	10
2. *Stenocypris* sp.	05	10	10	10	—	10	10	10	—	10	05	05
3. *Cyclocypris Globosa*	—	05	10	—	15	05	15	20	20	15	—	—
4. *Candocypria osborni*	05	10	—	05	05	10	10	—	20	—	10	05

During year 2000-2001 zooplanktons were represented by rotifera, cladocera, copepoda and ostracoda. Zooplankton in reservoir for both the years is shown in the form of circles (Fig. 11.36). Among zooplankton copepoda dominated, followed by cladocera, rotifera and ostracoda.

The seasonal variation in total zooplankton population during year 2000-2001 is presented in Fig. 11.37.

Rotifera accounted for about 7.09% during year 2000-01, and were represented by 8 species (Table 11.92). The highest density of rotifers was observed in the month of January, February and April. Throughout the summer months, rotifer population was maximum. However, during rainy and winter season, the rotifer population was comparatively less. The species *Branchionus diversicornis* and *Branchionus flacatus* were recorded for 8 months of the year. The peak period of rotifers was observed during month of January, February and April (80 organisms/litre) and it was minimum during month of October (10 organism/litre).

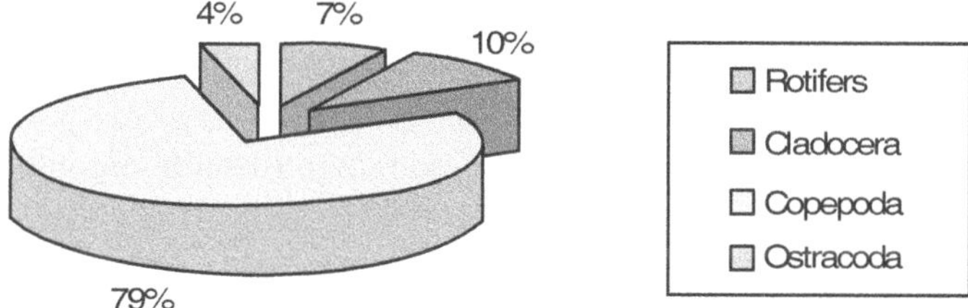

Fig. 11.36. The percentage of composition of zooplankton in Yeldari reservoir during 2000-01

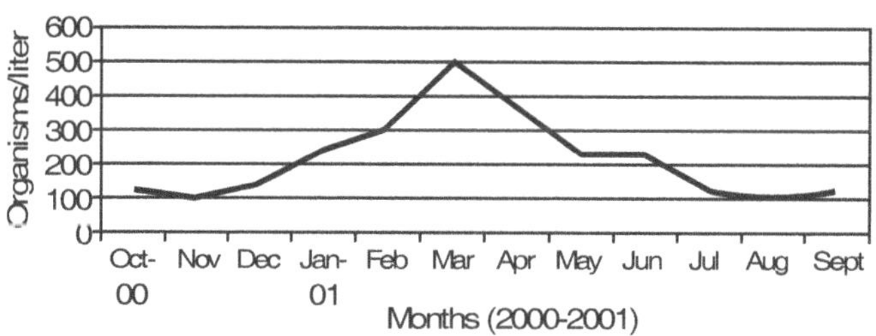

Fig. 11.37. Monthly fluctuations in zooplankton in Yeldari reservoir during year 2000-01

Table 11.92. Species composition of rotifera (density-organisms/litre) during year 2000-01

Species	Oct	Nov	Dec	Jan	Feb	Mar	Apr	May	Jun	Jul	Aug	Sept
1. *Branchionus calyflorus*	—	—	10	10	10	—	10	—	—	20	10	—
2. *B.diversicornis*	10	20	—	20	20	30	20	20	20	—	—	20
3. *B. flacatus*	—	—	10	20	20	10	20	20	20	—	10	—
4. *Euchlanis dilatata*	—	10	—	10	10	—	10	—	—	—	—	10
5. *Filina longiseta*	—	—	—	10	10	—	10	—	—	—	—	—
6. *Keratella tropica*	—	—	—	10	10	20	10	10	10	—	—	—
7. *Lecane bulla*	—	—	—	—	—	—	—	10	10	—	—	—
8. *Trichocera porellus*	—	—	—	—	—	—	—	—	—	—	—	—

Cladocerans were represented by 7 species and accounted for about 10.05% (Table 11.93). Ceriodaphnia cornuta was found for 11 months, followed by Indialona ganapati for 8 months (excluding the months of March, August, October and December). *Diaphanosoma excisum* present only during April which was totally absent during 1999-2000. *Diaphanosoma sarsi* during January and February, A*lona rectangula* and *Biapertura karna* during January and February, and *Moina micrura* during May, June and July. The maximum cladoceran population (210 organisms/litre) was recorded during April and minimum population (10 organisms/litre) during November.

Table 11.93. Species composition of cladocera (density-organisms/litre) during year 2000-01

Species	Oct	Nov	Dec	Jan	Feb	Mar	Apr	May	Jun	Jul	Aug	Sept
1. *Ceriodaphnia cornuta*	30	10	40	60	60	30	80	30	30	—	40	30
2. *Moina micrura*	—	—	—	—	—	—	—	10	10	—	—	—
3. *Diaphanosoma sarsi*	—	—	—	10	10	—	—	—	—	—	—	—
4. *Diaphanosoma excisum*	—	—	—	—	—	—	70	—	—	—	—	—
5. *Alona rectangular*	—	—	—	10	109	—	—	—	—	—	—	—
6. *Biapertura karna*	—	—	—	10	10	—	—	—	—	—	—	—
7. *Indialona ganapati*	—	—	—	10	10	20	60	10	10	20	—	20

Copepoda accounted for about 78.65% and were represented by 7 species (Table 11.94). They were dominated by *Cyclops viridis, Mesocyclops leukarti, Mesocyclops hyalinus* and *Nauplius larvae*. The population was more during summer and least during rainy season. The maximum was 1100 organisms/litre during April and minima was 100 organisms/liter during July. All the seven species of copepods were present during February, March and April. During May, July, August, September, October and November. *Phyllodiaptomus annae* was not seen in the major period of monsoon and winter. *Diaptomus marshianus* was not found in five months of the year 2001. While *Neodiaptomus lindbergi* was absent in winter season and two months of monsoon period. *Cyclops viridis* and *Nauplius larva* were found throughout the year, while *Mesocyclops leukarti* and *M. hyalinus* were recorded for eleven months.

Table 11.94. Species composition of copepoda (density-organisms/litre) during year 2000-01

Species	Oct	Nov	Dec	Jan	Feb	Mar	Apr	May	Jun	Jul	Aug	Sept
1. *Cyclops viridis*	50	80	50	40	90	100	300	400	20	20	40	40
2. *Mesocyclops leukarti*	60	50	40	30	70	65	350	250	—	30	60	30
3. *M. hyalinus*	80	30	10	30	90	85	250	250	30	20	50	—
4. *Diaptomus marshianus*	—	—	—	50	40	180	50	80	80	—	—	10
5. *Phyllodiaptomus annae*	—	—	20	30	10	200	40	—	40	—	—	—
6. *Neodiaptomus lindbergi*	—	—	—	—	50	70	70	20	40	—	—	10
7. *Nauplius larva*	80	30	40	30	90	340	40	80	90	30	30	40

Ostracoda came 4th in order of occurrence, and were represented by four species with summer maxima (Table 11.95). Their total contribution among zooplankton was about 4.19% (Fig. 11.36). *Cypris* sp. Dominated the reservoir. There was total absence of ostracods during September and October. During March and April all the 4 ostracods were found in the reservoir. During rainy season least number of ostracods were recorded.

Table 11.95. Species composition of ostracoda (density-organisms/litre) during year 2000-01

Species	Oct	Nov	Dec	Jan	Feb	Mar	Apr	May	Jun	Jul	Aug	Sept
1. *Cypris* sp.	—	20	20	20	10	20	10	25	10	10	—	—
2. *Stenocypris* sp.	—	—	10	05	—	10	10	10	05	—	10	—
3. *Cyclcypris globosa*	—	10	10	—	20	10	05	05	—	—	—	—
4. *Candocypria osborni*	—	—	—	05	10	10	05	—	20	10	—	—

The zooplankton population of two years (1999-2001) was represented by rotifera, cladocera, copepoda and osrtracoda at 8.92%, 11.47%, 73.95% and 5.64% of the total zooplankton population respectively.

Phytoplanktonic Diversity in Reservoir: The phytoplankton species occurred in the reservoir during year 1999-2001 is listed in Table 11.96 and the composition of phytoplankton in Fig. 11.38.

13 species of chlorophyceae, 10 of bacillariophyceae and 8 species of myxophyceae were recorded from the reservoir (Table 11.96). The phytoplankton were represented by myxophyceae, chlorophyceae, and bacillariophyceae at 39.38%, 57.91% and 2.70% of the total phytoplankton population respectively.

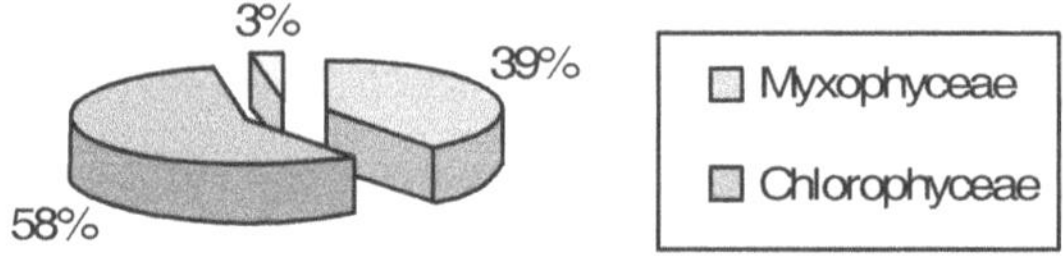

Fig. 11.38. Phytoplankton composition in Yeldari reservoir during 1999-2001

Table 11.96. List of phytoplankton inYeldari reservoir

Myxophyceae: *Anabaena* spp, *Arthrospira* spp, *Lygnbya majuscula, Microcystis areuginosa, Nostoc* spp., *Oscillatoria chlorina, Phormidium* sp.
Bacillariophyceae: *Cyclotella operculata, Cymbella turgida, Fragilaria* sp., *Gomphonema gracile, Melosira* sp., *Navicula mutica, Nitzchia* sp, *Pinnularia viridis, Synedra ulna.*
Chlorophyceae: *Eudorina* sp, *Pandorina morum, Volvox* sp., *Scenedesmus* sp., *Cosmarium microsporum, Ulothrix zonata, Microspora* sp, *Pediastrum duplex, Spirogyra margariata, Oedogonium* sp., *Chlorella vulgaris, Closterium* sp., *Cladophora* sp., *Stichococcus* sp.

Table 11.97. Phytoplankton composition in Yeldari reservoir during year 1999-2001

Groups	Range Units l^{-1}	Average
Myxophyceae	942 to 1630	1137
Chlorophyceae	1030 to 2430	1672
Bacillariophyceae	50 to 110	78

From chlorophyceae, *Ulothrix zonata, Scenedesmus* sp., *Closterium* sp., and *Pediastrum duplex* dominated the reservoir. The maximum population of chlorophyceae was recorded in March ($2430.l^{-1}$).

10 species from bacillariophyceae were identified. *Navicula* and *Cymbella* dominated the reservoir. Out of 10 species recorded, 9 species exhibited their presence during October and November and 9 in January 2000. Thus, these three months showed a maximum population of members of bacillariophyceae. The maximum population of bacillariophyceae during this year was recorded in October at $85.l^{-1}$.

Myxophyceae was represented by 8 species with dominance of *Oscillatoria, Merismopedia tennulssima, Anabaena* and *Arthruspira* spp. This group attained a highest peak in winter season, with maxima in November at $1630.l^{-1}$.

During year 2000-01, 14 species of cholrophyceae, 10 of bacillariophyceae and 7 species of myxophyceae were recorded from Yeldari reservoir.

Chlorophyceae was abundant in May with density of $2008.l^{-1}$. It was represented by new species (*Stichococcus* sp.) which was totally absent during previous year (1999-2000).

Bacillariophyceae was also more among the phytoplankton assemblage of the reservoir. Their population was maximum ($110.l^{-1}$) in October. It was mainly represented by *Fragilaria* sp, *Navicula* sp and *Nitzchia* sp.

Myxophyceae was richly represented in months of September (1248.l^{-1}), followed by November (1183.l $^{-1}$). The group was dominated by *Anabaena* spp., *Nostoc* spp., and *Merismopedia tenuissima.*

Among planktonic groups chlorophyceae dominated the reservoir at 48.66%, followed by myxophyceae (33.09%), copepoda (11.82%), bacillariophyceae (2.27%), rotifera (1.43), cladocera (1.83%), and ostracoda (0.90%) of the total planktonic population during year 1999-2001 (Table 11.98).

Table 11.98. Plankton composition in Yeldari reservoir during year 1999-2001

Groups	Range Unitsl^{-1}	Average (%)
Myxophyceae	942 to 1630	1137 (33.09)
Chlorophyceae	1030 to 2430	1672 (48.66)
Bacillariophyceae	50 to 110	78 (2.27)
PHYTOPLANKTON		2887 (84.03)
Rotifera	10 to 80	49 (1.43)
Cladocera	10 to 210	63 (1.83)
Copepoda	100 to 1100	406 (11.82)
Ostracoda	Nil to 50	31 (0.90)
ZOOPLANKTON		549 (15.97)
TOTAL PLANKTON		3436

Information of Ichthyofauna: Sakhare (2003) reported 29 species of fishes belonging to 20 genera falling in 4 orders have been identified (Table 11.99). Of the 4 orders, order cypriniformes dominated with 13 species falling under 10 genera of which genus *Puntius* is abundant with 4 species. Next in abundance is the order perciformes in which genus *Channa* is dominant with 3 species. In terms of species abundance genus Puntis tops the list. Out of 4 species occurring in the reservoir, *P. kolus,* and *P. sarana sarana* are predominant and occurred in all the areas. *Heteropneustes fossilis* and *Clarias batrachus* also occurred in all localities but in very less numbers. The fishes, which commonly occurred in reservoir water, are *Labeo rohita, Catla catla, Mystus* spp. and *Channa* spp. Among the three stations, first station was found to be rich in the availability of different species, since all the 29 species were recorded from this station.

Fishing tender and tender cost: The fishing tender of Yeldari reservoir was allotted to Purna Matsyavyvasai Sahakari Sanstha Maryadit, Sawali (Bu) by the District FisheriesDevelopment Officer, Parbhani (D.F.D.O. Parbhani) by the letter No.-4102, dated 04/10/2005 for the duration of 5 years *i.e.,* from July 2005 to June 2010 at the tender cost of Rs. 1, 41,290 per year.

Table 11.99. List of fishes recorded from catches of Yeldari reservoir

Order : Osetoglossiformes
Family : Notopteridae

1. *Notopterus notopterus* (Pallas)
2. *Notopterus chitala* (Hamilton-Buchanan)

Order : Cypriniformes
Family : Cyprinidae

3. *Catla catla* (Hamilton-Buchanan)
4. *Cirrhinus mrigala* (Hamilton-Buchanan)
5. *Labeo rohita* (Hamilton-Buchanan)
6. *Ctenopharyngodon idella* (Valenciennes)
7. *Cyprinus carpio communis* (Linnaeus)
8. *Hypopthalmicthys molitrix* (Valenciennes)
9. *Puntius sarana sarana* (Hamilton-Buchanan)
10. *Puntius sophore* (Hamilton-Buchanan)
11. *Puntius ticto* (Hamilton-Buchanan)
12. *Puntius kolus* (Sykes)
13. *Chela bacila* (Hamilton-Buchanan)
14. *Rohtee ogilbii* (Sykes)
15. *Chela bacila* (Hamilton-Buchanan)

Order : Siluriformes
Family : Siluridae

16. *Ompak bimaculatus* (Bloch)
17. *Wallago attu* (Schneider)

Family : Bagridae

18. *Mystus cavassius* (Hamilton-Buchanan)
19. *Mystus seenghala* (Sykes)

Family Clariidae

20. *Clarias batrachus* (Linnaeus)

Family : Heteropneustidae

21. *Heteropneustes fossilis* (Bloch)

Order : Perciformes
Family : Ambassidae

22. *Chanda nama* (Hamilton-Buchanan)
23. *Chanda ranga* (Hamilton-Buchanan)

Family Gobiidae

24. *Glossogobius giuris* (Hamilton-Buchanan)

Family : Channidae

25. *Channa striata* (Bloch)
26. *Channa marulius* (Hamilton-Buchanan)
27. *Channa gachua* (Hamilton-Buchanan)

Family : Mastacembelidae

28. *Macrognathus pancalus* (Hamilton-Buchanan)
29. *Mastacembelus armatus* (Lacepede)

Fishing Days and Closed Season: In Yeldari reservoir, fishing is carried out throughout the year. No 'closed season' has been declared by the State Fisheries Department or the society. However, on the suggestions of state fisheries department, the society has framed some rules for the conservation of fish. For example, catching of fish belonging to Indian major carps less than 1.5 kg in weight has been prohibited. Such types of fish if caught, it should be immediately released in the reservoir. This rule is not strictly followed and a large number of Indian major carps weighing less than 1.5 kg are caught.

Poaching, Fisheries Conservation and Escape of Fish from Reservoir: The strength of staff for checking of the poaching of fish from the reservoir is inadequate. There are only seven fisheries inspectors who besides performing their normal official duties are also entrusted with the duties of checking of poaching and implementation of the conservancy measures. As there are a large number of landing centers, a strength of seven inspectors is not sufficient to book the poachers and other fishery offenders under the Indian Penal Act for fish theft and/or under the Indian Fisheries Act, 1897, for violating fishery rules and regulations. It is worthwhile to state that during the present study, a good number of landing centers were visited and presence of many Indian Major Carps weighing less than 1.5 kg was observed in catches of the fishermen.

It has also been also found that there is a loss of fish, which escape from the sluice gates of the reservoir.

Problems of Aquatic weeds-vegetations/Removal of obstruction: Deforestation of standing trees in reservoir basin remained neglected. The submerged trees and shrubs remained permanent obstacles for fishing which cause loss or damage through the entanglement, prevent fishing operations and also provides fish with hideouts, this reduces the area available for the fishery.

The Crafts and Gears Used: Previously the society had purchased a motorboat fitted with 20 H.P. Ruston diesel engine. It had two insulted fish holds for transportation of fish catches in ice. But from the year 1986, this motorboat is not used because of heavy expenditure and other mechanical problems. At present, the society is not in a position to provide loans or subsidy for the purchase of craft, because the members have not returned the loans already taken. Now a days, the members purchase the craft individually without any financial assistance from co-operative society.

The craft used in the yeldari water is of primitive type, which is nothing but a platform of 6 × 3 feet size with a depth of 5 to 8 inches and is locally called as 'Nav'. It is prepared from the thermocoel and covered by a plastic covering. At present almost all crafts are of this type. The cost of one craft varies from 325 to 475 rupees in the local markets.

There are many drawbacks in such types of crafts such as: (1) They are not suitable to carry large nets which are heavy. (2) When fishermen get bumper catch it become difficult to navigate the craft with heavy weight of the catch.

Gears used for fishing are mostly surface gill nets. However, cast-nets, and rod and line are also used. Gill nets are with floats made up of soft wood, fixed at regular intervals whereas stone pieces or pebbles are also used as sinkers which are attached at intervals in the footrope. The nets are payed off in the evening and generally hauled up next morning. If sufficient catch is not obtained, then nets are payed at another location. The cost of nets varies according to quality of nylon used for their fabrication. Good quality nets are prepared from the Garware nylon which is considered as a good quality nylon yarn. The cost of such type of net ranges between Rs. 300 to 350 per kilogram of webbings, in Yeldari and Nanded markets. Apart from this, low quality nylon nets are also used which are popularly known as 'disco nets', the cost of which varies from Rs. 300 to 600 per kilogram of webbings. The quality of nylon yarn used for the preparation of this type of net is very poor. Rod and line is mostly used by fisher children who help their parents in improving their income. It is operated near shore of reservoir. Flourpaste, cockroaches, insects, earthworms or other small organisms are used as baits to attract and capture the fishes. It consists of a rod, nylon twine and hook. The nylon line is tied to a rod on one side and to the baited hook on the other side. The nylon is also provided with a wooden float. In certain cases, the hook is kept in position by attaching a stone to nylon twine.

Fish Seed Stocking: In the Yeldari reservoir the young stages of fishes mainly fry, semi-fingerling and fingerling commonly called as 'fish seed' were stocked by the fishing tender owners. The fish seed of Indian major carps namely *Catla catla, Labeo rohita, Cirrhina mrigala* and exotic carp species *Ctenopharyngodon idella,Cyprinus carpio* and *Hypophthalmichthys molitrix* were regularly stocked during the month of June to September every year.

According to Maharashtra State Government decision of Animal Husbandry, Dairy and Fisheries Department, G.R. No. Fishery Dept/1999/20/L.N./8/ADF-13 on dated 15th October 2001. The maximum number of fingerlings to be stocked in Yeldari reservoir of 6272.00 ha. average water spread area are 34.30 lakh. The fish seed was purchased from govt. fish seed hatcheries located at Siddheshwar reservoir and Bhategaon reservoir in Hingoli district and from Masoli fish seed hatchery located at Masoli reservoir near taluka Gangakhed in Parbhani district. The fish seed was also purchased from various private fish seed hatcheries and fish seed suppliers. The details of fish seed stocked in Yeldari reservoir is given in Table 11.100, 11.101 and 11.102.

Table 11.100. Fish seed stocking in Yeldari reservoir

Sr.No.	Year	Number of fish seed stocked	Fishing Tender Owner
1.	1998-99	20,00,000	Sawangi Mahalsa Matsyavyvasai Sahakari Sanstha Maryadit, Sawangi Mahalsa, Tq. Jintur Dist. Parbhani
2.	1999-00	4,17,500	—do—
3.	2000-01	Not Stocked	Cancellation of Lease Agreementbetween D.F.D.O. & Sawangi Fish Co-operative Society.
4.	2001-02	Not Stocked	Blank Year
5.	2002-03	Not Stocked	Blank Year
6.	1/1/2003 to 31/12/2003	16,00,000	International Meretek Pvt. Ltd., Nagpur Tq. Dist. Nagpur.

Contd...

Table 11.100: Contd...

7.	1/1/2004 to 31/12/2004	Not Stocked	International Meretek Pvt. Ltd., Nagpur Tq. Dist. Nagpur.
8.	2005-06	Not Stocked	Purna Matsyavyvasai Sahakari Sanstha Maryadit, Sawali (Bu)., Tq. Jintur Dist.Parbhani working from 4/10/05
9.	2006-07	2,54,13,199 fish seed 60,85,911 Prawn seed	—do—
10.	2007-08	Not Stocked	—do—

Table 11.101. Fish fingerling stocking and private agencies supplying fish seed to stock in to the reservoir

Sr.No.	Date of fish Seed stocked	Private fish seed Supplying Agencies	Fingerlings of fish Species stocked	Total no of fish seed stocked
1.	10/06/2006 and 16/06/2006	S.M. Fish Seed Suppliers, Mumbai.	IMC	1,92,36,745
2.	20/06/2006 and 30/07/2006	Pathak fish Suppliers Mumbai.	IMC	22,75,650
3.	13/09/2006 and 24/09/2006	Bhoiraj Fish Seed Production Centre, Gangakhed, District Parbhani.	IMC	3,30,000
4.	29/09/2006 and 05/10/2006	Sameer Fish Suppliers, Mumbai.	IMC	2,45,114
5.	08/10/2006 and 19/10/2006	Wanraj Enterprises fish Seed Suppliers, Mumbai	IMC	12,41,770
6.	27/10/2006	Shah Aqua India Fish Seed Suppliers, Mumbai	Silver Carp	2,83,480
	Total		**2,54,13,199**	

Table 11.102. Prawn seed stocking and private prawn seed supplying agencies

Sr.No.	Date of Prawn Seed stocked	Private Prawn Seed Supplying Agencies	Prawn Species Stocked	No. of Prawn Seed Stocked
1.	13/09/06 and 11/11/06	Surya Prawn Seed Suppliers, Mumbai.	*Macrobrachium rosenbergii*	11,51,220
2.	06/09/06 and 15/10/06	S.M. Prawn Seed Suppliers, Mumbai.	*Macrobrachium rosenbergi*	23,45,116

Contd...

Table 11.102: Contd...

3.	08/10/06 and 15/12/06	Wanraj Prawn Seed Suppliers, Mumbai.	*Macrobrachium rosenbergi*	15,64,927
4.	14/10/06 and 18/10/06	Chahare Prawn Seed Suppliers, Washim, Maharashtra.	*Macrobrachium rosenbergi*	7,72,800
5.	14/11/06	Bhagwati Fishing Corpation Ahmedabad, Gujarat.	*Macrobrachium rosenbergi*	41,275
6.	09/12/06 and 16/12/06	Samarth Aqua culture Seed Accessories, Mumbai.	*Macrobrachium rosenbergi*	2,12,573
	Total			**60,85,911**

Fig. 11.39. Fish seed stocking in Yeldari reservoir

New trends in fish seed stocking pattern by the Purna Fish Co-operative Society.

Purna Matsyavyvasai Sahakari Sanstha Maryadit, Sawali (Bu) had taken Yeldari reservoir on lease from the State Fisheries Department (DFDO, Parbhani) for the duration of five years from July 2005 to June 2010 for fishing. The Society had made an agreement with a private capital investing company in this sector. The Purna Matsyavyvasai Sahakari Sanstha Maryadit, Sawali (Bu) had made an agreement with Intellect Agri. Product Private Limited, Andheri (East), Mumbai, Maharashtra for fish seed stocking and purchase of fish and prawn catch from the reservoir on dated 27/02/06.

According to an agreement between the Fish Co-operative Society and the Intellect Agri. Product Private Limited Company, the company has to stock the fish fingerlings of Indian major carp species and prawn juveniles of *M. rosenbergi* species in Yeldari reservoir and have to purchase all the fish and prawn catch regularly from the fishermen involved in fishing in Yeldari reservoir and the Intellect Agri. Product Private Ltd. Company also has to give Rs. 5/kg for fishes and Rs. 10/kg for prawns as commission to the society.

According to agreement Intellect Agri. Products Pvt. Ltd. Company had purchased fish fingerlings and prawn juveniles from different private fish seed and prawn seed suppliers. The Purna Matsyavyvasai Sahakari Sanstha Maryadit, Sawali (Bu) (PMSSM, Sawli Bu) had kept the details of seed stocking of fish species and prawn species and the private fish seed and prawn seed supplying agencies shown in Tables 23.1, 23.2 and 23.3. The total no. of fingerlings stocked were 2,54,13,199 of which Indian Major Carp fingerlings were 2,51,29,719 and silver carp fingerlings were 2,83,480. The time duration required for fingerlings stocking was from June 2006 to October 2006. The total no. of prawn juveniles of *Macrobranchium rosenbergi* species stocked in Yeldari reservoir were 60, 85,911 number, this was the first time of such heavy stocking of prawn seed into the reservoir in the history of Yeldari reservoir fishery. The time duration required for the stocking of prawn seed was from Sept. 2006 to Dec. 2006 and fish seed was from June 2006-December 2006.

The fish seed and prawn seed were brought to the P.M.S.S.M. Sawli (Bu) Office from different places through containers loaded in trucks during the period of June 2006 to December 2006. The fish seed was released into the Yeldari reservoir only at one station *i.e.,* near the office of the P.M.S.S.M. Sawali (Bu) near the reservoir's earthen embankment at Yeldari camp.

For the fish seed stocking the P.M.S.S.M. Sawali (Bu) has adopted different method from the routine to release the fish seed into the reservoir. The purchased fish seed was brought in to the plastic bags filled with water and oxygen gas partially, and the bags were protected and packed in the plastic foldable tins. All the seed containing tins were loaded in to the large containers. Instead of release of fish seed by opening the fish seed bags, the fish seed bags were emptied into 6 inch diameter PVC plastic pipes connected in to the reservoir water from the embankment of reservoir. It was probably due to avoid the labour cost and to save the time. It was a successful method to stock the huge quantity of fish seed in crores of number for the large reservoirs like Yeldari.

In the year 2006-07, the flood gates of Yeldari reservoir were opened for 3 times during the period of August 2006 to October 2006 and were not opened during November 2006 to March 2007. (Source : Sub Divisional officer, Maintenance, Sub Division, Yeldari camp Tq. Jintur Dist. Parbhani.)

During the period of August 2006 when water was released for first time, it was considered that there may be loss of stocked fish seed from the reservoir. The fish and prawn seed was stocked in large quantity in to the reservoir to avoid the decrease in population of stocked fish seed and prawn seed loss through the released water from Yeldari reservoir in to Purna river basin, which later on enter in to the Siddeshwar reservoir.

Table 11.103. The fish catch at Yeldari reservoir From June 17, 2006 to May 31, 2008.

Sr.No.	Months and Year	Yeldari fish collection centre		Fish prawn collection centre		Khadki fish collection centre		Average fish catch/day of month in kg	Total fish catch of month in kg
		Fish catch/day of the month in kg.		Fish catch/day of the month in kg		Fish catch/day of the month in kg			
		Minimum	Maximum	Minimum	Maximum	Minimum	Maximum		
1.	Jun-06	—	—	—	—	—	—		
2.	Jul-06	30	155	—	—	—	—	92.5	2867.5
3.	Aug-06	28	944	—	—	—	—	486	15066
4.	Sep-06	179	1612	—	—	—	—	895.5	26895
5.	Oct-06	21	446	—	—	—	—	233.5	7238.5
6.	Nov-06	120	483	—	—	—	—	302.5	9045
7.	Dec-06	374	674	—	—	—	—	524	16244
8.	Jan-07	338	713	—	—	—	—	525.5	16290.5
9.	Feb-07	233	612	158	342	—	—	672.5	18830
10.	Mar-07	235	674	141	441	91	332	954	29667
11.	Apr-07	45	467	67	342	95	280	648	19440
12.	May-07	60	281	20	343	104	348	578	17918
13.	Jun-07	117	598	51	254	242	369	815.5	24465
14.	Jul-07	233	928	166	732	135	755	1474.5	45709.5
15.	Aug-07	233	1013	73	606	109	788	1411	43741
16.	Sep-07	334	785	130	495	313	902	1479.5	44385

Contd...

Table 11.103: Contd...

17.	Oct-07	386	1038	182	602	362	776	1673	51863
18.	Nov-07	381	1176	94	326	337	904	1609	48270
19.	Dec-07	439	1014	38	196	316	814	1408.5	43663.5
20.	Jan-08	329	1055	27	134	231	560	1168	36208
21.	Feb-08	714	1434	35	234	316	742	1788	50295
22.	Mar-08	208	1081	34	253	193	1107	1438	44578
23.	Apr-08	304	930	40	212	250	870	1303	39090
24.	May-08	210	880	70	250	210	850	1235	38285
	Total	5551	18993	1327	5762	3304	10397	22667	690054.5

Table 11.104. The Prawn catch at Yeldari reservoir From June 17, 2006 to May 31, 2008

Sr.No.	Months and Year	Yeldari prawn collection centre		Bamni prawn collection centre		Khadki prawn collection centre		Average fish prawn catch/day of month in kg	Total prawn catch of month in kg
		Prawn catch/day of the month in kg		Prawn catch/day of the month in kg		Prawn catch/day of the month in kg			
		Minimum	Maximum	Minimum	Maximum	Minimum	Maximum		
1.	Jun-06	—	—	—	—	—	—	—	—
2.	Jul-06	1.75	5.25	—	—	—	—	3.5	52.5
3.	Aug-06	2	11	—	—	—	—	6.5	100.75
4.	Sep-06	0.25	3	—	—	—	—	1.625	24.375
5.	Oct-06	0.25	2	—	—	—	—	1.125	17.4375
6.	Nov-06	0.25	4	—	—	—	—	2.125	31.875
7.	Dec-06	1.25	6	—	—	—	—	3.625	56.1875
8.	Jan-07	0.25	4	—	—	—	—	2.125	32.9375
9.	Feb-07	0.25	1.25	0.25	1			1.375	19.25
10.	Mar-07	0.25	5	0.25	1	0.25	0.75	3.75	58.125
11.	Apr-07	0.25	21	0.75	10	0.25	30.5	31.375	470.625
12.	May-07	3	26	4	233	19	123	204	3162
13.	Jun-07	69	386	15	73	49.75	101	346.875	5203.125
14.	Jul-07	87	199	34	104	44	139	303.5	4704.25
15.	Aug-07	52	187	21	62	36.5	113	235.75	3654.125

Contd...

Table 11.2: Contd...

17.	Oct-07	27	88	8	43	11	25	101	1565.5
18.	Nov-07	39	88	7	31	9	25	99.5	1492.5
19.	Dec-07	20	50	1	6	5	15	48.5	751.75
20.	Jan-08	14	44	0.25	5	4	16	41.625	645.187
21.	Feb-08	16	61	0.25	7	4	34	61.125	886.312
22.	Mar-08	16	62	2	29	4	33	73	1131.5
23.	Apr-08	16	60	2	25	4	30	68.5	1027.5
24.	May-08	16	50	0.5	10	4.5	30	55.5	860.25
	Total	438.75	1467.5	101.25	688	214.25	779.25	1844.5	28775.5615

It was found that, Intellect Agri Products Pvt. Ltd., Mumbai was not working properly according to the rules and conditions of the agreement with the Purna Fish Co-operative society. Therefore new agreement was finalized between Purna fish society and Aqua Fisheries and Agro Products Pvt. Ltd. Nariman Point, Mumbai on dated 15/02/2007 for the further period, the agreement was finalized by tender system, this agreement was with all legal deed process. It was an agreement only for the purchase of the catch from the reservoir and Cooperation in the fishery management process and the company has to pay the amount of stocked fish and prawn seed to the earlier company *i.e.* Intellect Agri Products Pvt. Ltd Company, Mumbai. Therefore earlier agreement with the Intellect Agri Products Pvt. Ltd., Mumbai was cancelled by the Purna Fish Co-operative Society.

Fish and Prawn Harvesting: On Yeldari reservoir there was 'no closed' day for fishing. Though P.M.S.S.M Sawli (Bu) had taken the fishing tender of Yeldari reservoir for the duration July 2005 to June 2010 on dated 4th October 2005. The reservoir was free for fishing from July, 2005 to June, 2006 to the Fishermen of villages located around the Yeldari reservoir; in this duration the fishermen engaged in fishing were about 250-300 in number from 34 different villages. Indian major carp species, the exotic carp species and *Macrobrachium rosenbergii* prawn species stocked in the Yeldari reservoir were harvested by using various kinds of nets called by various local traditional names.

Fish Marketing: During June 2005 to May 2006 the fish catch from the Yeldari reservoir was distributed and marketed by different methods. Some of the fishermen purchase the fish catch for Rs. 30 to 40 per kg. for fishes more than 01 kg. weight and for Rs. 20 to 30 per kg for fishes less than 1 kg. weight and they sale the fishes to middle man on commission at 5 to 10 Rs./kg for the large sized fishes and Small sized fishes. Most of the middlemen directly purchase the fishes from fishermen on Yeldari reservoir and sale the fishes in the surrounding market places at higher rate. Some of the fishermen and fisherwomen working on the reservoir collect their own fish catch and catch from other fishermen and sale it in the surrounding villages and market places.

The freshly collected fishes of Yeldari reservoir were sold in Yeldari-camp, Jintur, Selu, Parbhani, Risod, Sengaon, and Washim market. The freshly collected fishes were transported from Yeldari reservoir to different market places on bicycle, motor cycle or in four wheelers like tempo.

The fish preservation technique and facilities were not found during this study period (before June 2006). Recently a cold storage plant is being constructed near the reservoir at Yeldari campin December 2007. Visit to this cold storage plant for ots detail study was not permitted by the concerned authorities. The ownership of the cold storage plant was towards Aqua Fish products Pvt. Ltd., Mumbai in collaboration with Purna Fish Co-operative Society working at the Yeldari reservoir.

From June 2006, due to proper functioning of Purna Fish Co-operative Society, the fishery management of Yeldari reservoir was improved.

From June 2006, The society had banned the use of 'Zorli', or 'Wadap', 'Pandya' and 'Purai' type drag-bag nets and Phekjal type cast net because, by the use of these nets, the pre-mature young, small sized fishes may be harvested and it will affect negatively on the reservoir fish production. Due to the use of nets of very small mesh size of less than 1 cm, Purna society had banned the fishing during Jun, 2006 to July, 2006 in Yeldari reservoir by considering the breeding period of fishes, but in some villages located around the reservoir like Bamni, Khadki, Kini, Borkhedi the fishing was observed.

Table 11.105. Marketing of catch from Yeldari reservoir and wages (Year 2006 to 2008)

Sr.No	Fish/Prawn	Wages to fishermen	Purchase rate by Aqua fisheries and Agro Products Pvt Ltd. from the society	Purchase rate by middle man from Aqua fisheries and Agro Products Pvt Ltd	Purchase rate by retailers from middle man
1.	Local fishes, fishes below 1 kg	19 Rs./kg	24 Rs./kg	29 Rs./kg	31-34 Rs./kg
2.	Fish above 1 kg	24 Rs./kg	29 Rs./kg	34 Rs./kg	36-48 Rs./kg
3.	Prawns (Feb. to June 07)	150 Rs./kg	NA	230 Rs./kg	250-290 Rs./kg
4.	Prawns (July 2007 onwards)	70 Rs./kg	NA	230 Rs./kg	250-290 Rs./kg

In July, 2006 Purna society had established the first fish collection centre at Yeldari-camp. All catch obtained by the fishermen was collected at Yeldari camp fish collection centre up to December 2006. In this duration, with the help of 200 fishermen the fish fingerlings and prawn juveniles were stocked by the Purna society in Yeldari reservoir near Yeldari-camp fish collection centre, the fish seed contains Indian major carps, other local fish species and *M. rosenbergii* prawn. The prawn seed stocked in some irrigation reservoirs of Sengaon taluka in Hingoli district which reached in to the Yeldari reservoir through overflow.

It was became inconvenient to bring the fish catch at Yeldari-camp fish collection centre by the fishermen of different villages except from few villages like Yeldari-camp, Sawangi mahalsa, Murumkheda, Kini, Yeldari, Limbala tanda, Ambarwadi and Kawtha. Mean while fish catch of Kawtha village was weighed and brought to Yeldari-camp fish collection centre by using diesel engine boat or petrol engine speed boat by the Purna Fish co-operative society employees.

In January, 2007 the second centre for fish and prawn collection was started at village Bamni which was 12 km. away from Yeldari camp for the fishermen of Ambarwadi, Badnapur, Belkheda, Umrad, Saikheda, and Wazzar villages.

In February, 2007 the third centre for fish and prawn collection was started at Khadki, in Hingoli district, about 22 km away by road distance from Yeldari-camp, for the fishermen of villages Bhandari, Volgira, Khairi Ghumat, Pathonda, Borkhadi tanda, Dhotra, Sonsawangi, Khadki, Nansi, Bamni (ku), Dongaon, Salegaon, Ooty (Purna), Dhanora and Wazar of taluka Sengaon District Hingoli.

From February, 2007 onwards daily catch obtained by the fishermen was collected at 3 main fish collection centers established by the society. For the fish and prawn collection from Bamni and Khadki fish collection centers two jeeps with 5 to 6 employees from Intellect Agri Product Company in each jeep were involved. Regularly both jeeps along with employees go to Bamni fish collection centre and Khadki fish collection centre. These employees collect the fish and prawn catch from the fishermen and take the necessary records in their register of fish and prawn catch. They weigh the fish and prawn catch and pay the payment to the fishermen on the spot regularly for the catch. Later on these employees sort out the fish and prawn catch according to species and size of fishes and give it to middle man appointed by Intellect Agri Product Company. During August 2007 to November 2007 the prawn catch was not given to middle man and prawn catch was collected from Bamni and Khadki fish and prawn collection centre to Yeldari-camp fish and prawn collection centre. During this period the prawn was marketed to Mumbai, Kolkata and Amritsar fish markets. It was also reported that, the prawns of Yeldari reservoir were exported to China.

Yeldari-Camp Fish and Prawn Collection Centre

Yeldari-camp fish and prawn collection centre is present near the office of Purna Fish Co-operative society at Yeldari-camp. From July 2006 onwards. The daily fish and prawn catch from fishermen of villages Yeldari and Limbala tanda and Yeldari camp, Sawangi Mahalsa, Murumkheda, Kini and Kawtha villages were purchased and payment for the catch was given on the spot to the fishermen at this centre. The fishes and prawns were sorted according to species and size and weighed and given to middle-man regularly according to the agreement between Intellect Agri Product Company and middle-men.

The middle man sale the fishes and prawns on the spot to 5 to 6 retailers for Rs. 35 to 40 per kg for local fishes and IMC fish species less than 1 kg, where as IMC fish species more than 1 kg for Rs. 36 to 50 per kg. The remaining fish and prawn catch was sold to the traders at Parbhani. From Parbhani the fish and prawn catch was sold in Parbhani and surrounding markets and remaining fishes and prawns were transported to Kolkata, Mumbai and Amritsar by Railway transportation.

Bamni Fish and Prawn Collection Centre

At Bamni fish collection centre located near village Bamni, daily fish and prawn catch from the fishermen of Ambarwadi, Chaudharni, Badnapur, Bamni, kolpa, Sawangi (Bhambre), Belkheda, Umrad, Saikheda, and Wazar villages of were collected.The fishes and prawn catch were weighed and fish catch details were noted in the register maintained by the Society and payment for fishes and prawns were given on the spot by the Purna society regularly to the fishermen.

All the collected fishes and prawns were sorted and weighed by Purna Society and given to middle-man regularly according to the agreement between Intellect Agri Product Company and middle-men at the rates given in Table 3A.

The middle men sale the fishes and prawn to 5 - 6 retailers at rupees 35 to 40 per kg for local fishes and IMC fish species less than 1 kg, where as IMC fish species more than 1 kg at rupees 36 to 48 per kg. The remaining fish and prawn catch was sold to the trader at Parbhani. From Parbhani the fish and prawn catch was sold in Parbhani and surrounding market and remaining fishes and prawn were transported to Calcutta, Bombay, Amritsar, via Railway transportation.

Khadki Fish and Prawn Collection Centre

At Khadki fish collection centre daily fish and prawn catch from fishermen of villages Bhandari, Volgira, Khairi Ghumat, Pathonda, Borkhadi tanda, Dhotra, Sonsawangi, Khadki, Nansi, Bamni (Khu.), Dongaon, Salegaon, Ooty (Purna), Dhanora and Wazar of Sengaon taluka District Hingoli were collected. The fish catch from Bhandari, Volgira, Khairi Ghumat, Pathonda, and Borkhadi villages was collected at Bhandari and later on carried to Khadki fish and prawn collection centre. From each village, 1 to 4 fishermen regularly collect their own fish and prawns catch and catch from fishermen was already weighed and bring it on motorcycle to Khadki fish collection centre between 9 to 11 am. The fish and prawn catch of fishermen were weighed and fish catch details were noted in the register of the Purna Society and payment for fishes and prawns were given on the spot by the Purna society regularly. All the collected fishes and prawns were sorted and weighed by Purna Society and give it to middle-man regularly according to the agreement between Intellect Agri Product Company and middle-men at rate given in Table 3A.

The middle men collect all the fishes and prawn and preserve it in the containers with ice and they load the containers in mini truck or Jeep and carry to the Risod and Washim fish market. It is known from relevant source that, a very limited fishes were sold at Risod fish market @ Rs. 40 to 50 per kg. for local fish and @ Rs. 50 to 60 per kg for IMC fish species through agent or fish sellers. Most of the fishes and almost all prawns were sold to the fish traders at Washim @ Rs. 25/kg for local fish and @ 40 Rs./kg for IMC species and @ Rs. 250/kg for prawn. From Washim, the fishes were transported in local market, and the prawns were marketed at Akola, Nagpur and Mumbai markets.

Fisher Communities: Around the Yeldari reservoir there are 34 villages located closer to reservoir water. In these 34 villages the fisher communities of different caste and tribes are present. During the present study period the fisher population was recorded and it was 964. From almost all the 34 villages there are 16 villages included in Jintur taluka of Parbhani District where as 28 villages included in Sengaon taluka of Hingoli District located on other side of reservoir. The fisher population recorded from 16 villages of Taluka Jintur Dist. Parbhani was 526 and from 18 villages included in Tq. Sengaon Dist. Hingoli was 438. The population status of fisher communities distributed around Yeldari Reservoir is given in the Tables 11.103 and 11.104. Out of 964 fishermen , fulltime active fishermen number was 524. The fulltime active fishermen population recorded from 16 villages included in Tq. Jintur Dist. Parbhani was 343 and from 18 villages included in Tq. Sengaon Dist. Hingoli were 181.

Table 11.106. Fishermen Population Status of Yeldari Reservoir (Year 2006-08)

Sr.No.	Village Name	Taluka of District Parbhaniand Hingoli.	Fishermen population	Fisher Community Tribes and Caste and their population number.
1.	Yeldari camp	Jintur	85	Bhoi-51, Boudh-30, Telgu-1, Muslim-03
2.	Sawangi Mahalsa	Jintur	46	Banjara-03, Bhoi-09, Boudh-20, Holar-02, Muslim12,
3.	Murumkheda	Jintur	08	Bhoi-02, Boudh-06,
4.	Kini	Jintur	69	Bhoi-12, Boudh-15, Chambar-36, Hatkar-01,Vanjari-05
5.	Ambarwadi	Jintur	56	Andh-2, Boudh-11, Chambar-11, Matang-23, Vanjari-02, Waddar-07
6.	Kawtha	Jintur	76	Andh-48, Bhoi-03, Boudh-12, Hatkar-08, Matang-05,
7.	Badnapur	Jintur	07	Bhoi-07
8.	Chaudharni	Jintur	21	Boudh-17, Hatkar-02, Muslim-02,
9.	Bamni	Jintur	32	Bhoi-06, Baudh-03, Chambar-02, Dhanger-02, Hatkar-01, Koli-01, Matang-10, Muslim-7,
10.	Kolpa	Jintur	37	Bhoi-06, Boudh-22, Hatkar-07, Koli-02
11.	Kumbhephal	Jintur	09	Boudh-09
12.	Sawangi Bhambre	Jintur	43	Bhoi-09, Boudh-11, Koli-21, Muslim-02
13.	Umrad	Jintur	23	Boudh-23
14.	Bekheda	Jintur	03	Boudh-03
15.	Saikheda	Jintur	03	Boudh-03
16.	Wazar (Kh)	Jintur	08	Boudh-02, Rajput-05, Muslim-01
17.	Yeldari	Sengaon	10	Hatkar-10

Contd...

Table 11.106: Contd...

18.	Limbala-Tanda	Sengaon	23	Banjara-23
19.	Bhandari	Sengaon	32	Andh-01, Banjara-05, Bhoi-03, Boudh-05, Hatkar-01, Muslim-17
20.	Khairi-Ghumat	Sengaon	13	Banjara-05, Hatkar-06, Muslim-02
21.	Holgira	Sengaon	07	Banjara-05, Hatkar-02
22.	Borkhadi	Sengaon	39	Banjara-26, Boudh-12, Hatkar-01
23.	Dhotra	Sengaon	49	Boudh-41, Hatkar-07, Maratha-01
24.	Pathonda	Sengaon	26	Andh-11, Banjara-04, Boudh-10, Hatkar-01
25.	Son Sawangi	Sengaon	25	Banjara-22, Hatkar-03
26.	Khadki	Sengaon	43	Banjara-21, Boudh-21, Hatkar-01
27.	Bamni (Kh)	Sengaon	38	Banjara-31, Hatkar-03, Muslim-04
28.	Nansi	Sengaon	27	Banjara-02, Boudh-25
29.	Dongergaon	Sengaon	43	Banjara-21, Bhoi-16, Boudh-06
30.	Ooty (Purna)	Sengaon	20	Bhoi-08, Boudh-12,
31.	Salegaon	Sengaon	13	Boudh-11, Muslim-01, Vanjara-01
32.	Dhanora	Sengaon	20	Bhoi-08, Boudh-06, Hatkar-04, Maratha-02
33.	Pimpri	Sengaon	06	Boudh-06
34.	Barda	Sengaon	04	Boudh-04

Total Fishermen Population = 964.

Table 11.107. Full-time Fishermen Population Status of Yeldari Reservoir (Year 2006-08)

Sr.No.	Village Name	Taluka	Full time Fishermen population	Fisher Community Tribes and Caste and their population
1.	Yeldari camp	Jintur	76	Bhoi-45, Boudh-26, Muslim-03, Telgu-1
2.	Sawangi Mahalsa	Jintur	38	Banjara-02, Bhoi-07, Boudh-18, Holar-01, Muslim-10,
3.	Murumkheda	Jintur	04	Bhoi-01, Boudh-03.

Contd...

Table 11.107: Contd...

4.	Kini	Jintur	45	Bhoi-08, Boudh-11, Chambar-24, Hatkar-01, Vanjari-01
5.	Ambarwadi	Jintur	38	Boudh-08, Chambar-10, Matang-13, Vanjari-02, Waddar-05
6.	Kawtha	Jintur	63	Andh-41, Bhoi-02, Boudh-09, Hatkar-06, Matang-05
7.	Badnapur	Jintur	04	Bhoi-04
8.	Chaudharni	Jintur	12	Boudh-09, Hatkar-01, Muslim-02
9.	Bamni	Jintur	12	Bhoi-02, Boudh-02, Dhangar-01, Matang-05, Muslim-02,
10.	Kolpa	Jintur	17	Bhoi-04, Boudh-09, Hatkar-03, Koli-01
11.	Kumbhephal	Jintur	04	Boudh-04
12.	Sawangi Bhambre	Jintur	16	Bhoi-04, Boudh-02, Koli-09, Muslim-01
13.	Umrad	Jintur	11	Boudh-11
14.	Belkheda	Jintur	02	Boudh-02
15.	Saikheda	Jintur	01	Boudh-01
16.	Wazar (Kh)	Jintur	10	Hatkar-10
17.	Yeldari	Sengaon	06	Hatkar-06
18.	Limbala-Tanda	Sengaon	09	Banjara-09
19.	Bhandari	Sengaon	20	Andh-01, Banjara-01, Bhoi-03, Boudh-04, Hatkar-01, Muslim-10
20.	Khairi-Ghumat	Sengaon	07	Banjara-02, Hatkar-04, Muslim-02
21.	Holgira	Sengaon	04	Banjara-03, Hatkar-01
22.	Borkhadi	Sengaon	16	Banjara-08, Boudh-08
23.	Dhotra	Sengaon	26	Boudh-24, Hatkar-01, Maratha-01
24.	Pathonda	Sengaon	10	Andh-05, Banjara-02, Boudh-02, Hatkar-1
25.	Son Sawangi	Sengaon	11	Banjara-09, Hatkar-02
26.	Khadki	Sengaon	15	Banjara-10, Boudh-05
27.	Bamni (Kh)	Sengaon	19	Banjara-15, Hatkar-02, Muslim-02

Contd...

Table 11.107: Contd...

28.	Nansi	Sengaon	07	Banjar-01, Boudh-06
29.	Dongergaon	Sengaon	09	Banjara-06, Bhoi-02, Boudh-01
30.	Ooty (Purna)	Sengaon	07	Bhoi-03, Boudh-04
31.	Salegaon	Sengaon	01	Boudh-01
32.	Dhanora	Sengaon	13	Bhoi-07, Boudh-03, Maratha-02, Hatkar-01
33.	Pimpri	Sengaon	—	—
34.	Barda	Sengaon	—	—

Total Full Time Fishermen Population=533

Caste and Tribe

The fisher population present in 34 villages belongs to different caste and tribes. The fishermen caste and tribes were Andh (Adiwasi), Banjara, Bhoi, Chambhar, Dhangar, Hatkar, Holar, Koli, Maratha, Matang, Muslim, Rajput, Telgu, Vanjari and Waddar. Details of active fishermen caste and tribe population is given in the Table 11.108.

Table 11.108. Caste and tribe population of fisher communities in the villages around Yeldari reservoir

Sr.No	Caste and tribe of fisher Population	Fisher population of 16 villages of Tq. Jintur Dist. Parbhani	Fisher population of 18 villages of Tq.Sengaon Dist. Hingoli	Total Fisher population	Percentage of population
1	Andh	50	12	62	6.43
2	Banjara	03	165	168	17.42
3	Bhoi	105	35	140	14.52
4	Boudh	187	159	346	35.89
5	Chambar	49	—	49	5.08
6	Dhangar	02	—	02	0.20
7	Hatkar	19	39	58	6.01
8	Holar	02	—	02	0.20
9	Koli	24	—	24	2.48
10	Maratha	—	03	03	0.311
11	Matang	38	—	38	3.94
12	Muslim	27	24	51	5.29
13	Rajput	05	—	05	0.51

Contd...

Table 11.108: Contd...

14	Telgu	01	—	01	0.10
15	Vanjari	07	01	08	0.82
16	Waddar	07	—	07	0.72
17	Total	526	438	964	

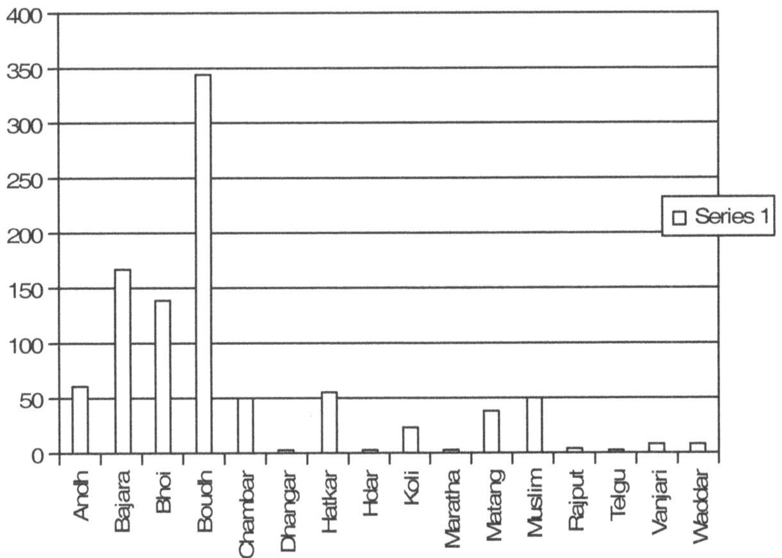

Fig. 11.40. Caste and tribe Fisher population number involved in the Yeldari reservoir fishery (Year 2006-08)

Involvement of Fisherwomen in Fishing and Other Activities

It was found that the fisher women of various fisher caste and tribe are involved in the fishing activities along with their husband. At Yeldari reservoir, 8 to 10 fisher women were involved in actual fishing activities; they operate the thermocoel rafts, gill-net and also assist the fishermen in the operation of large nets like Zorli' and 'Pandya'. All fisherwomen were active swimmers. The clothing of the fisherwomen were saries, locally it was called as 'Lugda'.

Majority of fisherwomen were involved in household activities, but at some places like Yeldari-camp, Kini, Kawtha, and Ambarwadi, Bamni (Bu) of Tq. Jintur Dist. Parbhani and Bhandari, Kawtha, Bamni of Tq. Sengaon Dist. Hingoli the fisher women belongs to caste Boudh, Bhoi, Chambar, and Banjara were involved in fish marketing of their own catch and the marketing of purchased catch.

There was no remarkable involvement of fisherwomen in the Yeldari reservoir fishery but they play an important role to assist their husband in the fishery activities, but whatever fisherwomen population was actively working and involved in actual fishing activities, their activities were really brave in such a vast spread and deep reservoir, it was beyond the imagination of common man.

Housing

Around the Yeldari reservoir there are few villages around the peripheral region of the reservoir like Kini, Kawtha, and Sawangi – Mahalsa having 100% population of fisher communities; in these villages, there are separate colonies of Boudh, Muslim, Chambar, and Bhoi. The type of house of fisher communities distributed in various villages around Yeldari reservoir are of simple type, constructed with the stone, bricks, clay and concrete. Maximum houses are of tin protection, some are having concrete slabs and others are with hut like structure. All housing system is in small plot of 1500 to 2000 sq. ft. area.

The fisher communities built their temporary hut like structure of polythene and bamboo at the site of fishing away from their village on the coast of reservoir. These huts are temporary, easy to construct and dismental. The temporary huts constructed during peak fishing season. In these temporary huts sometimes entire family of fishers take the rest or only fishermen use this hut as a shelter while the rest of family members remain in their villages.

Wages and Income

The main source of income for the fisher communities working at the Yeldari reservoir can be explained as:

Situation before June 2006: There was no fixed and assured income for the fisher community because, the Yeldari reservoir fishery management was very poor before year 2006, in this period the fishing contract was toward Meretek Company, Nagpur. Some time the reservoir remain open hence there was no ban on the use of small mesh size nets and mosquito nets for fishing, similarity the fish seed stocking by fishing tender owner companies, co-operative societies and Government agencies was inadequate as compared to the vast area of reservoir.

The fisher community's income was indefinite and poor, it was up to Rs. 50 to 60 per day per fisherman.

Situation after June 2006: After year 2006, the Purna society had taken the charge of fishing at Yeldari reservoir by tender system and flourished as a modern fishery co-operative society with new trends of reservoir fishery management. The society had stocked nearly 2.5 crores fish seed of Indian major carps and 60 lakhs juveniles of prawn species *Macrobrachium rosenbergii.* The society has also established a separate patrolling and vigilances squad on the land and in the water to control fish poaching, theft and control on the use of mosquito nets. This unit was having jeeps, petrol speed boat, binoculars, mobile sets. The result of this management and control on the young fish catch, theft of fish and fish marketing regulation finally resulted in the development of large sized fishes in the reservoir. This results in increase in the rate of fish and prawn catch to increase the wages to the fisher communities up 100-150 per days per person as assured income source in the peak fishing season. Due to this increase in wages the fishermen population involvement found increased in fishing activities after year 2006.

Educational Facilities and Educational Status

Marathi is the main language of communication and learning in the villages present around the Yeldari reservoir.Information about educational facilities and educational status of active fishermen, fisherwomen and their children was collected by survey and personal interviews of fisher communities belongs to village Yeldari camp, Sawngi mahalsa, Kini, Kawtha, Ambarwadi, Bamni and Khadki. The educational facilities present in the villages present around Yeldari reservoir are of moderate type. In most of the villages Marathi medium primary schools are present only at Yeldari-camp, Ambarwadi, and at village Bamni high school is present, where as Junior collage and sinner collages are not present in any village present around Yeldari reservoir. For junior collage and senior collage education is available at town places like Jintur, Sengaon, and Parbhani etc. As the fishermen are engaged in fishing through out the day, there was not much attention and awareness about education to their children was found in all the fisher communities . The children of fisher communities belong to different caste and tribe go to the primary school. The drop out percentage of children from schools is comparatively more in fisher community than other communities present in the villages present around the Yeldari reservoir. To study the educational status 408 active fishermen from above mentioned villages were interviewed, Out of 408 active fishermen 164 *i.e.* 40.19% active fishermen were illiterate and 245 *i.e.* 49.80% active fishermen were literate. Out of 245 literate active fishermen 184 active fishermen *i.e.* 73.77 % active fishermen had taken the primary education only and most of them had taken the education up to 2nd standard only, 48 active fishermen were educated up to high school, 11 *i.e.* 4.50% fishermen were educated up to 11th and 12th class and only 3 *i.e.* 1.22% active fishermen were graduate. (Table 11.80) The educational status in fisherwomen is very worst. out of 293 fisherwomen from above mentioned village, it was found that 237 *i.e.* 80.88 % fisher women were illiterate and 56 *i.e.* 19.12% fisherwomen were literate and educated up to primary school. (Table 11.81). Out of 177 fisher community children 41 *i.e.* 23.13% children were illiterate and 136 *i.e.* 76.83% children were literate. Out of 136 literate children 113 *i.e.* 83.08% child were learning in primary school, 13, *i.e.* 9.55% children were learning in high school and 10 *i.e.* 7.35% children were learning at 11th and 12th level. Out of 41 illiterate children most are girls. (Table 11.109).

Table 11.109. Educational status of active fishermen from some villages around Yeldari reservoir

Sr.No.	Name of the village	Fisher population	Illiterate	Literate			
				1st to 7th	8th to 10th	11th to 12th	Graduation
1.	Yeldari camp	56	45	11	—	—	—
2.	Sawangi Mahalsa	28	19	09	—	—	—

Contd...

Table 11.109: Contd...

3.	Kini	46	40	06	—	—	—
4.	Kawtha	59	45	14	—	—	—
5.	Ambarwadi	47	41	6	—	—	—
6.	Bamni	21	16	05	—	—	—
7.	Khadki	36	31	05	—	—	—
		293	**237**	**56**			

Table 11.110. Educational status of fisherwomen from some villages around Yeldari reservoir

Sr.No.	Name of the village	Fisher population	Illiterate	Literate			
				1st to 7th	8th to 10th	11th to 12th	Graduation
1.	Yeldari camp	85	48	15	13	06	02
2.	Sawangi Mahalsa	46	24	13	8	—	—
3.	Kini	69	18	39	10	02	—
4.	Kawtha	76	24	45	07	—	—
5.	Ambarwadi	56	19	29	05	02	01
6.	Bamni	33	16	14	02	01	—
7.	Khadki	43	15	25	03	—	—
	TOTAL	408	164	180	48	11	3

Table 11.111. Educational status of children of fisher communities from some villages around Yeldari reservoir

Sr.No.	Name of the village	Fisher population	Illiterate	Literate			
				1st to 7th	8th to 10th	11th to 12th	Graduation
1	Yeldari camp	30	08	18	02	02	—
2	Sawangi Mahalsa	22	06	12	02	02	—
3	Kini	35	10	16	04	05	—
4	Kawtha	27	05	18	03	01	—
5	Ambarwadi	27	04	21	02	—	—
6	Bamni	12	—	12	—	—	—
7	Khadki	24	08	16	—	—	—
		177	41	113	13	10	—

Modern Facilities

The fisher community were having colour television set in their home, 10 to 20% fisher were having personal mobile phone sets. It was observed that the use of motorcycle by the fishermen was found increased after year 2006. The motorcycles were used in fish transportation from site of catch to various fish and prawn collection centers of Purna society. At Yeldari-camp, Kini, Kawtha and Bhandari villages the use of motorcycle by fishermen was found increased.

Problems of Aquatic Weeds, Obstacles and Hide Out

It was observed that, there were no obstacles of weeds and tree boulders acting as obstruction for gill net fishing but the fishermen are well acquainted with the problem and information of these problematic areas of fishing was found transferred from senior fishermen to new comers. Hence this doesn't affect the fishing process in major.

The Yeldari reservoir fishery remains open through out year and no closing day of fishing hence there is continuous movement of the fishermen and their fishing activities in the reservoir, hence no any development of the weed species found in the reservoir.

Recently it was observed that the area near the concrete embankment of the Yeldari reservoir, acting as fish hide out because there is ban on fishing activity in 200 meter area from the embankment gates of the reservoir. This act as a safe place for fishes to hide out. This fishing restriction was from the Irrigation department to avoid any accident and planned mishap of damage of the concrete embankment like bomb explosion of the embankment of the reservoir.

If, with planned, secured and monitored fishing permission are given to the trained fishermen or if the irrigation department it self involved in fishing process, then there is a chance of getting huge fish catch from this thick populated area. It was found that, there are huge shoals of large fishes swimming near the embankment of the reservoir, which can be observed even from the concret platform of the reservoir.

Flood and Water Release from Flood Gates

During heavy rainy season, it is essential to release the water beyond the storage capacity of reservoir, similarly for the irrigation to agriculture the water from Yeldari reservoir is released in the down stream Purna river basin and the water reach into Siddheshwar reservoir in the downstream area, from where the water is released through the canal to agriculture, through this system of water release from Yeldari reservoir the stocked young fish and prawn juveniles, large grown fishes and prawns lost because there is no any prevention or protection to prevent this loss. Recently purna society identified this problem and diploid active fishermen for the catch of lost fishes and prawn in the downstream purna river basin.

To compensate the problem of fish loss from the reservoir, the Purna society decided to stock heavy number of fish seed in lakhs.

There is a chance of prevention of fish and prawn loss from the reservoir by constructing a well designed filter in the river basin in a suitable area.

Success of the Yeldari Reservoir Fishery and Future Prospects

The Yeldari reservoir fishery has been studied up to some extent by Sakhare (2001) especially for the physico-chemical characters of reservoir water and discussed about some aspects of fishery management and concluded that the Yeldari reservoir has good fishery potential and suggested regular studies on various aspects for the fishery development.

Through this detailed study as a Yeldari reservoir case study, the real problems of fishery management, role of co-operative society and status of fisher communities, Adoption of modern techniques of reservoir fishery management by Purna society etc are studied in detail as above and present status of Yeldari reservoir fishery is explained.

The striking change in the increase in fish production from the reservoir and increase in wages and income to fisher community was seen due to the efforts of Purna Matsyavyvasai Sahakari Sanstha Maryadit, Sawali (Bu).

The society had identified the fact that, it is necessary to stock the fish and prawn seed in the ratio of available area. The society had also employed a special team of workers as guards specially for patrolling around the reservoir for the prevention of theft, poaching and control on netting of fish and prawn from the reservoir.

If the fish co-operative societies or companies with some other new successful management plans and strategies are permitted in the Yeldari reservoir fishery then there is a better chance of Yeldari reservoir improvement in terms of increase in fish production, socio-economic upliftment of fisher communities, employment generation to trap the untamed huge fishery potential of Yeldari reservoir in future, which will help to improve the inland fishery production of this region.

MASOLI RESERVOIR

The Masoli reservoir is with total catchment area 281.07 Sq. Km. It was constructed on the Masoli river near village Isad in Gangakhed Taluka of Parbhani district. It is located at the latitude of 18°54′ 10″ N and longitude of 76°45′ 05″ E. The area receives average annual rainfall 788mm. The length of earthen embankment is 987 meters. The gross storage capacity of the reservoir is 34.08 million cusec meters. Its dead storage capacity is 6.94 million cu mt. The reservoir was constructed for the irrigation purpose. It irrigates near about 2591 hectares of agricultural land. Only one outlet canal of 15 km length emerges from the reservoir for irrigation. The reservoir is also a main source of water for Gangakhed town.

The reservoir is rich in flora and fauna in and around, as there is no industry on both sides as well as in catchment area of the reservoir; hence it is totally free from pollution load.

The state fisheries department recommended recently stocking of the reservoir with 4.56 lakh nos of fingerlings per year, giving an average estimated annual yield of around 137 tonnes. Government has already constructed a Chinese type of hatchery in the vicinity of the reservoir, for producing the required number of seed to meet the demand in the Parbhani district. The reservoir is regularly stocked and harvested by a private contractor. No ecological studies have so far been made on reservoir. There is no record maintained to know the fish production. However, the fishermen themselves disposed of the catches by selling to the merchants. The preliminary observations on the fish fauna revealed that the reservoir supports a good fishery of Indian major carps, gill nets fabricated out of synthetic twine are operated throughout the year. The nets are without foot rope and have thermocole floats. During the summer months drag nets are also operated. The boats used are improvised out of vehicle tyres tubes and thermocole platform.

The gears used for fishery are mainly gill nets and cast nets. About 50 fishermen are involved in fishing. A small amount of catch is utilized for household's consumption. The catch is sold in the market of Gangakhed, Nanded and Parbhani.

Water Quality: Kadam *et al* (2006) studied the water quality of Masoli reservoir (Table 11.112) and concluded that water is suitable for fisheries.

Table 11.112. Physico-chemical characteristics of Masoli reservoir

Parameters	Station A	Station B	Station C
Water Temperature (°C)	30.4	30. 1	30.5
Humidity %	43	43	42
Dissolved Oxygen (mg/l)	12.3	12.6	11.2
pH	8. 1	8.2	8.0
Turbidity (NTU)	130	140	155
Chlorides (mg/l)	16.9	18.3	14.2
Alkalinity (mg/l)	124	136	112
Total Hardness (mg/l)	105	109	100
Calcium (mg/l)	69	67	65
Magnesium (mg/l)	8.78	10.24	8.54
Sulphate (mg/l)	19	21	20
Total dissolved solids (mg/l)	260	240	280
Biochemical oxygen demand (mg/)	3.74	3.92	3.27
Chemical oxygen demand (mg/l)	3.61	3. 11	3.47

Ecological Aspects: In Masoli reservoir, species belonging to chlorophyceae, euglenophyceae and dinophyceae represented as true planktonic forms whereas the diatoms exhibited a mixed population consisting mostly the benthos species that are detached from substratum (Kadam *et al.* 2006). The blue green algae also show a similar picture like diatoms. The phytoplankton algal count per ml area at different stations is depicted in Table 11.113.

Table 11.113. Phytoplankton algal count per ml in the area.

Type of phytoplankton	Season		
	Summer	Monsoon	Winter
	Station 'A'		
Chlorophyceae	1400	1200	1100
Bacillariophyceae	1500	850	700
Cynophyceae	650	950	500
Euglenophyceae	800	450	150
Dinophycaeae	700	350	100
	Station 'B'		
Chlorophyceae	1250	950	800
Bacillariophyceae	1400	750	650
Cynophyceae	400	750	450
Euglenophyceae	900	650	350
Dinophyceae	400	375	160
	Station 'C'		
Chlorophyceae	1100	950	700
Bacillariophyceae	1200	750	600
Cynophyceae	700	800	400
Euglenophyceae	750	450	250
Dinophyceae	**200**	**300**	**140**

Table 11.114. Macrophytic community of Masoli reservoir

Life form	Name of species	Class	Family
Floating	*Salvinia molesta*	Fern	Salviniaceae
	Eichhornia crassipes	Monocot	Pontederiaceae
	Lemna sp.	Monocot	Lemnaceae
	Pistia sp.	Monocot	Araceae
	Potamegaton sp.	Monocot	Potamegatonaceae
	Nelumbo mucifera	Dicot	Nymphaceae
Submerged	*Marsilea quadrifolia*	Fern	Narasiliaceae
	Naja sp.	Monocot	Najadaceae

Contd...

Table 11.114: Contd...

	Ceratophylum sp.	Dicot	Ceratophyllaceae
	Utricularia exoleta	Dicot	Lentibulariaceae
Emergent	*Trapa* sp.	Dicot	Hydrocaryaceae
	Cyperus sp.	Monocot	Cyperuaceae
	Marsilea quadrifolia	Fern	Narasiliaceae
Marginal	*Ipomea cornea*	Dicot	Convolvulaceae
	Cyperus sp.	Nionocot	Cyperaceae
	Marsilea quadrifolia	Fern	Narasiliaceae

The submerged weeds under collection include *Marsilea* sp. *Hydrilla verticellata, vallisneria spiralis, Najas sp., Cerotyphyllum sp.* and *Utricularia exoleta*. Submerged aquatic weeds show vegetative growth during monsoon and flourished during post monsoon months. Hydrilla sp. showed profuse growth during summer. The growth of submerged plants during post monsoon months might have been due to high mineral concentration and better light condition. During the present investigation maximum population of *Hydrilla, Najas, Potamegaton* and *Trapa* were observed.The zooplankton diversity in reservoir is depicted in Table 11.115.

Table 11.115. Zooplankton diversity

Copepoda: *Cyclop* sp., *Nauplii, Mesocyclops leucarati, Clanoid copepods, Diaptomous* sp. *and Neodiaptomous* sp.
Cladocera: *Daphnia carinata, Ceriodophnia* sp., *Monia dubia, Bosmina* sp. *and Simocephalus* sp.
Rotifers: *Filinia terminalis, Filina longiset, Kerattela tropia, Kerotella cochlearis, Testudinell* sp., *Asplancha sp., Branchionus caudatos, B. rubens, B. forticula, B. calyciflarus, B. angularis, B. falcutus, Horella* sp.
Ostracoda: *Cypris and Meta cypris.*

For the fisheries activities, the reservoir was under the control of District fisheries Department.Fish seed of *Labeo rohita, cattla catla* and *cirhina mrigala* was stocked. Just after one year of stocking, the fishing activitiy was carried out (Chavan, 2006). Total annual fish catch during the year 2001 (June) to 2002 (June) and July 2002 to July 2003 from the reservoir is shown in Table 11.116.

Table 11.116. Fish catch of Masoli reservoir

Sr.No.	Fishes	Fish catch (kg / ha)	
		Year 2001-02	Year 2002-03
1.	IMC	10	12
2.	Cat fishes	08	07
3.	Weed fishes	10	11
4.	Mastacembalus armatus	05	08

IMC - Indian major carps – *Catla catla, cirhina mrigala, Labeo rohita*
Catfishes – *Wallago attu, Mystus blaker, Mystus seenghala, H. fossilis*
Weed fishes – *Puntius ticto, P. sarana, P. hexacticus, Chela phulo, Chela bakaila*

KARPARA RESERVOIR

The Karpara reservoir is constructed across the river Karpara at Niwali in Jintur taluka of Parbhani District. The reservoir was formed for irrigation by constructing dam on the river Karpara. The reservoir has the water spread area of 551 ha. The district fisheries department has recommended stocking of the reservoir with 5.45 lakh nos of fingerlings for obtaining an yield of 163.65 tonnes. The group wise composition of landing in recent years shows that about 50 per cent of catch comprises local minors and minnows. These two groups mainly constitute small sized fishes of low economic value. The average market price of these groups is less than half of the value of major carps and local majors. The fish catches of low value and under sized fishes from this reservoir also show that fishing is not properly managed leading to the depletion of the stock of quality fish groups. Conservation measures such as mesh size regulation are not strictly followed which is evident from the landings of under sized fishes.

The fishing craft is in the form of rafts, which are prepared by tying together 14 sealed empty kerosene tins. Paddles having blades at both the ends propel this craft. Fishermen sit at the centre of raft before it is launched. The vehicle tyre and tubes are also used as crafts.

PALAS-NILEGAON RESERVOIR

The reservoir is located within the geographical coordinates of 76°10′150″ longitude E and 17°40′45″ latitude N. he reservoir came into existence with the impounding of river Bori near Pune-Hyderabad national highway at village Bhabulgaon of Osmanabad district. The water-spread area at full reservoir level is 206 hectare. Length and height of earthen dam are 1074 m and 14.95 m respectively. Out of identified 28 fish species *Catla catla, Mastacembelus armatus, Labeo rohita, Cirrhnus mrigala, Mystus seenghala, Channa* sp and *Puntius sarana* were predominant. The species like *Wallago attu, Chela phulo, Notopterussp, Heteropneustes fossilis, Clarias batrrachus, Ctenopharyngodon idella, Cyprinus carpio, Hypophthalmichtys molitrix, Ompak bimaculatus, and Rohtee cotio* are rarely recorded. The Indian major carps form the major fishery in this reservoir. Their production gradually increased with the increasing of stocking density. Exotic fishes though stocked along the carps however, could not survive and grow in the reservoir due to which they contribute very negligible fishery. Phytoplankton belonging to three groups *i.e.*, myxophyceae, chlorophyceae and bacillarophyceae and zooplanktons belonging to rotifera, cladocera, copepoda, and ostracoda were identified. The plankton showed summer peak coinciding with the mean depth, influx and outflow of water. Rotifera, copepoda, ostracoda, myxophyceae and chlorophyceae contributed to the summer peak of planktonic density. The reservoir is leased to fishermen's cooperative society. Every year co-operative society gives fishing rights to contractor on a fixed amount. The members of the society never fish the reservoir. The water transparency varied between 73-117 cm, temperature between 21 – 34°C. The pH and dissolved oxygen were in the range of 7.2 – 8.6 and 5.3 - 10.8 mg/l respectively. The free carbon dioxide ranged between 7.2 to

8.3 mg/l and total alkalinity from 168.3 to 393.2 mg/l. The total hardness and total dissolved solids were in the range of 62 to 188.6 and 103 to 299.4 mg/l.

JAWALGAON RESERVOIR

Jawalgaon reservoir was built in Solapur district by impounding the river Nagzari irrigates an area of 4,451 ha. The length and height of dam is 1230 m and 21.71 m respectively. Species like *P. kolus, P. sarana, P. sophore, P. ticto, Chanda ranga* that are smaller in size and are larvivorous in nature. Some fishes are of medicinal value and some are useful as bait. Based on the interviews with the local fishermen, the per day catch per fisherman is estimated to be approximately 2 kg only and per hectare fish yield from the reservoir is about 37.43 kg/ha/yr. One contractor does the fishing without any restriction on mesh size. This resulted in indiscriminate exploitation of stocked fishes much below the desirable size.

The water temperature ranged from 19 to 31°C, pH found to be 6.4 to 8.5. Total-dissolved solids ranged from 113 to 233 mg/l. Transparency was found to be 27 to 69 cm. The dissolved oxygen and free carbon-dioxide content were between 2 to 6.2 and nil to 1.8 mg/l. respectively. Total alkalinity and chlorides fluctuated from 94.28 to 183 mg/l. The results revealed that the water of reservoir is suitable for fisheries.

BORI RESERVOIR

Bori reservoir in Osmanabad district is an artificial impoundment constructed across the river Bori, supplying drinking water to Tuljapur, Naldurg and Shri Tuljabhavani sugar factory. The total water spread area of reservoir is 746 ha. The reservoir is situated near Naldurg town close to the Tuljapur-Naldurg road. The reservoir is leased to fishermen's co-operative society. The reservoir has faunistic diversity of 20 fish species; the overall percentage of Indian major carps is low. In Bori reservoir natural recruitment of major carps is either absent or poor due to nonavailability of suitable breeding grounds. The reservoir needs regular stocking support of major carp fingerlings. The reservoir is likely to give more fish yield than the present fish yield (approximately 13 kg/ha/yr). It may be possible only with regular stocking of major carp fingerlings @ 1000 nos/ha/yr. The preliminary study of planktons indicates three major classes of phytoplankton, which includes chlorophyceae, cyanophyceae, and bacillariophyceae. Maximum phytoplanktons were recorded during month of April and minimum during October. While zooplankton commomprised the organisms belonging to cladocera, rotifera, copepoda and ostracoda. Copepoda population was dominated among the zooplanktons. Maximum density of zooplankton was registered during month of May.

The water temperature ranged from 19 to 27°C. The minimum temperature was recorded in November, while maximum in the month of April. pH was in the range of 6.9 to 7.0; the minimum and maximum values were recorded in months of April and December respectively. Transparency ranged from 40-288 cm, indicating low values during august and high in the month of March. Dissolved oxygen ranged from 5.2 to 9.12 mg/l. Its maximum and minima were recorded in April and October respectively. Carbon-dioxide ranged between Nil to 2.0 mg/l. Total alkalinity range was observed in between 78 to152 mg/l.

HANGARGA RESERVOIR

It supplies drinking water to Tuljapur town, the historical town in Maharashtra in its religious, architectural and secular splendor. Goddess Shri Tuljabhavani is the presiding deity of the town. Hangarga reservoir is with water spread area of 180 ha. The reservoir is leased to one private party. The gill nets in use are with very small mesh size leading to an indiscriminate harvest. As a result, the fishes were not allowed to reach optimum size for the harvest. The major fishes currently encountered in order of abundance are *W.attu, Mystus* sp., *O. bimaculatus, Channa* spp., and *Puntius* spp. Due to constant increase in population of above fishes, this reservoir may be categorized as 'catfish reservoir'. Reservoir was not stocked with seed of Indian major carps in past. Hence the Indian major carps contribute very negligible fishery. The yield from reservoir has been estimated at 29 kg/ha/yr. The most unfavorable feature of the reservoir is the lack of sufficient water during summer days, which result in the fish kill. Such type of condition was observed during the summer period of 2003 and 2004.

Water temperature in the reservoir ranged from 17 to 29°C. During monsoon period water was somewhat muddy and brownish in colour, the secchi disk transparency ranged between 14 to 21 cm. Dissolved oxygen varies from 3.8 to 7.2 mg/l. Free carbon dioxide in reservoir water was recorded at negligible level indicating the good water quality. pH of Hangarga reservoir was alkaline (8 to 8.2). The total dissolved solids values were in the range of 195 to 255 mg/l. The chlorides ranged between 21.27 to 32 mg/l. The water samples showed total hardness from 94 to 106 mg/l. The total alkalinity of reservoir water was in the range of 54 to 103 mg/l.

RAMDARA RESERVOIR

In Ramdara reservoir of Tuljapur (Maharashtra) tilapia was stocked with Indian major carps. After few years of stocking tilapia has adversely affected the indigenous fishes and nearly eliminated the Indian major carps from the catch. The performance of tilapia in Ramdara has been discouraging mainly because of early maturity, continuous breeding, overpopulation and dwarfing of the species. The middlemen purchase the fish from the fishermen. There is no sale of fish by fishermen directly to consumers.

DHOM RESERVOIR

The dam was constructed across the river Krishna near Wai in Satara district .It is an earthfill and gravity dam The capacity of dam is 13.50 TMC. The dam also have Hydroelectric Project of 2 mw. The reservoir is about 20 kms in length. It is an hydroelectric project with captacity of 4 mg electricity generation capacity. The main purpose of the dam is to supply the water to agriculture,industries and for the drinking to Wai, Pachgani, Mahababaleshwar and the surrounding villages on the bank of reservoir. The salient features of the reservoir are depicted in Table 11.117.

Table 11.117. Salient features of Dhom Reservoir in Satara district

1. Name of reservoir	Dhom
2. Location	Wai
3. Year of construction	1976
4. River	Krishna
5. Type	Earthfill-gravity dam
6. Height of dam	160 feet
7. Length of dam	8130 feet
8. Volume	6335 km^3
9. Capacity of dam	331,100 km^3
10. Surface area	2498 km^2

The reservoir is regularly stocked by the seed of Indian major carps. The details of the seed stocking and fish production in the reservoir are depicted in Table 11.118.

Table 11.118. Fisheries of Dhom Reservoir

No.	Year	Seed Stocking (lakhs)	Fish Production (kg)	Fish Production (kg/ha/yr)
1.	2007-08	2.50 Sr.	13736	
2.	2008-09	4.00	23178.50	
3.	2009-10	10.00	32520	
4.	2010-11	13.725	38203	
5.	2011-12	4.403	40061	

KANHER RESERVOIR

It is an earthfill and gravity dam constructed across river Wenna near Satara. The dam is situated in Northwest of Medha Tahsil of Satara district. It is located at latitude 1°744″ 16°02″ North and longitude 73°53″ 43°10″ East. The height of the dam above lowest foundation is 165.2 feet while the length is about 6411 feet.The volume content is 6308 km^3 with gross capacity of 286,000 km^3. The main purpose of dam is to supply water for drinking,domestic purpose and irrigation as well as fishing practices are carried out under the supervision of district fishery development office, Satara. The morphometry of the reservoir is presented in Table 11.119.

Table 11.119. Morphometry of Kanher Reservoir in Satara district

1. Name of reservoir	Kanher
2. Location	Satara
3. Year of construction	1986
4. River	Wenna
5. Type	Earthfill-gravity dam
6. Height of dam	165.2 feet
7. Length of dam	6411 feet
8. Volume	6308 km^3
9. Capacity of dam	271,680 km^3
10. Surface area	18.63 km^2
11. Designed spillway capacity	3203 m^3/sec

Table 11.120. Yearswise seed stocking and fish production in Kanher reservoir

S.No.	Year	Seed Stocking	Fish Production (kg)	Fish Production (kg/ha/yr)
1.	2007-08	5.00	76.50	
2.	2008-09	3.00	12590	
3.	2009-10	5.50	74960	
4.	2010-11	NA	60000	
5.	2011-12	9.56	NA	

Pawar and Pawar (2012) reported the occurrence of 50 fish species belonging to 5 orders from the Kanher reservoir. The fishes belonging to order cyprinifirmes were dominant with 40 species to be followed by fishes of order siluriformes with 5 species, while order perciformes was with 3 species only. The orders like synbranchiformes and anguilliformes were represented by one species each. They concluded that 10.32% fish species were carnivorous 60.43% omnivorous and 38.02% herbivorous.

Pawar and Sonawane (2011) studied water quality of Kanher dam and concluded that the most of the parameters were within normal range which indicates better quality of reservoir water.

KOYANA (SHIVAJISAGAR) RESERVOIR

Koyana is one of the major tributaries of Krishna river system. Both the river and the tributary originate from Mahabaleshwar plateau, the famous hill resort of Maharashtra state, located in western ghats (locally called Sahyadri). They flow in opposite directions at their origin, Krishna to east and Koyana to west; both then turn southwards to meet at Karad, to form the famous confluence 'the preetisangam'. Both get the benefit of high rainfall of 6182 mm/yr at Mahabaleshwar, during south-west monsoon.

The Koyna Dam is one of the largest dams in Maharashtra constructed on Koyna River. It is located in Koyna Nagar of Satara district, nestled in the Western Ghats on the state highway between Chiplun and Karad. The main purpose of dam is to provide hydroelectricity with some irrigation in neighbouring areas. Today the Koyna Hydroelectric Project is the largest completed hydroelectric power plant in India having a total installed capacity of 1,920 MW. Due to its electricity generating potential Koyna river is considered as the 'life line of Maharashtra'.The spillway of the dam is located at the center. It has 6 radial gates. The dam plays a vital role of flood controlling in monsoon season. The catchment area dams the Koyna river and forms the Shivajisagar Lake which is approximately 50 km (31 mi) in length. It is one of the largest civil engineering projects commissioned after Indian independence. The Koyna hydro-electric project is run by the Maharashtra State Electricity Board.

Table 11.121. Morphometric features of Shivajisagar Reservoir

1. Location	Koyananagar
2. Coordinates	17°242 063 N73°452 083 E17°242 063 N 736452 083 E
3. Type of dam	Concrete dam
4. Consruction began	1956
5. Catchment Area	871.78 sq.km
6. Maximum submerged Area	11535 ha.
7. Minimum submerged Area	1435 ha.
8. Maximum depth of water column	78.8 m
9. Minimum depth of water column	30.5 m
10. Nature of submergence	Deep valley
11. Rainfall at catchment area	6182 mm/yr
12. Length of dam	807 m
13. Height of dam	80.35 m
14. Gross storage capacity	2797.4 mcm
15. Dead storage capacity	707.90 mcm
16. Maximum depth of dead storage	30.5 m
17. Number of radial gates	6 tender gates
18. Purpose of dam	Electricity generation
19. Number of generators	12
20. Generation capacity	1920 MW

The Shivajisagar reservoir came into existence with the first impoundment in 1961. The reservoir was constructed at Helwak village in Patan tahsil of Satara district. Shivajisagar dam was constructed on Koyana river in a deep valley, in between two mountain ranges, running parallel to eachother, over a distance of 65 km, right from Koyananagar (Helwak), the actual dam site, to Tapola ad beyond, upstream. By virtue of 6182 mm/yr rainfall in the vast catchment bowl of 871.78 sq.km; huge quantity of 2797.4 mcm of water accumulates in the reservoir, just within a span of two early months of south-west monsoon, July and August.

The former seasonal stretch of the narrow and shallow Koyana river had no sizable fish production prior to the first impoundment of the Shivajisagar, which came into existence, consequent to the construction of dam. At present, about 65.7 mt of fish is harvested every year (Valsangkar, 1993). The per hectare yield in relation to maximum water spread area is 5.69 kg only. Important fish species represented in the catches are depicted in Table 11.122.

Table 11.122. Fish fauna of Shivajisagar reservoir

Sr.No.	Scientific Name	Local name
1.	*Tor khudree* (Sykes)	Khadas
2.	*Tor mussullah* (Sykes)	Mashil
3.	*Puntius sarana* (Hamilton)	Pituli
4.	*Puntius kolus* (Day)	Kolashi
5.	*Puntius dobsoni* (Day)	Panghat
6.	*Puntius amphibious* (Valenciennes)	Khavali
7.	*Puntius sahyadriensis* (Silas)	Phakati
8.	*Rohtee ogilbii* (Sykes)	Gudani
9.	*Osteobrama vigorsii* (Sykes)	Phek
10.	*Labeo calbasu* (Hamilton)	Kanoshi
11.	*Labeo sindensis* (Hamilton)	Tambati
12.	*Labeo rohita* (Hamilton)	Rohu
13.	*Catla catla* (Hamilton)	Catla
14.	*Cirrhinus mrigala* (Hamilton)	Mrigal
15.	*Hypophthalmichthys molitrix* (Valenciennes)	Chandera
16.	*Cyprinus carpio* (Linnaeus)	Cyprinus
17.	*Ompak bimaculatus* (Bloch)	Wonj
18.	*Wallago attu* (Bloch)	Shivada
19.	*Mystus vittatus* (Bloch)	Shengal
20.	*Mystus seenghala* (Sykes)	Shengal
21.	*Channa marulius* (Hamilton)	Maral
22.	*Salmostoma* spp. (Hamilton)	Alkui

The major carp availability in commercial catches continues to be negligible. Ever since 1988, on an average 4.16 lakh of major carp fingerlings are being stocked at Tapola, 45 km upstream in the Shivagisagar reservoir, every year. So far their significance in the fish catches is not noteworthy. One remarkable feature of this reservoir is that 20% of the fish catches from the reservoir consists of mahseer alone. This indicates that mahseer can adopt itself, though on a small scale, to changed habitat, if the essential factors such as running streams, rocks-gravel or lined poole or puddles and undisturbed environment are available. The mahseer is represented by two species *i.e., Tor mussullah* and *Tor khudree.* They are indigenous to the Krishna river system.Mahseer in the Shivajisagar reservoir have established themselves in the lacustrine conditions, and are breeding naturally in the upstream areas, towards shallower regions of the reservoir, when the climatic conditions like low temperature (<24°C), fresh influx of rain waters with high dissolved oxygen and fluviatile nature of the spawning grounds etc are conducive.

The official record of fishing permits issued to the individual members of the four fishermens co-operative societies, for undertaking fishing in Shivajisagar (Koyana) reservoir, indicated that o an average 73 fishermen are actively engaged in fishing in the reservoir.

The fishing period is restricted to 180 days/year during March to August. There is no machinery to collect daily data of fish landings. In the absence of it, estimates are made by taking into consideration the number of authentic fishermen holding fishing permits, the fishing period of 180 days, and average fish catch of 5 kg/person/day. This gives production of 65.7 mt per season.

Readymade nylon gillnets (entailing nets), below 10 cm stretched mesh size and drift nets having mesh size less than 5 cm (stretched) are being used by the local population. About 50 fishing craft made locally, by using inferior wood and technology are plying in Shivajisagar for fishing operations. These country crafts are operated by groups of 10-12 fishermen. They cannot reach distant fishing sites on their own. As such, aluminium boats discarded by the mercantile marine ships have been purchased by the co-operative societies. These powerful and stable motorized boats are employed in towing small country craft, for reaching distant fishing grounds, and returning to the landing sites. Four such motorized boats, serving as motherships, are in operation in the reservoir. Recently, about 15 outboard engines (Yamaha) have been introduced at Tapola, the distal landing centre in the reservoir. This reform has improved the lot of fishermen considerably. They are operating their boats independently and are covering larger distant areas. Consequently the average fish catch per individual has gone to 10 kg/day, in this area (Valsangkar, 1993). Fishermen in Shivajisagar reservoir are non-traditional. They belong to higher caste, scheduled caste ad nomadic tribes. This non-traditional fishermen population has adopted fishing as vocation, equal to the formation of Shivajisagar reservoir. Basically they were agriculturists, and labourers in forest areas. They lack knowledge about fish, fish behaviour, gear and craft. Fishing is done in groups of twelve fishermen.

GIRNA RESERVOIR

Situated near Malegaon town in Nashik district, the Girna dam, erected in 1970 across the rivers Girna and Panzan at latitude 20°-28′N and longitude 74° – 43′E. It irrigates at about 56545.6 ha of land. It is a hydroelectric project. The total capacity of electricity production is 7000 kw. The reservoir has an effective waterspread of 2500 ha at a height of 1318 feet from mean sea level. The reservoir is fed by perennial Girna river and seasonal Panzon river. The reservoir is free from pollutants and weed infestation. The salient features of reservoir are depicted in Table 11.123.

Table 11.123. Salient features of Girna Reservoir of Nasik district

1. Year of construction	1970
2. River	Girna and Panzan
3. Location	Malegaon
4. Rainfall (mm)	1120
5. Length of dam (m)	963.1
6. Maximum height (m)	43
7. Catchment Area (sq.km)	4279
8. Waterspread area (ha)	5420
9. Irrigation potential (ha)	56545
10. Water temperature (°C)	18.5 to 31
11. Transparency (cm)	29 to 56
12. Total dissolved solids (mg/l)	87 to 99
13. pH	7 to 8.5
14. Total Alkalinity (mg/l)	80 to 118
15. Dissolved oxygen (mg/l)	4.3 to 7.5
16. Free carbon-dioxide (mg/l)	Nil to 2
17. Chlorides (mg/l)	10 to 12.3
18. Total hardness (mg/l)	63.8 to 95

(Source : Sakhare and Joshi, 2004)

The different types of gears used in Girna reservoir are Mahajal, Dori, Chaal, Disco and Gill nets. The seed of major carps and common carp were stocked at fry stage. Reservoir was under stocked in the past (Valsangkar, 1987). Khalid and Siddiqui (1990) reported the reservoir fishery potential of Nashik district in Maharashtra with special reference to Girna reservoir fisheries. Girna reservoir harbours a substantial population of catfishes. This could be the reason for having no apparent relationship between stocking and production of fish in Girna reservoir (Khalid and Siddiqui, 1990). A fisherman earns an average income of

Rs. 1000 to 1500 per month. Construction of dam attracted many fishermen families to migrate from nearby districts and towns to reservoir site.

The commercially important species in the reservoir are *Cyprinus carpio, Catla Catla, Labeo rohita, Labeo calbasu, Mystus* spp., *O.bimaculatus, M. arnatus, Wallago attu, Channa* spp., and *Puntius* spp.

Consturction of dam attracted may fishermen families to migrate from nearby districts and towns to reservoir site.There are nearly 35 bheel (tribal) families, totally devoted to fishing.

MAMDAPUR RESERVOIR

The reservoir came into existence with the impounding of Nilkanth Nalla in Barshi taluka of Solapur district of Maharashtra. The morphometric features are depicted in Table 11.124.

Table 11.124. Morphometric features of Mamdapur reservoir in Solapur district

1. River	Nilkanth Nalla
2. Location	Barshi
3. Year of Construction	1983
4. Rainfall (mm)	592
5. Length of dam (m)	10.20
6. Height of dam (m)	14.50
7. Water spread area (ha)	100.40
8. Irrigation potential (ha)	326

Sakhare (2012) studied the water quality of Mamdapur reservoir in relation to fisheries (Table 11.125). The minimum water temperature was recorded in November 2010 (23°C) and the maximum in April 2011 (37°C).

Table 11.125. Water quality of Mamdapur reservoir, Maharashtra

Parameters	Range
1. Water temperature (°C)	23 – 37
2. pH	7.8 – 9.2
3. Transparency (cm)	100 – 145
4. Total dissolved solids (mg/l)	90 – 310
5. Chlorides (mg/l)	42 – 59.17
6. Total hardness (mg/)	100 – 197
7. Dissolved oxygen (mg/l)	5.2 – 8.6
8. Total alkalinity (mg/l)	193 – 289
9. Free carbon-dioxide (mg/l)	Nil – 2.5

Transparency is a physical variable significant to production. During present study secchi disc transparency ranged from 100 to 145 cm. The minimum transparency values were recorded in July 2010 while the maximum in April 2011. The low values recorded during rainy season may be due to the heavy rains and winds of high velocity. Similarly high values during winter and summer period can be attributed to low moderate wind velocity.

The pH is affected not only by levels of carbon dioxide but also by other organic and inorganic components of water. Further, any alteration in water pH is accompanied by changes in other physico-chemical parameters. In the present study pH ranged from 7.8 to 9.2. The minimum values were recorded in May 2011 while the maximum in January 2011. In case of Mamdapur reservoir the average pH value during the study period was observed to be 8.5 which indicated that the reservoir is a medium productive.

The total dissolved solids were high in summer followed by winter and monsoon. In lotic sector it was maximum in December and minimum in July.

The dissolved oxygen is of great importance to all living organisms. It dissolves in water due to direct diffusion from air and/or photosynthetic activity of autotrophs. The diffusion of oxygen is dependent on temperature of water, its salinity, total dissolved solids, and movements of water etc. During the study period, the dissolved oxygen ranged from 5.2 to 8.6 mg/l. The high dissolved oxygen was found during monsoon, while lower values were recorded in summer.

Free carbon-dioxide is an extremely necessary constituent in an aquatic environment. Free carbon dioxide in water occurs due to respiration of aquatic biota, decomposition of organic matters and also due to infiltration through the soil. Carbon dioxide is an important component of the buffer system, which influences carbonate and bicarbonate concentrations in water. Free carbon dioxide generally varied between nil and 2.5 mg/l. It was not detected in January 2011 and April 2011.

Alkalinity of any water is mainly due to carbonates, bicarbonates and hydroxide. It is an index of nutrient status in a water body. The total alkalinity in Mamdapur reservoir ranged from 193 to 289 mg/l. The availability of carbon dioxide for primary production is related to alkalinity.

The term hardness is frequently used as the assessment of the water quality. This is governed by the concentration of calcium and magnesium salts, which combined largely with bicarbonate, carbonate, sulphates, chlorides and others anions of mineral acids. Desirable level of total hardness for fish culture generally fall within the range of 20 to 300 mg/l. Based on the average hardness level, Mamdapur reservoir can be considered suitable for fisheries. The maximum total hardness (197 mg/l) was found in May 2011, its minima (100 mg/l) was recorded in October 2010. The higher values of total hardness were recorded during summer and lower values were recorded in the winter season.

HARNI (KATGAON) RESERVOIR

Harni (Katgaon) reservoir is situated near Katgaon village in Tuijapur tahsil of Osmanabad district and is connected by an approach road of the length of 6.5

km from Solapur-Hyderabad-National Highway No. 9. This approach road takes off near the Khanapur stage on the National Highway. The project is constructed across Harni river, a tributary of river Bori, having catchment area of 190.624 km^2 (73.60 sq. miles) at the project site. The head works of the project were started in June 1961 and completed in June 1964. This project envisages construction of an earthen dam and includes other works such as ogee spillway with its appurtenant works, head regulator on right flank and canals on left and right flanks taking off from the same head sluice.

Table 11.126. The salient features of Harni (Katgaon) reservoir

1. Name of River	Harni
2. Location	Katgaon village
3. Type	Earthfill
4. Sub basin number	19
5. Water Spread Area (ha)	239
6. Year of Construction	1964
7. Height of dam (m)	16.55
8. Length of dam (m)	3059
9. Length of ogee spillway	276.025 meters
10. Maximum flood discharge	58,490 cusecs
11. Reservoir area (10^3m^2)	317
12. Effective storage capacity (10^3m^3)	11180
13. Gross storage capacity (10^3m^3)	13580
14. Live storage capacity (Mcum)	11.170
15. Carry Over capacity (Mcum)	1.410
16. Catchment Area (km^2)	190.624
17. Designed spillway capacity (m^3/sec)	1647
18. Length of irrigation channel Left flank canal (km) Right flank canal (km)	 27.355 14.481
19. Irrigation potential (ha)	1,660
20. Purpose	Irrigation

Fig. 11.41. Harni (Katgaon) reservoir of Osmanabad district

Fig. 11.42. A view of Harni (Katgaon) reservoir from embankment

No village was submerged under this reservoir. The villages such as Katgaon, Khanapur and Darshnal of Tuljapur and Tandulwadi and Musti of Solapur taluka receive benefit due to the irrigation facilities under this reservoir. The salient features of Harni (Katgaon) reservoir are depicted in Table 11.92.

The climate of reservoir area is primarily dry. The rainy season starts from mid-June and remains till end of September. With onset of the rainy season there is an appreciable drop in temperature. From October to November the climate is humid. From mid-November to January it is winter. From February to March the climate is dry and from April to June it is summer. May is the hottest month. The average rainfall of area is 730 mm. Though the area is thought to belong to assured rainfall zone the area has experienced moderate, severe and acute drought conditions for more than 30% of the last 90 years. Hence the entire area comes under the category of drought area. The relative humidity during the afternoons being 20%. During the southeast monsoon season, the humidity is as high as 80% in the mornings and 60% in the afternoons.

Fishermen's Co-operative Society: The fishing rights of the reservoir are given on lease to the co-operative society named 'The Jay Ambika Matsya Vyvsaik Sahakari Sanstha Limited'. The society was established on 22/03/1978 under Maharashtra Co-operative Act with registration number of OSM/RSR-552 Date-22/03/1978. The initial capital was Rs.480/- only.

At the time of registration there were a total of 49 members. The number of member has increased by about 2 times during the last 23 years. In the year 1987-88, the total number of members were 49. In 2001-02 it increased to 51, further and in subsequent years finally the total number of members become 57 till January 2010. The members of the co-operative society belong to different castes and religions. These are the non-traditional fishermen like muslim (Islam), dhangar (bandi) and lingayat.

Socio-Economic Study: Jetithor (2011) studied socio-economic status of fishermen. All 57 members are not full time fishermen. All of them are part time fishermen. The study area is socio-economically backward. It is observed that 62.07% of respondents were illiterate. General opinion of the communities is that education does not really help, expect, perhaps in getting a government job. The members of the community feel that there is need for an educational system that would import those enabling skills in their chosen profession and in areas such as management and accounting.

The socio-economic study was made in different villages at the periphery of the reservoir. A study has been made on the existing co-operative society depending on the reservoir and their constraints with a view to identifying the probable solution to augment the fish production from the reservoir in a sustainable manner. Most of the resources from which the society members earn their livelihood have been studied. The earnings of a livelihood by fishers in these areas is a grave challenge to be met.

The per capita annual income from fisheries range from Rs. 9000 to 37000/ .50% of the fishermen reported an income in the range of 10,000 to 17,000 per annum. 27% fishermen get income from fisheries in the range of 18,000 to 22,000/

-, and only 23% of the members get annual income in the range of 23000 to 37000/-. This range of variation in earnings indicates unreliability of fishing as a dependable occupation.

Agricultural labour and general labour are the two other occupations that these fishermen take for the purpose of gainful employment. Both these activities provide more working days and earnings than does the fisheries. It appears that fishing is only a minor activity compared to other economic actitivites.Women seem to have more working days in agriculture and other areas, than do men. From the point of view of the sources of income, fisheries in Harni (Katgaon) reservoir provide only 24% of the family income. General and agriculture labour provide 76% of the family income. It should be noted that most people, both men and women, shift from one type of job to another based on season. Both agriculture and fisheries provide work during the season. When the season is over, people shift to general labour, wherever available.

All the respondents were members of the co-operative society and the study shows that they were also active in the banking transactions. Marriage, house construction, acquisition of milch animals, cycles and fishing equipments constitute the major purposes for which bank loans are availed. It can also be seen from this study that 8.75 of the respondents possessed *Pucca* house and some of them possessed bicycle, black and white television. Radio, cow and land. The study revealed that only 60% fishermen have their own nets. 62.5% inhabitants do not have ownership rights on a agricultural land through which they can earn a living. The socio-economic status of fishermen is presented in Table 3.1.

From the above socio-economic analysis, it can be seen that the fishermen are rightly termed as the poorest among the inhabitants around the reservoir.

Lease Amount: The fishing activities in Harni (Katgaon) reservoir were undertaken immediately after its formation in year 1964. According to the lease policy introduced by the Maharashtra government; the fisheries department leases the fishing rights of the Harni (Katgaon) reservoir to the co-operative society for fishing on contract basis every year. During 2007-08, the lease amount was Rs.24000/- on which the society was allowed to exploit fish from the Harni (Katgaon) reservoir. During 2008-09, the lease was Rs. 24800/-.

Seed Stocking: The society generally purchased fish seed from the Chandni fish hatchery, which is a government hatchery constructed near Paranda (in Osmanabad district), about 70 kms from the reservoir site. Details of fish seed stocked in the reservoir and stocking rate/ha for two years is shown in Table 11.127.

Table 11.127. Details of fish seed stocking in Harni (Katgaon) reservoir

Year	Catla	Rohu	Mrigal	Silver carp	Grass carp	Common carp	Total /ha	Stocking
2007-08	200,000	100,000	100,000	75000	100,000	100,000	6,75,000	2824
2008-09	60,000	40,000	30,000	—	10,000	20,000	1,60,000	669

During year 2007-08, the society purchased 6,75,000 fish seed, which gave a per hectare stocking density of 2824 fingerlings. In second year (2008-09) society could purchase 1,60,000 seed, which gave per he stocking density of 669 fingerlings. But recently during year 2009-10 the reservoir was not stocked with seed. Usually the seed was purchased in the form of fry, and it was grown up to the fingerling stage in ponds constructed in the vicinity of reservoir. The fingerling stage is then stocked in reservoir.

Seed of major carps (*Catla catla,Labeo rohita,Cirrhinus mrigala*) and *Hypophthalmichthys molitrix, Ctenopharyngodon idella* and *Cyprinus carpio* were stocked in the reservoir. Silver carp was not stocked in year 2008-09. It was due to the non-availability of seed. Among the varieties, catla seed was stocked in large numbers when compared to the seed of others. The details of seed stocking are depicted in Table 11.124 and Fig. 11.43. The stocking rate/ha ranged from 669 (year 2008-09) to 2824 (year 2007-08), with an average of 1747 fingerlings/ ha/yr.

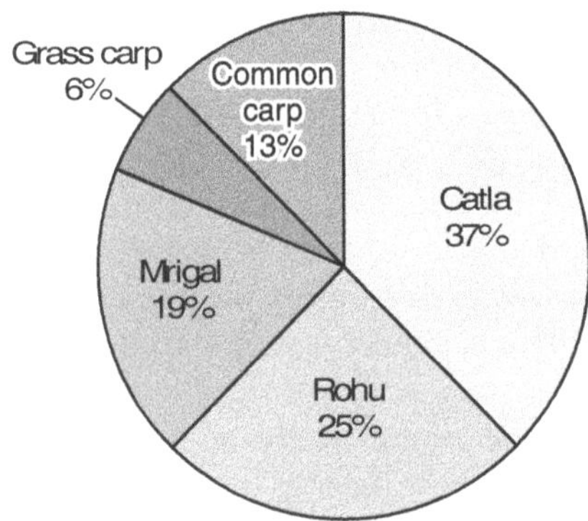

Fig. 11.43. Seed composition during year 2008-09

Auto stocking was observed in Harni (Katgaon) reservoir. Carps have been observed to breed due to the availability of certain congenital factors. These factors vary from season to season depending upon the on-set of monsoon, quantum of rainfall, volume of water entering the reservoir etc. In reservoir Common carp (*Cyprinus carpio*) breed three times in a single year.

Fish Fauna: A total of 37 fish species belonging to 13 families have been recorded from the reservoir (Table 11.128). Family cyprinidae contributed the largest number of species (17), followed by family bagridae and channidae (3 each). Family notopteridae, siluridae, mastaembelidae and ambassidae represented by two species each, while family cobitidae, claridae, heteropneustidae, mugilidae, anabantidae and gobiidae were represented by one species each.

Table 11.128. Ichthyofauna of Harni (Katgaon) reservoir in Osmanabad district

Class : Pisces
Subclass : Teleostomi
Order : Clupeiformes
Suborder : Notopteroidei
Family : Notopteridae

1. *Notopterus notopterus* (Pallas)
2. *Notopterus chitala* (Ham.)

Order : Cypriniformes
Suborder : Cyprinoidei
Family : Cyprinidae

3. *Chela phulo* (Ham.)
4. *Catla catla* (Ham.)
5. *Labeo rohita* (Ham.)
6. *Labeo boggut* (Sykes)
7. *Labeo fimbriatus* (Bloch)
8. *Cirrhinus mrigala* (Ham.)
9. *Cyprinus carpio* (Linn.)
10. *Hypophthalmichthys molitrix* (Sykes)
11. *Ctenopharyngodon idella* (Ham.)
12. *Amblypharyngodon mola* (Ham.)
13. *Discognathus modestus* (Ham.)
14. *Osteobrama cotio* (Ham.)
15. *Puntius sarana sarana* (Ham.)
16. *Puntius ticto ticto* (Ham.)
17. *Puntius chola* (Ham.)
18. *Puntius sophore* (Ham.)
19. *Rasbora daniconius* (Ham.)

Family : Cobitidae

20. *Nemacheilus botia* (Ham.)

Order : Siluriformes
Family : Bagridae

21. *Mystus cavasius* (Ham.)
22. *Mystus seenghala* (Sykes)
23. *Mystus vittatus* (Bloch)

Family : Claridae

24. *Clarias batrachus* (Linn.)

Family : Heterpneustidae

25. *Heteropnesutes fossilis* (Bloch)

Family : Siluridae

26. *Wallago attu* (Bloch and Schneider)
27. *Ompak bimaculatus* (Bloch)

Order : Mugiliformes **Family : Mugilidae** 28. *Mugil cephalus* (Linn.)
Order : Channiformes **Family : Channidae** 29. *Channa gachua* (Ham.) 30. *Channa striatus* (Bloch) 31. *Channa marulius* (Ham.)

Fish Catch: The fishery is exploited by fishermen who have been organized from village Katgaon into a co-operative society. The details of fish catch from the reservoir for the period of 2007-08 to 2008-09 is shown in Table 11.126. The maximum fish catch was recorded during year 2007-08. It gave an average fish production of 83.68 kg/ha/yr. The minimum production was during 2008-09 at 9600 kg, giving an average fish production of 40.16 kg/ha/yr. During the study period the catch of carps showed the higher catches than those of local fishes (Table 11.129).

Table 11.129. Total fish catch and per hectare fish prduction

Year	Local fishes (kg)	Carps	Total catch	Fish production /Ha
2007-08	5000	15,000	20,000	83.68
2008-09	2400	7,200	9600	40.16

The stocked landings dominated over the unstocked species. Important fishery comprised fishes such as *Catla catla, Labeo rohita, Cirrhinus mrigala, Hypophthalmichthys molitrix, Cyprinus carpio,* and *Channa* spp. A rare appearance of silver carp in the landings indicates that fish is not suitable in Harni (Katgaon) reservoir.

Fishing Days and Closed Season: In Harni (Katgaon) reservoir fishing is carried out throughout the year. No closed season has been declared by the district fisheries department. However, on the suggestions of district fisheries department, the society has framed some rules for the conservation of fish. For example, catching of fish belonging to Indian major carps less than 1 kg has been prohibited. Such types of fish if caught, it should be immediately released in the reservoir. This rule is not followed by fishermen members.

Overexploitation and Improper Fishing System: The over exploitation of the reservoir plays a major role in extinction of the species. The co-operative society deliberately issues the license to any number of fishermen. Many fishermen from Solapur district migrate to this area during rainy season and catch huge quantity of fish. This has adversely affected the livelihood of permanent local fishermen. Over exploitation of the reservoir for fishery has resulted in excessive mortality and reduction in effective population size of the fish. Monsoon is the breeding season for most of the fishes and the fishing activity is at peak during this season.

Conservation of Breeding Grounds: There is an urgent need to breed some important fish species in reservoir for maintaining the stock of fish species. Side by side the breeding grounds should be restored scientifically. Otherwise, the productivity will not sustain for a long time; and after a period ranging from one to several years, the fish yield will definitely decline to an alarming level. Several autobreeding grounds had been found to be lost because of indiscriminate fish catching by fishermen, loss of eggs, spawns etc. of different fish species due to the use of mosquito net by fishermen of the reservoir. District Fisheries Department has taken various steps for building up awareness among the fishermen against the use of mosquito nets conservation of brooders etc. through publicity and awareness camps. Besdies, several training programmes were also organized by fisheries department for capacity building among the fishers depending on the reservoir.

Restriction to Net Usage: In Harni (Katgaon) reservoir fishing is carried out throughout the year. No 'closed season' has been declared by the district fisheries department or the co-operative society. Fishing activity is at its peak during the monsoon season. A majority of the total fish catch is during the monsoon season. Since monsoon is the breeding season for most of the fishes, it is advisable to ban fishing of native fishes. This can be achieved by restricting the net sizes being used by the fishermen. Only large sized gill nets should be allowed during monsoon, which are useful to catch introduced fishes. Even after the monsoon season, the present restriction on minimum net size should be strictly followed so as to catch only mature fishes.

Permanently Stopping Migrating Fishermen: Fishing licenses should be issued to permanent fishermen residing near the reservoir and moderate reservoir who are solely dependent on the reservoir for their livelihood. The present status of fisheries in the reservoir and moderate reservoir productivity indicates the depletion of fish resource. In order to reduce the fishing pressure, it is advisable to avoid migratory fishermen from fishing. Since their fishing period is monsoon, large quantities of brooders are destroyed by overfishing.

Fishing Gears and Crafts: The major fishing gear in reservoir is surface gillnet. The net is made of nylon twine. Presently, the plastic twine (mono-filament) nets are also being used in the reservoir. Dragnets are also used especially for catching catfishes and weed fishes. The other fishing gears such as cast nets, long lines and traps are also operated but the catches are insignificant. The presence of underwater obstacles restricts the use of active gear in reservoir and the choice is often limited to passive gear such as gillnets.

The thermocoel platform of 6 × 3.5 feet size is used as fishing craft. It is covered by a plastic covering. This craft is not suitable to carry large nets which are heavy. The wooden boats are operated only during monsoon season. These boats are taken on lease from fishermen of nearby reservoir. Coracle, a saucer shaped country craft, is also used in Harni (Katgaon) reservoir. It is made of a split bamboo frame, covered with buffalo skin. Apart from being simple and inexpensive, coracle is durable. It is also a versatile craft used for laying and lifting of nets, besides navigation and transport of fish and other material. The members of cooperative society purchase the craft individually without any financial assistance from co-operative society or fisheries department.

Temperature (°C): Seasonal variation in air and water temperature are presented in Tables 11.130, 11.131, 11.132 and 11.133. During 2007-08, in the summer the mean air temperature values recorded were 36.05, 37.05 and 37.2°C at stations I, II and III respectively. While its minimum value in summer was 32.2°C at station II in May 2007, maximum was 41°C at station III in April 2008. The corresponding mean values in 2008-09 were 36.3, 37.3 and 34.75°C at stations I, II and III respectively. Its minimum (35°C) was recorded at station II and III in February and maximum value (40°C) was noticed at station I in April 2009.

Table 11.130. Monthly values of Air temperature (°C) (Year 2007-08)

Stn	Oct	Nov	Dec	Jan	Feb	Mar	Apr	May	Jun	July	Aug	Sept
I	27	27	30	31	34	39	37	35.4	32	34	29	30.2
II	23	27.4	28	33	37	40.2	40	32.2	33.2	30	30.5	33.3
III	25.5	32	30	31.5	35	36	41	36.8	30	33	34	29.8

Table 11.131: Monthly values of Air temperature (°C) (Year 2008-09)

Stn	Oct	Nov	Dec	Jan	Feb	Mar	Apr	May	June	July	Aug	Sept
I	26.2	27	25	29.4	29	36.9	40	39.4	35.6	34.2	29.4	29.2
II	25	27.8	29.2	30	35	38.8	39	35.5	33.8	29.9	30.2	29.4
III	27	29.2	26.4	30.2	36	37.5	39.4	38.8	32	30.8	28.7	28.6

In monsoon the average air temperature values were 31.3, 31.5 and 31.7°C at stations I, II and III respectively. During this period its minimum (30°C) was noticed in June at station III. The corresponding average values in 2008-09 were 31.3, 31.5 and 31.7°C at stations I, II and II respectively. While its minimum (28.6°C) was seen in the September at station III, maximum (35.6°C) was recorded at station II in June.

In winter the average air temperature was 27.5, 27.8 and 29.7°C in II, III sampling stations and I respectively. While its minimum (23°C) was recorded at station II in October, maximum (33°C) was found at station II. Its corresponding values during 2008-09 were 27.5, 27.8 and 29.4°C at stations I, II and III respectively. Its maximum value recorded was 30.2°C in January at station III, while minimum value was 25°C at station II in October 2008.

During the study period water temperature ranged between 29°C – 32°C in 2007-08 in summer, minimum temperature (29°C) was recorded in March at II and maximum temperature (32°C) at I in May. The corresponding values in 2008-09 was noticed in the range of 28 – 32°C. Its minimum value (28°C) was recorded in March at station III, while its maximum (32°C) was recorded at stations III and II in April and May.

Table 11.132. Monthly values of Water temperature (°C) (Year 2007-08)

Stn	Oct	Nov	Dec	Jan	Feb	Mar	Apr	May	Jun	July	Aug	Sept
I	17	18.2	19	20	31	29	30	32	28	23	23	26
II	17	18.1	20.2	21	30.8	29	32	31	29	22	24	25
III	18	20	18	20	31	31.8	32	32	29.5	24	23	26

Table 11.133. Monthly values of Water temperature (°C) (Year 2008-09)

Stn	Oct	Nov	Dec	Jan	Feb	Mar	Apr	May	Jun	July	Aug	Sept
I	17	17	18	19.5	30	29	30	31.8	32	23	24	25
II	17.5	18	19	18	29.1	29	31	32	30	24	24.5	25.6
III	18	17.5	19	18	29.8	28	32	31	31	23	24	26

In monsoon the water temperature ranged between 22 to 35°C. The mean water temperature values were 25°C, 25°C and 25.6°C at stations I, II and III respectively. While its minimum (22°C) was found at station II in July, maximum value (29.5°C) was recorded at station III in June 2008. The corresponding mean values recorded during the monsoon 2008-09 were 26, 26.2 and 26°C at stations I, II and III respectively. During this period temperature ranged between 23°C to 32°C. Its maximum value (32°C) was recorded in June at station I, June, while minimum value (23°C) at station I&III in August.

During 2007-08 winter the mean water temperature recorded was 18.55, 19.7 and 19°C in stations I, II and III respectively. The water temperature ranged between 17°C to 21°C. While minimum (17°C) was noticed in October at station I&II, maximum temperature (21°C) was found at station II in January. During the second year (2008-09), the mean water temperature found was 17.8, 18.1 and 18.1°C at stations I, II and III respectively. While its minimum (17°C) was found at I in October and November, maximum value (19°C) was recorded in month of December at station II and III.

pH (Potentia Hydrogeni): Seasonal variations in pH at different spots in Harni (Katgaon) reservoir are presented in Tables 11.134 and 11.135. In first year summer 2008, the mean pH values of 8.1, 8.2 and 8.2 were recorded at stations I, II and III respectively. The corresponding mean summer values in second year were 8.2, 8.1 and 8.1. During first year of investigation, the minimum pH of 8 was recorded at station I in April and 7.9 was at station II in April 2009, while the maximum recorded was 8.4 at station II in February 2008 and 8.4 at station I&II in March and 8.4 at station III in April 2009.

Table 11.134. Monthly values of pH (Year 2007-08)

Stn	Oct	Nov	Dec	Jan	Feb	Mar	apr	May	Jun	July	Aug	Sept
I	8.0	8.1	7.7	7.5	8.1	8.4	8.0	8.0	6.8	7.1	7.4	7.2
II	7.6	7.9	7.4	7.3	8.4	8.4	8.0	8.2	7.3	6.9	7.5	7.2
III	7.7	7.3	8.1	7.2	8.2	7.9	8.4	8.3	6.5	7.0	6.9	7.1

Table 11.135. Monthly values of pH (Year 2008-09)

Stn	Oct	Nov	Dec	Jan	Feb	Mar	Apr	May	Jun	July	Aug	Sept
I	8.0	8.4	7.5	7.6	8.2	8.5	8.0	8.1	6.9	7.0	7.3	7.1
II	7.8	7.7	7.2	7.5	8.3	8.1	7.9	8.2	7.1	6.8	7.2	6.9
III	7.6	7.2	7.3	8.0	7.9	8.4	8.1	8.0	7.1	7.5	7.1	7.0

In first year summer 2008, the mean pH values of 8.1, 8.2 and 8.2 were recorded at stations I, II and III respectively. The corresponding mean summer values in second year were 8.2, 8.1 and 8.1. During first year of investigation, the minimum pH of 8 was recorded at station I in April and 7.9 was at station II in April 2009, while the maximum recorded was 8.4 at station II in February 2008 and 8.4 at station I and II in March and 8.4 at station III in April 2009.

In monsoon 2008, the mean pH values of 7.1, 7.1. and 6.8 were recorded at stations I, II and III respectively. In monsoon 2009 they were 7.0, 7.0 and 7.1 at I, II and III respectively. During this season a maximum pH of 7.5 was found at stations II in August 2008. While the minimum pH of 6.5 was recorded at station II in June 2008 and 6.9 at station I in June 2009.

In winter the mean pH values at three spots was 7.8, 7.5 and 7.5 in 2007-08 and corresponding values in 2008-09 were 7.8, 7.5, and 7.5. During this period the maximum of 8.4 at station I in 2008, while a maximum value of 8.1 was recorded at station I and station III in in November and December 2007.

Transparency (cms): The transparency values at three sampling stations of the Harni (Katgaon) reservoir for the period of two years are presented in Tables 11.136 and 11.137.

The mean transparency values during summer 2008 were 106.2, 108.5 and 108.5 in stations I, II and III respectively. While its minimum value (90) was noticed at station I in February, maximum value (127) was recorded in May at station II and III. The corresponding mean values in 2009 were 106, 107.5 and 107.2 at stations I, II and III respectively, while its minimum (91) was recorded at station I in February and maximum (124) was noticed at station II in May 2009.

Table 11.136. Monthly values of Transparency (cms) (Year 2007-08)

Stn	Oct	Nov	Dec	Jan	Feb	Mar	Apr	May	Jun	July	Aug	Sept
I	82	85	90	97	91	97	114	122	66	50	54	52
II	83	84	88	98	92	98	116	124	69	48	52	50
III	82	84	85	100	92	96	118	123	67	49	53	51

In monsoon of year 2008, the mean transparency values recorded at I, II and III were 54.25, 54.55 and 54. In 2009 the corresponding values were 55.5, 54.7 and 55. In year 2008, minimum value (39) was recorded at station III in September and its maximum (66) was found at station II in June. During the year 2009 the minimum value (48) was noticed at station II in July, while maximum value (69) was noticed at station II in June.

Table 11.137. Monthly values of transparency (cms) (Year 2008-09)

Stn	Oct	Nov	Dec	Jan	Feb	Mar	Apr	May	Jun	July	Aug	Sept
I	82	83	89	90	90	115	115	125	64	50	53	50
II	82	84	85	100	92	117	117	127	66	49	52	51
III	80	75	88	100	91	118	118	127	65	59	53	39

During year 2008 winter season, the mean transparency values at three stations of study were 87.25, 87.75 and 88.5, while in second year, the corresponding values were 88.5, 88.2 and 87.7 at stations I , II and III respectively. During first year winter the maximum (100) was recorded at stations II and III in January 2008. During second year winter, the minimum of 82 was recorded at station I and II in October and maximum of 100 in January 2009 at station III was recorded.

Total Dissolved Solids (mg/l): The values of total dissolved solids (TDS) expressed in mg/l for the two years period are presented in Table 11.138 and 11.139.

During year 2008, the mean seasonal value at the three stations were 214.5, 214 and 214 in summer, 103.7, 100.2 and 103.5 in monsoon and 114.5, 114 and 114.5 in winter. The corresponding values during 2009 were 217, 217.5 and 213 in summer, 108.5, 108.9 and 108.5 in monsoon, and 122.5, 122 and 122.5 in winter.

Table 11.138. Monthly values of Total Dissolved Solids (mg/l) (Year 2007-08)

Stn	Oct	Nov	Dec	Jan	Feb	Mar	Apr	May	Jun	July	Aug	Sept
I	110	122	130	128	190	260	198	220	103	117	111	103
II	115	120	123	130	180	292	198	200	104	118	114	99.7
III	114	119	127	130	199	198	222	235	106	110	116	102

Table 11.139: Monthly values of Total Dissolved Solids (mg/l) (Year 2008-09)

Stn	Oct	Nov	Dec	Jan	Feb	Mar	Apr	May	Jun	July	Aug	Sept
I	105	116	115	128	195	280	198	185	98	103	106	108
II	100	115	120	121	180	296	190	190	95	105	112	89
III	98	118	122	120	192	280	196	188	110	198	96	90

The seasonal minimum values in the first year summer were 180 at station II in February and in monsoon it was 90 at station III in September and in winter it was 98 at station III in October. In the second year of the investigation, a minimum of 180 at station II in February (summer), 102 at station III in September (monsoon) and 114 at station III in October (winter) were recorded.

The seasonal maximum values in the first year of studies were 296 at station II in March (summer), 198 at station III in July (monsoon), 128 at I in January (winter). In the second year the maximum values of 292 at station II in March (summer), 118 at station II in July (monsoon) and 130 at station II and III in January (winter) were recorded.

The observation on TDS for two years (2007-08 and 2008-09) clearly indicate that the total dissolved solids values were high in summers followed by winter and monsoon.

Dissolved Oxygen (mg/l): Dissolved oxygen contents of water samples from three stations along the Harni (Katgaon) reservoir are computed in Tables 11.140 and 11.141.

During the first year (2007-08) of the study the dissolved oxygen ranged between 4.1 to 9.8 mg/l and seasonal dissolved oxygen values at the three stations (I, II and III) were 4.5, 4.5 and 4.4 (summer), 9.1, 9.3 and 9.2 (monsoon) and 7.9, 7.9 and 7.9 (winter) respectively. In summer the minimum value (4.1) was observed at station III in April and maximum of (4.9) was found at station II in March.In winter minimum (6) was found at station II in January ,while maximum (9.2) was found at station II in October.

Table 11.140. Monthly values of Dissolved oxygen (mg/l) (Year 2007-08)

Stn	Oct	Nov	Dec	Jan	Feb	Mar	Apr	May	Jun	July	Aug	Sept
I	9.1	8.8	8.1	5.8	4.3	4.8	4.5	4.4	8.3	9.1	9.3	9.8
II	9.2	8.3	8.3	6.0	4.4	4.9	4.3	4.6	8.4	9.4	9.6	9.8
III	9.1	8.2	8.1	6.3	4.2	4.7	4.1	4.7	8.5	9.3	9.4	9.7

Table 11.141. Monthly values of Dissolved oxygen (mg/l) (Year 2008-09)

Stn	Oct	Nov	Dec	Jan	Feb	Mar	Apr	May	Jun	July	Aug	Sept
I	9.7	8.7	8.4	6.9	6.1	5.7	4.9	4.4	8.2	9.2	9.5	9.8
II	9.3	8.4	8.3	6.8	6.2	4.8	4.6	4.2	8.5	9.3	9.4	9.7
III	9.4	8.6	8.6	6.4	6.3	5.1	4.2	4.3	8.3	9.2	9.3	9.6

During the second year (2008-09) of the study, the dissolved oxygen ranged from 4.2 to 9.8. In summer the seasonal mean values of 5.2, 4.9 and 4.9 were found at stations I, II and III respectively, while its minimum (4.2) was found at station II in May. In monsoon the mean values were found to be 9.1, 9.2 and 9.1 at three sampling stations. The minimum (8.2) value was found at station I in June, while maximum value (9.8) was found at station I in September. In winter the mean dissolved oxygen values were 8.4, 8.2 and 8.4 at stations I , II and III respectively. While its minimum (6.4) was recorded at station III in January, maximum (9.7) was recorded at station I in October.

Phenolphthalein Alkalinity (mg/l): Monthly values of phenolphthalein alkalinity are presented in Tables 11.142 and 11.143. The seasonal mean values of phenolphthalein alkalinity during 2007-2008 summer were 26.8,27 and 27.1 mg/l at stations I, II and III respectively. Its minimum (25.2) was found at station II in February, while its maximum (27.9) was found at station II in May. The corresponding mean values in summer of second year (2008-09) were 24.8, 25.6

and 25 at stations I, II and III respectively. While its minimum value (23) was found in February at station I in February 2009.

Table 11.142. Monthly values of Phenolphthalein Alkalinity (mg/l) (Year 2007-08)

Stn	Oct	Nov	Dec	Jan	Feb	Mar	Apr	May	Jun	July	Aug	Sept
I	30.9	31.1	31.2	31.8	25.4	27	27.5	27.4	27.9	28	29.2	30
II	30.4	30.9	30.9	31.4	25.2	27.4	27.8	27.9	28	28.4	29.5	30.4
III	30.8	30.4	31	30	26	27.8	27.1	27.6	28.1	28.6	29.9	30.6

Table 11.143. Monthly values of Phenolphthalein Alkalinity (mg/l) (Year 2008-09)

Stn	Oct	Nov	Dec	Jan	Feb	Mar	Apr	May	Jun	July	Aug	sept
I	30.8	28.7	31.6	31	23	24	25.6	26.8	27.9	28.1	29	30
II	31.8	31.4	31.4	23.5	23.5	24.8	25.9	28.4	27.6	28.6	29.4	30.1
III	31.6	31.2	31	30	23.2	24.9	25.2	27	28	28.9	29.5	30.2

During monsoon (2007-08), the mean values in three stations were 28.7,29 and 29.3.While its minimum (27.9) was found at station I in June, its maximum (30.6)was recorded in September at station III. The corresponding mean values in 2008-09 was 28.75, 28.9 and 29.1 at stations I, II and III. The minimum (27.6) was found in June at station II, while its maximum (30.2) was recorded at III in September.

During winter (2007-08) the mean values were 31.2, 30.9 and 30.5 at station I, II and III respectively. While its minimum (30) was found at III in January, and maximum (31.8) was noticed at I in January. The corresponding mean values in the second year (2008-09) of the present study were 30.5, 29.5 and 30.9 mg/l at station I, II and III. While its minimum value (23.5) was noticed during January at station II and maximum value (31.8) in October at station I.

Total Alkalinity (mg/l): Monthly values of total alkalinity are presented in Tables 11.144 and 11.145. During the year 2007-08, the station-wise mean values of the total alkalinity in summer were 214.7, 215.8 and 214 mg/l at station I, II and III respectively. Its minimum (189) at III in May and maximum (248.2) were found at station III in February. In monsoon, the mean values were 142.5,142.2 and 144.2 mg/l. The minimum (112) was observed in June at station II and maximum (187) was found at station II in September. In winter the mean values were 272.02,272.8 and 273.3, minimum (261) at station I in October and maximum (278) were found at station II and I in December and January respectively.

Table 11.144. Monthly values of Total Alkalinity (mg/l) (Year 2007-08)

Stn	Oct	Nov	Dec	Jan	Feb	Mar	Apr	May	Jun	July	Aug	Sept
I	261	272.1	277	278	246	220	204	189	116	132	140	182
II	265	172.2	278	276	248.2	215	203	197	112	128	142	187
III	264	276.2	276	277	245	223	199	189	118	130	148	185

Table 11.145. Monthly values of Total Alkalinity (mg/l) (Year 2008-09)

Stn	Oct	Nov	Dec	Jan	Feb	Mar	Apr	May	Jun	July	Aug	Sept
I	261	265.1	270	275.1	251	225	209	194	116	130	140	188
II	260.1	267.5	271.8	276.3	252.2	220	208	192	117	123	148	189
III	266.2	265.2	273	271.4	250	228	204	194	119	129	146	185

During the year 2008-09, the mean values of total alkalinity were 219.7, 219.05 and 219 in summer, 143, 144.2 and 144.7 in monsoon, and 267.8, 268.6 and 268.9 in winter at stations I, II and III respectively. In summer, the minimum value (192) was noticed at station II in May, while the maximum value (252.2) was found at station II in February. In monsoon minimum value (123 was found at station II in July and maximum (189) at station II in September. In winter the minimum value (261) was found at station I in October, while maximum (276.3) was found at station II in January.

During the two years study period higher total alkalinity values were observed in winter followed by summer and monsoon.

Free Carbon Dioxide (mg/l): Station wise monthly values of free carbon dioxide are given in Tables 11.146 and 11.147.

In the first year (2007-08), the free carbon dioxide ranged from Nil to 0.7 mg/l. It was found to be absent for 11 months. It was recorded at station II in December 2007.

Table 11.146. Monthly values of Carbon dioxide (mg/l) (Year 2007-08)

Stn	Oct	Nov	Dec	Jan	Feb	Mar	Apr	May	Jun	July	Aug	Sept
I	—	—	—	—	—	—	—	—	—	—	—	-
II	—	—	0.7	—	—	—	—	—	—	—	—	—
III	—	—	—	—	—	—	—	—	—	—	—	—

Table 11.147. Monthly values of Carbon dioxide (mg/l) (Year 2008-09)

Stn	Oct	Nov	Dec	Jan	Feb	Mar	Apr	May	Jun	July	Aug	Sept
I	0.3	—	—	—	—	—	—	—	—	—	—	—
II	—	—	—	0.4	—	—	—	—	—	—	—	—
III	0.2	—	—	—	—	—	—	—	—	—	—	—

In second year (2008-09), the free carbon dioxide was totally absent from November to December 2008 and February to September 2009.The maximum carbon dioxide (0.4 mg/l) was recorded in January 2009 at station II.

Chlorides (mg/l): Monthly chloride values of water samples from the three stations (I, II and III) are tabulated in Tables 11.148 and 11.149.

Monthly chloride values at three sampling stations in the first year (2007-08) were 89.2, 89.4 and 89.6 in summer, 34.5, 34.5 and 34.7 in monsoon and 60.8, 61.4 and 61.4 in winter.

Table 11.148. Monthly values of Chlorides (mg/l) (Year 2007-08)

Stn	Oct	Nov	Dec	Jan	Feb	Mar	Apr	May	Jun	July	Aug	Sept
I	46.6	58.1	66.2	72.4	81.2	86.1	93.2	96.4	33.5	34.2	33.8	36.8
II	46.8	58.4	67.1	73.6	81.3	88.2	92.8	96.2	33.6	34.1	33.9	36.4
III	47.8	58.9	66.8	72.2	82.2	87.4	92.7	96.1	33.8	34.4	34.2	36.3

Table 11.149. Monthly values of Chlorides (mg/l) (Year 2008-09)

Stn	Oct	Nov	Dec	Jan	Feb	Mar	Apr	May	Jun	July	Aug	Sept
I	45.6	52.1	63.4	70.0	80.4	91.1	96.0	96.2	35.2	36.8	39.6	41.6
II	45.4	53.2	64.8	70.2	80.1	90.0	96.2	96.4	37.5	38.3	39.4	41.2
III	46.0	53.6	64.6	70.4	80.8	91.6	96.1	96.9	36.6	38.2	39.8	41.7

In 2008-09, the mean values of chloride at the stations I, II and III were 91.9, 90.6 and 91.3 in summer, 38.3, 39.3 and 39 in monsoon, and 57.7, 58.4 and 58.6 in winter. During this period, the seasonal maximum and minimum values were 96.9 at spot III in May and 80.1 at station II in February (summer), 41.7 at III in September and 35.2 at station I in June (monsoon) and maximum at 70.4 at station III in January in and minimum at 45.4 at II in October (Winter).

To summarise, during the investigation the study higher values of chlorides were recorded in summer and lower in rainy season.

Total Hardness (mg/l): Tables 11.150 and 11.151 presents station wise and month-wise values of Total Hardness for two-year study period.

Table 11.150. Monthly values of Total hardness (mg/l) (Year 2007-08)

Stn	Oct	Nov	Dec	Jan	Feb	Mar	Apr	May	Jun	July	Aug	Sept
I	117	99	133.4	132.4	154	208	206	204	125	130.1	137.4	139.2
II	128	96	132.2	133.6	152.4	210	209	202	124	130.8	141	152
III	113	111	132.4	133.8	156	199	208	210	124.5	130.9	136.2	132.4

Table 11.151. Monthly values of Total hardness (mg/l) (Year 2008-09)

Stn	Oct	Nov	Dec	Jan	Feb	Mar	Apr	May	Jun	July	Aug	Sept
I	114.6	120	130	138	184.8	190.1	191.2	207.8	131	133	135	132.8
II	115	121	129	138.4	187.4	190.4	192	208	131.5	132	138	133.4
III	114.4	121.2	128.8	139	185.2	189	190	209	131.8	133.4	137	132

During the first year (2007-08) the seasonal mean total hardness values at 3 stations (I, II and III) were 193.3, 193.3 and 193.2 (summer), 132.9, 136.9 and 132.9 (monsoon) and 120.4, 122.4 and 122.5 (winter). During this period minimum value (152.4) was recorded at station II in February and maximum (210) was found at station III in May (summer). Similarly minimum values (124) were found at station II in June and maximum (152) at station II in September (monsoon), and minimum (96) was recorded at station III in November and maximum 133.8 was found at station III in January (winter).

During summer of this year while the maximum total hardness (209) was found at station III in May, its minimum (153) was recorded at station I in February. In monsoon the maximum total hardness (138) was found at station II in August and minimum (131) was recorded at station I in June and in winter the maximum (139) was found at station III in January, minimum (114.4) was recorded at station III in October.

The higher values of total hardness were recorded during summer and lower values were recorded during the winter season.

Zooplankton Diversity: During 2007-08, rotifer dominated the zooplankton population (40.80%), followed by copepoda (23.74%), cladocera and ostracoda (12.92% each) and protozoa (9.7%). The percentage composition of zooplankton during year 2007-08 is presented in Fig. 11.44. Total 34 different species of zooplankton were identified (Table 11.152). The maximum zooplankton population was observed during summer and minimum during winter season (Fig. 11.44).

Fig. 11.44. The percentage of composition of zooplankton in Harni (Katgaon) reservoir during 2007-08

Table 11.152. List of zooplankton in Harni (Katgaon) reservoir

Protozoans
Diffugia spp.
Arcella spp.
Vorticella spp.
Opercularia spp.
Paramecium caudatum
Colpidium sp.
Rotifera
Branchionus flacatus
Brachionus calyciflorus
Brachionus diversicornis

Filina longiseta *Keratella tropica* *Keretella quadrata* *Lecane bulla* *Trichocera porecelus* *Trichocera longiseta*
Cladocera *Indialona ganapati* *Moina micrura* *Diaphanosma sarsi* *Alona rectangular* *Biapertura karna* *Ceriodaphnia cornuta* *Diaphanosoma excisum*
Copepoda *Diaptomus marshianus* *Phylladiaptomus anus annae* *Nedodiaptomus linddbergi* *Nauplius larva* *Mesocyclops leukarti* *Mesocyclops hyalinus* *Cyclops viridis*
Ostracoda *Stenocypris* sp. *Cypris obensa* *Stenocypris* sp. *Cyclocypria globosa* *Candocypria osborni* *Cyprinotus* sps.

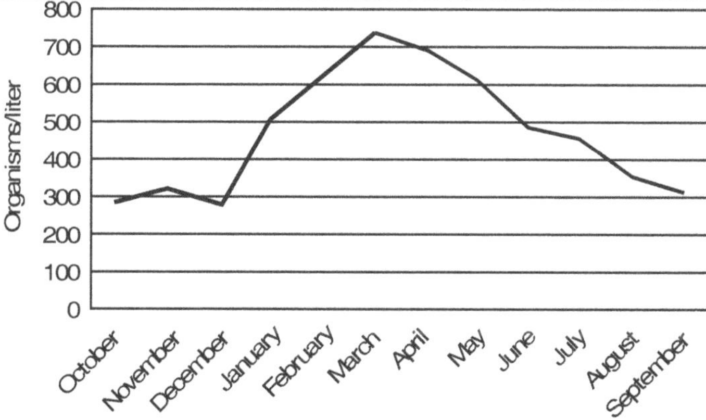

Fig. 11.45. Monthly fluctuations in zooplankton in Harni (Katgaon) rservoir during year 2007-08

Protozoans were high in rainy season. Rotifers, cladocerans, copepods as well as ostracods showed the summer maxima. The protozoans, ostracods, rotifers and copepods showed minima in winter season, while cladocerans were observed less in rainy season.

Protozoans were represented by 6 species (Table 11.153). The species *Paramecium caudatum* and *Colpidium* sps. were observed in 10 months, while *Opercularia* sps. and *Arcella* sps. were absent in rainy summer season respectively. *Difflugia* and *Vortocella* sps. were absent for the months of summer (February and March). The maximum protozoan population was recorded in August 2008 (73 organisms /litre) and minimum population in the month of November 2007 (17 organisms/litre).

Table 11.153. Composition of protozoans (density: organisms/liter) during year 2007-08

	Oct	Nov	Dec	Jan	Feb	Mar	Apr	May	Jun	Jul	Aug	Sept
Difflugia sps.	20	03	09	03	–	–	05	02	03	18	30	21
Arcella sps.	12	05	07	06	–	–	–	–	10	15	16	12
Voritcella sps.	08	02	02	04	–	–	04	–	05	13	12	13
Opercularia sps.	09	–	03	13	02	06	10	16	–	–	–	–
Paramecium, caudatum	07	03	02	20	18	30	16	18	11	18	13	03
Colpidium sps.	06	04	04	08	04	09	05	07	08	04	02	05

Rotifers were represented by 9 species (Table 11.154) and species *Brachionus flacatus,Brachionus diversicornis* and *Keratella tropica* recorded throughout the study period, while *Lecane bulla* was accounted for 8 months only. The peak of rotifers was observed during months of May 2008 (264 organisms/liter) and it was minimum in the month of October (100 organisms/liter).

Table 11.154. Composition of rotifera (density: organisms/liter) during year 2007-08

	Oct	Nov	Dec	Jan	Feb	Mar	Apr	May	Jun	Jul	Aug	Sept
Brachionus falcatus	24	23	32	38	51	46	60	70	63	50	42	38
Brachionus calyciflorus	12	38	–	22	16	21	18	26	30	25	14	24
Brachionus diversicornis	22	22	16	29	38	49	50	40	17	26	15	22
Filinia longiseta	–	19	19	12	19	18	23	20	15	–	05	–

Contd...

Table 11.154: Contd...

Keratella tropica	21	16	12	10	22	30	24	21	34	27	16	25
Keratella quadrata	11	12	14	16	14	12	22	30	21	22	—	-
Lecane bulla	–	18	18	21	32	29	22	-	19	13	-	-
Trichocera porcelus	10	–	13	24	20	24	22	29	28	20	-	-
Trichocera longiseta	–	**14**	**19**	**20**	**21**	**27**	**20**	**28**	**25**	**19**	**20**	**24**

Cladocera were represented by 6 species (Table 11.155). The species *Indialona ganapati, Diaphanosoma sarsi, Ceriodaphnia cornuta* and *Diaphanosoma excisum* were recorded throughout the investigation, while *Alona rectanguila* was totally absent in rainy season. *Moina micrura* and *Biapertura karna* were absent in two months of winter season. The maximum cladoceran population (126 organisms/litre) was recorded during March and minimum population (32 organisms/litre) during November.

Table 11.155. Composition of cladocera (density: organisms/litre) during year 2007-08

	Oct	Nov	Dec	Jan	Feb	Mar	Apr	May	June	July	Aug	Sept
Indialona gurupuli	07	02	07	16	12	19	22	07	07	06	10	07
Moina micrura	06	–	–	18	22	28	18	08	09	09	11	10
Diaphanosoma excisum	09	04	03	14	15	10	04	07	08	04	01	02
Diaphanosoma sarsi	08	09	09	10	14	20	16	05	02	04	06	04
Alona rectangular	04	08	06	17	21	18	13	10	–	–	–	–
Biapertura karna	–	–	07	15	18	22	08	06	06	13	07	06
Ceriodaphnia cornuta	07	09	02	09	16	09	09	04	03	03	02	05

Copepoda was represented by 7 species (Table 11.156). The maxima was 207 organisms/liter during March and minima was 54 organisms/liter during December.*Mesocyclops sps* and *Neodiaptomus lindbergi* were accounted throughout the year while *Diaptomus marshianus* was absent in three months of rainy season (July to September). *Nauplius larva* and *Cyclops viridis* were absent in October and December 2007.

Table 11.156. Composition of copepoda (density: organisms/litre) during year 2007-08

	Oct	Nov	Dec	Jan	Feb	Mar	Apr	May	Jun	Jul	Aug	Sept
Diaptomus marshianus	13	15	09	12	18	21	30	22	14	–	–	–
Phyllodiaptomus annae	–	08	05	08	17	20	23	26	16	16	18	–
Neodiaptomus lindbergi	20	12	13	13	27	27	40	22	13	13	08	14
Nauplius larva	–	07	–	15	18	30	25	28	28	11	17	07
Mesocyclops leukarti	11	26	14	20	25	41	31	21	15	23	12	12
Mesocyclops hyalinus	14	08	13	32	42	50	28	18	19	09	16	20
Cyclops viridis	–	**11**	–	**07**	**11**	**18**	**08**	**29**	**10**	**08**	**07**	**06**

Ostracoda was represented by 5 species (Table 11.157). The highest density of ostracods was observed in April (112 organisms/liter), while lowest density was recorded in November 2007 (21 organisms/liter).

During second year of investigation (2008-09) zooplankton were represented by protozoa, rotifera, cladocera, copepod and ostracoda. The percentage composition of zooplankton during year 2008-09 is presented in Fig. 11.39. The monthly variation in total zooplankton population during year 2008-09 is shown in Fig. 11.46.

Table 11.157. Composition of ostracoda (density: organisms/litre) during year 2007-08

	Oct	Nov	Dec	Jan	Feb	Mar	Apr	May	Jun	Jul	Aug	Sept
Stenocypris sps.	–	–	–	12	20	16	27	21	10	15	20	28
Cypris obensa	10	06	10	07	19	18	30	19	09	10	14	03
Cyclocypris globosa	12	07	04	18	22	23	23	22	21	17	13	07
Candocypria osborni	–	03	08	13	19	20	17	17	10	12	–	05
Cyprinotus sps.	06	05	06	08	10	13	15	10	12	13	10	07

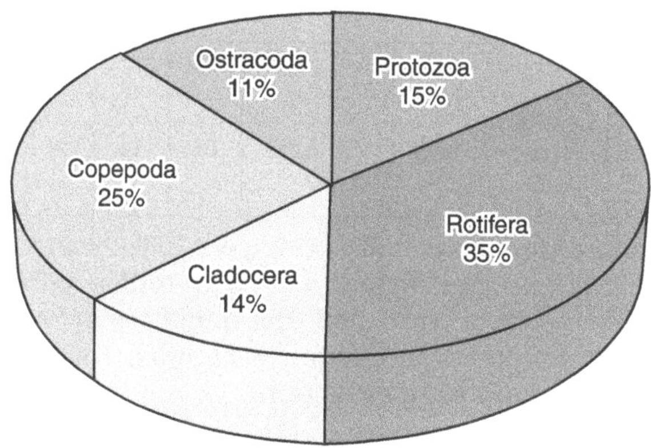

Fig. 11.46. The percentage of composition of zooplankton Harni (Katgaon) reservoir during 2008-09

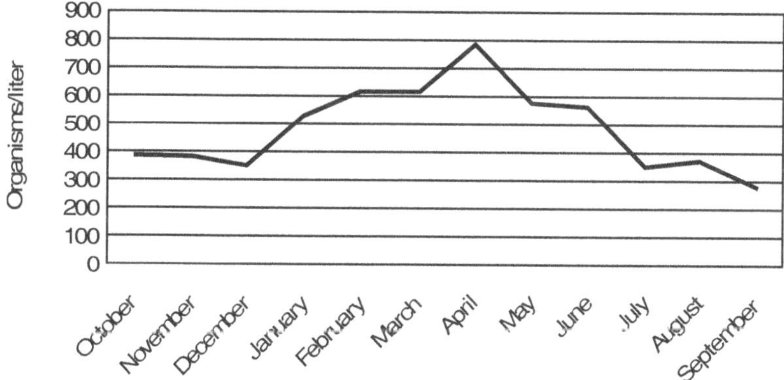

Fig. 11.47. Monthly fluctuations in zooplankton in Harni (Katgaon) reservoir during year 2008-09

Protozoa was represented by 6 species (Table 11.158). The maxima was 95 organisms/liter during October and minima was 54 organisms/litre during January 2009.*Vorticella* sps. were absent throughout the summer and *Cospidium* sps. were present only in 7 months. *Difflugia* sps. and *Arcella* sps. were also absent in three months of summer.

Table 11.158. Composition of protozoans (density: organisms/litre) during year 2008-09

	Oct	Nov	Dec	Jan	Feb	Mar	Apr	May	Jun	Jul	Aug	Sept
Difflugia sps.	23	12	08	09	–	–	–	15	20	24	28	20
Arcella sps.	20	08	20	10	13	–	–	–	13	21	17	30
Voritcella sps.	19	14	08	06	–	–	–	–	10	19	26	18
Opercularia sps.	12	11	12	03	14	09	27	20	18	–	–	–

Contd...

Table 11.158: Contd...

Paramecium, caudatum	21	16	10	08	23	38	32	23	21	13	-	-
***Colpidium* sps.**	–	–	–	**08**	**13**	**15**	**06**	**16**	–	–	**06**	**04**

Rotifera accounted for about 35% during year 2008-09 and were represented by 9 species (Table 11.159). The highest density of rotifers (278 organisms/litre) was recorded in the month of April 2009 and minimum in September 2009 (102 organisms/litre). Throughout the summer months, rotifer population was maximum. It was minimum in rainy season.

Table 11.159. Composition of rotifera (density: organisms/litre) during year 2008-09

	Oct	Nov	Dec	Jan	Feb	Mar	Apr	May	Jun	Jul	Aug	Sept
Brachionus falcatus	28	24	13	35	33	42	56	34	17	21	30	29
Brachionus calyciflorus	–	–	15	17	30	19	26	30	32	15	26	18
Brachionus diversicornis	–	08	19	26	28	39	30	27	30	18	21	24
Filinia longiseta	–	20	–	23	10	08	17	11	16	22	–	–
Keratella tropica	26	25	17	29	25	27	33	25	26	17	25	20
Keratella quadrata	20	15	13	09	14	22	28	29	12	–	–	–
Lecane bulla	21	17	09	20	22	15	30	26	20	14	–	14
Trichocera porcelus	20	15	20	18	14	18	21	06	12	09	–	–
Trichocera longiseta	22	18	21	28	20	28	29	10	22	23	28	–

Cladocerans accounted for about 14% and were represented by 7 species (Table 11.160). The maximum population was recorded in March (131 organisms/liter) and minimum population in the month of November (37 organisms/litre). Cladocerans were maximum throughout the summer season. Out of 7 species, 5 were present throughout the year, while *Alona rectangular* was absent in rainy season. *Biapertura karna* was not seen in October 2008.

Table 11.160. Composition of cladocera (density: organisms/litre) during year 2008-09

	Oct	Nov	Dec	Jan	Feb	Mar	Apr	May	Jun	Jul	Aug	Sept
Indialona ganapati	08	04	09	17	13	17	25	08	07	08	12	09
Moina micrura	09	10	09	19	20	30	21	14	10	08	13	14
Diaphanosoma excisum	08	03	03	16	18	10	03	04	10	05	02	03
Diaphanosoma sarsi	09	03	03	12	13	18	12	07	03	06	05	03
Alona rectangular	08	06	07	15	25	19	17	12	–	–	–	–
Biapertura karna	–	05	04	17	20	28	11	03	08	14	07	03
Ceriodaphnia cornuta	**04**	**06**	**04**	**08**	**14**	**19**	**06**	**04**	**02**	**03**	**03**	**07**

Copepoda was accounted for about 25% and were represented by 7 species (Table 11.161), and *Neodiaptomus lindbergi* was absent throughout the study period. *Diaptomus marshianus* and *Mesoccyclops hyalinus* were present for 8 and 9 months respectively. *Cyclops viridis, Mesocyclops leukarti, Nauplius larva* and *Phylladiaptomus annac* were present in 11 months. The peak period of copepods was observed during month of April (255 organisms/litre) and it was minimum during the month of September 2009 (26 organisms/litre).

Table 11.161. Composition of copepoda (density: organisms/liter) during year 2008-09

	Oct	Novr	Dec	Jan	Feb	Mar	Apr	May	Jun	Jul	Aug	Sept
Diaptomus marshianus	–	22	10	10	38	30	21	14	28	–	–	–
Phyllodiaptomus annae	–	14	10	21	18	08	20	10	30	12	15	18
Neodiaptomus lindbergi	20	10	20	18	10	20	35	28	26	20	18	08
Nauplius larva	08	08	14	13	19	12	22	15	18	08	20	–
Mesocyclops leukarti	16	12	06	18	32	42	70	40	27	16	22	–
Mesocyclops hyalinus	12	24	18	20	40	10	65	35	22	–	–	–
Cyclops viridis	**06**	**09**	**15**	**16**	**16**	**15**	**22**	**20**	**20**	**11**	**11**	–

Ostracoda was represented by 5 species (Table 11.162). *Cyclocypris globosa* and *Cyprinotus* sps were present for 12 months. *Stenocypris* sps were absent for 3 months of winter season and two months of rainy season. *Cypris obensa* and *Candocyprias osbornii* were absent in two months of rainy season. Throughout the summer month, ostracod population was maximum. It was minimum in rainy season.

Table 11.162. Composition of ostracoda (density: organisms/liter) during year 2008-09

	Oct	Nov	Dee	Jan	Feb	Mar	Apr	May	Jun	Jul	Aug	Sept
Stenocypris Sps.	–	–	–	06	12	13	30	25	18	10	–	–
Cypris obensa	07	11	05	10	14	16	22	16	22	–	–	11
Cyclocypris globosa	12	10	08	15	13	12	18	10	06	08	10	16
Candocypria osborni	13	09	09	13	11	10	13	28	20	–	08	–
***Cyprinotus* sps**	**12**	**08**	**11**	**12**	**08**	**10**	**15**	**16**	**12**	**06**	**04**	**08**

Phytoplankton Diversity: 10 species of chlorophyceae, 5 species of cyanophyceae, 6 species of bacillariophyceae and 3 species of euglenophyceae were recorded from the Harni (Katgaon) reservoir (Table 11.163).

Table 11.163. Phytoplankton diversity in Harni (Katgaon) reservoir

Chlorophyceae: *Ulothrix zonata* *Zygnema* sp *Pediastrum duplex,* *Pediastrum simplex,* *Scendesmus armatus,* *Oedogonium patulum,* *Ankistrodesmus falcatus,* *Chlorella valgoris,* *Cosmarium contractum,* *Closterium limneticum*
Cyanophyceae: *Oscillatoria chlorine* *Oscillatoria limnosa* *Anabaena constricta* *Merismopedia punctata* *Microcystis aerugenose*

Contd...

Table 11.163: Contd...

Bacillariophyceae: *Navicula gracilis* *Navicula viridula* *Nitzschia subtilis* *Bacillaria paradoxa* *Diatoms vuloare* *Synedra affinis*
Euglenophyceae: *Eugelna stellata,* *Euglena viridis* *Euglena pisciformis*

From chlorophyceae *Chlorella valgaris, Pediastrum* spp and *Scendesmus armatus* dominated the reservoir. The maximum population of chlorophyceae was recorded in April (1360.l^{-1}).

From cyanophyceae *Oscillatoria* spp and *Microcystis aeurogenose* dominated the reservoir. This group attained a highest peak in summer, with maxima in May at 2010.l^{-1}.

Among bacillariophyceae *Navicula* spp., *Synedra affinis* and *Bacillaria paradoxa* dominated the reservoir. Their population was maximum (470.l^{-1}) in July. The maximum population of bacillariophyceae in rainy season may be due to the inflow of water inside the reservoir.

Euglenophyceae was represented by 3 species. *Eugela* spp. increased in number in winter season. The maximum population of euglenophyceae was recorded in December (540.l^{-1}).

Table 11.164. Plankton groups and their composition in Harni (Katgaon) reservoir during year 2007-09

Groups	Range (Units l^{-1})	Average (%)
Chlorophyceae	370 to 1360	1060 (34.99)
Cyanophyceae	420 to 2010	910 (30.04)
Bacillariophyceae	350 to 470	300 (9.90)
Euglenophyceae	215 to 540	280 (9.24)
PHYTOPLANKTON		2550 (84.19)
Protozoans	17 to 95	58 (1.91)
Rotifers	100 to 270	182 (6.00)
Cladocerans	32 to 131	64(2.11)
Copepods	26 to 255	117(3.86)
Ostracods	21 to 112	58 (1.91)
ZOOPLANKTON		479 (15.81)
TOTAL ZOOPLANKTON		3029

TERNA RESERVOIR

It is located in Osmanabad taluka near Terna Sugar Factory. It was constructed in 1970 across the river Terna. The total length of dam is 2651 meters. The basin of the reservoir is erratic, undulating and strewn with boulders of various dimensions. These boulders pose a lot of difficulties in operation of fishing nets. The stocked Indian major carps firmly established in the reservoir.

Fig. 11.48. Terna Reservoir

The natural breeding and auto stocking of major carps also observed in the reservoir. The major groups of fishes seen in the landings of the reservoir are major carps. In Terna reservoir major carps constitute almost 65% of the total landings followed by catfishes (25%) and miscellaneous species (10%). This shows that the major share of the catch comprises quality fish, which fetch good market price. Gill net and hook and lines are common gears used for fishing. The details of seed stocking and fish production is depicted in Table 11.165. The fish catch varied from 10110 kg (year 2011-12) to 29190 kg (2008-09).

Table 11.165. Seed stocking and fish production in Terna reservoir

Year	Seed stocked	Fish production (kg)
2007-8	12.50	24710
2008-09	20.00	29190
2009-10	10.00	23000
2010-11	10.00	12150
2011-12	10.00	10110

The craft used in fishing is of primitive type, which is nothing but a platform of 6 × 3 ft size with a depth of 4-10 inches and is locally called as 'Nav'. This is prepared from the thermocoel and covered by a plastic covering. Such type of craft is not suitable to carry large nets, having heavy weight. When fishermen get large quantities of fish catches, it become difficult to carry them to the landing centers. It is observed that *Catla catla, Labeo rohita and Mystus* spp. were consistently dominant throughout the study period.

Fig. 11.49. Craft used for fishing in Terna reservoir

Fig. 11.50. Problem of submerged vegetation

CHANDANI RESERVOIR

It was constructed across river Chandani in 1965 for purpose of irrigation. The government fish hatchery was established adjacent to the reservoir. The total water spread area of the reservoir is 340 hectares. The length of reservoir is 1920 meters with height of 17.18 meters.

Fig. 11.51. A view of Chandani Reservoir

The main fishing gear is gill nets operated from a thermocole raft. Rod and line and cast nets are also in use. The fishery of Chandani reservoir consisted mainly of indigenous species.Though some stocking of major carps has been done during early years, its impact on the fishery was negligible. The dominant species during present investigation were *Mystus* spp., *Channa* spp. and *Mastacembelus armatus.* Since the reservoir harbor different species of predatory fishes, it is desirable to stock reservoir with fingerlings of 100-200 mm in size.

The craft used for fishing similar to like that of Terna reservoir. The fishermen were also observed to rely on another kind of improvised materials. They showed considerable ingenuity in fabricating makeshift rafts out of discarded old rubber tubes. A woolen platform is placed over the rubber tube and tied tightly with rope. It is used for hook and line operation and also for setting and hauling of gill-nets. There was no strict mesh regulation imposed for the nets operated for fishing.

Fig. 11.52. Craft used for fishing in Chandani reservoir

SINA KOLEGAON RESERVOIR

It is newly constructed shallow reservoir in Paranda taluka. Reservoir is very close to Sonari village. The reservoir has an opportunity to provide water to agricultural land as well as to Bhairavnath Sugar Factory. The wooden boats are commonly used for fishing. Reservoir was stocked by fry stage. The minimum stocking was done in 2009-10 @ 314 seed/ha ,while the maximum stocking was @ 628 seed/ha in year 2011-12. The fish production varied from 18 kg/ha/yr to 90 kg/ha/yr.

Table 11.166. Seed stocking and fish production in Sina Kolegaon reservoir

Year	Seed stocked	Seed stocking /ha	Fish production (kg)	Fish production /ha
2009-10	700,000	314	40000	18
2010-11	13,90,000	624	200000	90
2011-12	14,00,000	628	135000	61

Details of seed stocking in the reservoir is shown in Table 11.166. The highest stocking of reservoir with 14,00,000 seed was done in 2011-12, which gave a per ha stocking density 628 fry. The minimum stocking of 7,00,000 seed was done in 2009-10 which gave per ha stocking density 314 fry. The details of the fish production from the reservoir is shown in Table 11.166. The study of the table revealed that the maximum fish production for the last five years (from 2007-08 to 2011-12) was recorded at 2,00,000 kg during 2010-11. It gave an average fish production of 89.76 kg/ha/yr. The minimum production was during 2009-10 at 40,000 kg, giving an average production of 17.95 kg/ha/yr. The fish catches are mainly constituted by *Catla catla, Labeo rohita, Mystus* spp., *Channa* spp., *Mastacembelus* sp. and *Notopterus* spp.

Fig. 11.53. Sluice gates of Sina Kolegaon reservoir

Fig. 11.54. Shallowness of Sina Kolegaon Reservoir

SANGMESHWAR RESERVOIR

It was impounded in 1977 across in Bhum taluka. The water spread area is 448 ha. The reservoir harbours a variety of fish species but the fishery was significantly dominated by *Catla catla*. The major carnivores (although present in sparse density) were *Channa* spp., *Mystus* spp. and *Wallago attu*. The annual fish landings ranged from a low of around 11.16 kg/ha/yr (year 2008-09) to a high of 66.96 kg/ha/yr (year 2007-08). Consistently high catches were recorded during September-October followed by a sharp decline thereafter. The gears used in reservoir are surface gill-nets, cast nets and hook and lines. The crafts used are similar to that of Terna and Chandani reservoir.

Table 11.167. Stocking and fish production in Sangmeshwar Reservoir

Year	Seed stocked	Fish production (kg)
2007-8	9,94,000	30,000
2008-09	5,00,000	5000
2009-10	1,00,000	10400
2010-11	10,00,000	10,000
2011-12	10,00,000	10,000

KHASAPUR RESERVOIR

The water spread area of reservoir is 330 ha at FRL. It was constructed in 1988 near Khaspaur village in Paranda taluka of Osmanabad district. About 22 fishermen associated with co-operative society are earning their livelihood from the fishing in reservoir. The fishery is exploited by licensed fishers whose numbers are regulated by the district fisheries department. The main fishing gear is gill nets operated from a coracle. Rod and line and cast nets are also in use. All the licensees are members of the fishermen's co-operative society and their catches are handed over to the co-operative society at prices fixed each year. After making local sales, the major part of the catches are handed over to the middlemen.

The common species in the catches are *Catla catla, Labeo rohita, Cyprinus carpio, Wallago attu, Mastacembelus armatus, Channa* spp., *Puntius* spp, and *Mystus* spp. Except for *C. catla, L. rohita* and *C. carpio*, which are introduced, all other species are native to the reservoir ecosystem. The percentage of carps in total catch varied from 14 to 20%. There are two visible peaks in the catches, one in August-September and the other in January-February, the former coinciding with the increasing phase of the reservoir level and the latter with the decreasing phase. Generally, in summer season the catches are poor. It is due to low fishing effort, Details on fish seed and fish catch are not available, but after visiting this reservoir, it was concluded that the reservoir is an example of a productive but poorly managed reservoir. The potential for production is great but due to poor management and lack of planning, it has become a loss making reservoir for district fisheries department. The reservoir needs regular stocking with quality fish seed of carps.

BENITURA RESERVOIR

The reservoir was constructed in 1994 across river Benitura in Omerga taluka. The water spread area of reservoir is 306 ha. The height and length is 13.48 m and 1780 m with gross storage capacity of 12810 $10^3 m^3$. From 2007-08 to 2010-11 reservoir was regularly stocked by 2,50,000 seed per year, while recently in 2011-12 it was stocked by 4,00,000 seed of Indian major carps. The fish catch varied from 31 kg/ha/yr (2007-08) to 262 kg/ha/yr (2009-10). During recent year the fish catch was 55740 kg *i.e.*, 182 kg/ha/yr.

Table 11.168. Seed stocking and fish production in Sina Kolegaon reservoir

Year	Seed stocked	Seed stocking /ha	Fish production (kg)	Fish production /ha
2007-8	2,50,000	817	9370	31
2008-09	2,50,000	817	39340	129
2009-10	2,50,000	817	80200	262
2010-11	2,50,000	817	36300	119
2011-12	4,00,000	1307	55740	182

POWAI LAKE

The water spread area of lake is 210 ha with mean depth of 3.2 meters. Powai lake is sitautaed at L19908′N and L72°54′E, 27 kms away in the North-East of Mumbai city. Now a days due to high discharge of waste waters, building raw materials, cement and most of the heavy metals like lead of vehicles, silt load from the surrounding area due to quarrying work, it has lost its diverse nature of fish population (Salaskar and Yeragi, 2004).

Powai Lake is situated 55 meters above the mean sea level is known as 'Anglers Paradise'. It is meant exclusively for angling and sports. It is about 27 km away from the north east of Mumbai city. The lake environment is well protected by Dr. Babasaheb Ambedkar Udyan and Indian Institute of Technology on the eastern side. This lake came into existence in the year 1891, when Mumbai Municipality got constructed a masonry dam of 10 meter height between two hillocks across Powai basin to conserve the rain water for drinking purpose, which later commonly known as Powai Lake,since it impounded in Powai area. The Lake is Powai Lake is 2.10 km^2 and has a maximum depth of 6.1 meter and minimum depth of 4.3 meters. It has total water spread area of 210 hectares. Rainwater is the main source of the lake to maintain water level.from the very beginning, its water was not meant for human consumption, nevertheless, even today; it is used by industrial areas including big industry like Larsen and Turbo. The lake is extensively used for washing and bathing activities and exhibits characteristics of a eutrophicated ecosystem. Average rainfall at Powai is about 2540 mm, and the lake overflows for about for 60 days each year. The lake is leased out Maharashtra state Angling Association (MSAA), Mumbai for angling in addition to conservation. It was registered in year 1955 under the society's registration Act 1860.

Powai Lake is supported by 37 species (Kulkarni, 1947 and Amore 1955). Bhagat (1977) has listed 32 species while Singh Kohli (1995) has listed 10 main fish species in the lake. The fish fauna of Powai lake includes *Catla catla, Labeo rohita, Cirrhinus mrigala, Labeo calbasu,* and exotic fishes such as *Tilapia mossambica,Osphronemus goramy, Hypophthalmichthys molitrix, Tor tor, Ctenopharyngodon idella,* and *Cyprius carpio.* The species wise catches in lake are depicted in Table 11.169.

Table 11.169. Species-wise catches in Powai Lake of Mumbai

Name of fish	Annual catch in kg (1995)	Percentage by weight (1995)	Annual catch in kg (1996)	% by weight (1996)
Labeo rohita	473.50	19.79	343.13	13.28
O. goramy	3.00	0.13	2.50	0.10
Cirrhinus mrigala	355.75	14.87	341.50	13.22
Catla catla	442.93	18.52	511.88	19.81
Labeo calbasu	2.25	0.09	1.50	0.06
Hypophthalm-ichthys molitrix	899.50	37.60	870.38	33.68
Tor spp.	2.00	0.08	15.50	0.60
Cteopharyngodon idella	21.38	0.89	45.88	1.78
Hybrid varieties	191.88	8.02	451.83	17.48
TOTAL	2392.18	100	2583.90	100

The average per day catch in year 1950 was 22 kg through angling (Annual Report, Department of Fishery, Mumbai). It has gone down to 8 kg per day in 1989 (Singh Kohli,1995). Salaskar and Yeragi (2004) found that the average per day catch was 6.5539 kg per day in 1995 and 7.079 kg per day in 1996. Thus, there is a drastic depletion in the per day catch of fish. This may be due to the heavy siltation and high discharge of waste waters into the lake.

Powai Lake has a fairly rich fauna of fishes, but nowaday due to high discharge of waste waters, building raw materials, cement and most of the heavy metals like lead of vehicles, silt load from the surrounding area due to quarrying work, it has lost its diverse nature of fish population (Salaskar and Yeragi, 2004). The fish productivity of the lake is 44.724 kg/ha/yr in 1995 and 45.6376 kg/ha/yr in year 1996.

Over the years, real estate development around the lake and ersosion from the adjacent hills has reduced the total area of water spread and the water depth.Domestic sewage from nearby settlements, particularly slums entered the lake directly and solid wastes are dumped on its shores. Aquatic weeds such as

Ipomoea and water hyacinth grow luxuriantly over the lake causing a serious problem.

Recent Efforts of the Brihanmumbai Municipal Corporation (BMC) and the Government of Maharashtra to improve water quality through bioremediation and aeration resulted in the lake shifting from hypertrophic to mesotrophic condition, and reduction in the sludge at the lake bottom.The list of fishes commonly recorded in Powai Lake are depicted in Table 11.167.

Table 11.170. Fishes commonly recorded in Powai Lake of Mumbai

1. *Catla catla*
2. *Labeo rohita*
3. *Labeo calbasu*
4. *Cirrhina mrigala*
5. *Tor khudree*
6. *Osphronemus goramy*
7. *Hypophthalmichtys molitrix*
8. *Cyprinus carpio*
9. *Ctenopharyngodon idella*
10. *Tilapia mossambica*
11. *Mystus* spp.
12. *Clarius batrachus*
13. *Oxygaster bacaila*
14. *Danio malbaricus*
15. *Rasbora daniconius*
16. *Aplocheilus lineatus*
17. *Barbus filamentasis*

In the fish catch of Powai Lake *Labeo rohita* dominates in number but as far as weight *Hypophthalmichthys molitrix* dominates. The percentages of fishes in number by weight is *Labeo rohita* (19.79%) and 13.28%, *Osphroemus goramy* (Beehead) 0.13 and 0.10%, *Cirrhina mrigala* 14.87% and 13.28%, *Catla catla* 18.52% and 19.81%, *Labeo calbasu* 0.09 and 0.06%, *Hypophthalmichthys molitrix* 37.60 and 33.68, *Tor tor* 0.08% and 0.60%, *Cyprinus carpio* 8.02% and 17.48% and *Ctenopharyngodon idella* 21.38% and 1.78% in year of 1995 and 1996 (Salasakar and Yeragi, 2004).

The percentage of fishes in number by catch number is Labeo *rohita* 29.53% and 17.85%, *Osphraemus goramy* 0.14% to 0.163%, *Cirrhina mrigala* 20.82% and 26.85%, *Catla catla* 10.12% and 7.813%, *Labeo calbasu* 0.21 % and 0.10%, *Hypohthalmichthys molitrix* 24.53 % and 17.79%, *Tor tor* 0.28% and 0.21%, *Cyprinus carpio* 13.12% and 27.67% and *Ctenopharyngodon idella* 1.21% and 1.51% in the year of 1995-96.

The Lake has a large number of crocodiles. The death of crocodile due to ingestion of chemicals discharged from pipelines flowing into the lake was noticed in 1999.

Salaskar ad Yeragi (2004) studied primary productivity of Powai Lake during the year 1995-96. The gross primary productivity values varied between 600 mg/c/m^3/day (December, 1995) to 4324 mg/c/m^3/day (September, 1996).Likewise the net production values ranged from 225.20 mg/c/m^3/day (June, 1996) to 3337 mg/c/m^3/day (September, 1996). The major peaks of net production coincided with those of gross production. Both gross and net primary productivity were high during monsoon. During this time nutrient load in the lake is also found to be highest (Salaskar and Yeragi, 2004).

Many non-governmental organizations (NGOs) areworking independently or in groups to protect water quality of Powai Lake including the Maharashtra State Angling Association (MSAA), Nushad Ali Sarovar Samardhini (NASS), and Bombay Natural History Society (BNHS) and Young Environmentalists Programme. MSAA have also published a book 'Powai-The Anglers Paradise' to create public interest in Powai Lake and MSAA activities. MSAA stock 2 to 5 Lakh fish fingerlings per year. MSAA members catch about 5000-6000 fish. The bulk of the fish are stolen by poachers but fortunately many prize specimens still live at Powai because MSAA release back many of the fish which are caught as per association rules. This NGO have also introduced mahseer seed in Powai Lake with the help of Tata Hydropower Project, Lonavala (Maharashtra). MSAA approached the state and central governments with a request to protect the Powai Lake. In response to representation of MSAA, in 2001 Ministry of Environment and Forest, Government of India, New Delhi, included Powai Lake in the National Lake Conservation Mission with allocation of Rs. 26 crores to the Bombay Municipal Corporation (BMC) for conservation and management of the lake.

MAKHMALI LAKE

Game and Salaskar (2007) studied environmental impact of macrophytes on Makhmali Lake of Thane and recorded 5 species of macrophytes. During most of the year surface of the lake, especially in the shallow areas, is found to be covered with water hyacinth (*Eichhornia crasspies*) which was found to be dominating throughout the year.Other macrophytes found in the lake are *Lemna polyrhiza, Hydrilla verticillata, Azolla pinnata* and *Ipomea aquatia*. These aquatic plants cover more than 40% of the lake water area reducing the shore lie as well as obstructing accessibility. *Eichhornia crassipes* grew maximum during winter and summer and flowering was observed in June and September.

REFERENCES

- Agarwal, S.C.1990. Limnology. A.P.H. Publishing Corporation, New Delhi, p. 150.
- Amore, D.L. 1955. "Powai" Hindu Kitab Ltd. Publisher, Mumbai.
- Bhadane, R.S. 2012 a. Studies of monthly variations in DO, BOD and COD parameters of Gangapur dam water at Nashik, Maharashtra. *Ecology and Fisheries*, 5(1) : 47-54.
- Bhadane, R.S. 2012b. Zooplankton of dynamics of Gangapur at Nashik. *Ecology and Fisheries*, 5(2) : 51-56.
- Bhagat, M.J. 1977. Ecology ad Sport Fishery of Freshwater Lake, M.Sc. Thesis, University of Mumbai.

- Bhalerao, S.N., 2012. Study of fish diversity and water quality at Kasar Sai dam, Hinjewadi, Pune, MS, India, *International Research Journal of Biological Sciences. 1(4) : 51-55,* Paper available online at www.isca.in.
- Bharambe, C.M. and Patil, Bharathi, 2013. Ecology of Nalganga reservoir (Buldana distirict, Maharashtra) in reference to physico-chemical status.In:Emerging Trends in fisheries and Aquaculture, Proceednigs of the National Conference on Emerging Trends in Fisheries and Aquaculture (Eds. V.B. Sakhare and B.Vasanthkumar), Astral International Private Limited, New Delhi, pp. 207-217.
- Boyd, C.E., 1982. Water quality management of pond fish culture. Elsevier Publication, pp 3-18.
- Chavan, S.P., 2006. An evaluation of fisheries potential of Masoli reservoir in Parbhani district of Maharashtra. In : Ecology of Lakes and Reservoirs (Ed. V.B. Sakhare), Daya Publishing House, Delhi, pp188-189.
- CIFRI, 1997. Ecology and Fisheries of Bhatghar reservoir, Bulletin Number 73, Special Publication on Occasion of Golden Jubilee Celebrations of CIFRI, Barrackpore.
- Das, Manas Kumar and Das, R.K., 1995. Fish diseases in India-A review. Environment and Ecology, 533-541.
- Das, R.K. 1996. Monitoring of water quality, its importance in disease control. Paper presented in Nat. Workshop on fish and prawn disease, epizootics and quarantine adoption in India. October 9, 1996. CIFRI, pp 51-55.
- Deshmukh, U.S., 2001. Ecological studies of chhatri lake, Amravati; with special reference to planktons and productivity. Ph. D. thesis, Amravathi University, Amravathi.
- Gaike, P.P., Patil, P.V. and Shejule, K.B., 2011. Hydro-biological study of Dahipal dam Dist. Jalna (MS) India. *Science Research Reporter*. 1(3) : 170-172.
- Game, A.S. and Salaskar, P.B., 2007. Environmental impact of macrophytes on Makhmali lake, Thane, Maharashtra. *J.Aqua.Biol*. 22(2) : 203-204.
- Harney, N.V., sitre, S.R., Wadhave, N.S. and Nasare, P.N., 2010. Checklist of birds of Ghodpeth reservoir of Bhadrawati tahsil in Chandrapur district of Maharashtra. *Ecology and Fisheries*. 3(1) : 127-133.
- Jaiswal, D.P. and Ahirrao, K.D., 2012. Ichthyodiversity of the Rangavali Dam, Navapur, District Nandurbar, Maharashtra State. *Journal of research in Biology*. 3 : 241-245
- Jayabhaye, U.M. and Madlapure, V.R., 2006. Studies on zooplankton diversity inParola dam, Hingoli, Maharashtra, India. *J.Aqua.Biol*. 21 (2) : 67-71.
- Jayabhaye, U.M. and Madlapure, V.R and Malviya, M.K., 2006. Study of fish diversity in the Parola dam near Hingoli, Hingoli district, Maharashtra, India. *J.Aqua.Biol*. 21(2) : 65-66.
- Jayabhaye, U.M., Madlapure, V.R. and Salve, B.S. 2007. Phytoplankton diversity of Parola dam, Hingoli, Maharashtra. *J. Aqua. Biol*. 22 (2) : 27-32.
- Jetithor, S.G. 2011. Studies on some aspects of ecology and fisheries of Harni (Katgaon) reservoir in Osmanabad district, Maharashtra, Ph.D. thesis, Dr. Babasaheb Ambedkar Marathwada University, Aurangabad.
- Jhingran, A.G., 1990. Recent advances in the reservoir fisheries management in India. In Sena De Silva (Ed.) Reservoir fisheries of Asia. Proceedings of the 2nd Asian reservoir fisheries workshop held in Hangzhov peoples Republic of China 15-19 October, 1990.
- Jhingran Arun, G. and Sugunan, V.V., 1990. General guidelines and planning criteria for small reservoir fisheries management. In Reservoir Fisheries in India (Jhingran

A.G.and V.K. Unnithan Eds.) pp. 1-8. Asian Fisheries Society, Indian Branch, Mangalore, INDIA.

- Kadam, S.U., Gaikwad, J.M. and Babar Md., 2006. Water quality and ecological studies of Masoli reservoir in Parbhni district, Maharashtra. In : Ecology of Lakes and Reservoirs (Ed.V.B.Sakhare), Daya Publishing House, Delhi, pp. 163-175.
- Kalbande, Subhangi, Telkhade, Pravin and Zade, Suresh, 2012. Fish diversity of Rawanwadi Lake of Bhandara district, Maharashtra, India. *Abhinav Journal.* 2(2) : 30-33, paper available on wwww.abhinavjournal.com.
- Khalid, M.A., Siddiqui, M.S., 1990. Reservoir fishery potential of Nasik district in Maharashtra with special reference to Girna dam reservoir fisheries, pp. 97-101. In : In Reservoir Fisheries in India (Jhingran A.G. and V.K. Unnithan Eds.) pp.1-8. Asian Fisheries Society, Indian Branch, Mangalore, INDIA.
- Koli, K.B. and Muley, D.V., 2012. Study of zooplankton parameters in Tulshi reservoir of Kolhapur district (M.S.) India. *E-International Scientfic Research Journal,* 4(1) : 38-46.
- Kulkarni, A.N., Kanwate, V.S. and Deshpande, V.D., 2006. Checklist of birds of Shikachiwadi reservoir Dist. Nanded, Maharashtra. *J.Aqua.Biol.* 21 (1) : 80-85.
- Kulkarni, C.V., 1947. Note on freshwater fishes of Bombay ad Salsette Islads, *J. Bombay Natural History Society,* 47 (2) : 319-326.
- Maharashtra Pollution Control Board and Central Institute of Fisheries Education. 2011. Final Report on Assessment of Riverine fisheries and Linking with water quality restoration programme-River Godavari in Maharashtra.
- Manjare, S. A. Vhanalakar, S.A. and Muley, D. V., 2010. Water quality assessment of Vadagon tank of Kolhpur (Mharashtra), with special reference to zooplankton. International Journal of Advanced Biotechnology and Research. 1(2) : 91-95. Paper available on http://www.bipublication.com
- Meshram, Manisha and Nasare, P.N., 2011. Phytoplakton diversity of Futala lake in Nagpur, Maharashtra.*Ecology ad Fisheries.* 4 (1) : 67-70.
- Maris, D.F. 1966. A total alkalinity atlas for marine lake waters. *Limnol. Oceanogr.* 11 : 68-72.
- Mc Kee, J.E. and Wolf, H.V. 1963. Water quality criteria, 2nd edition. State of California, State water quality control Board, Sacramento, Publication 3-A, 548 Moyle, J.B. 1945. Some chemical factors influencing the distribution of aquatic plants in Minnesota. Amer. Mild. Natur. 34 : 402-420.
- Mule, D.V. and Patil, I.M., 2006. A study of water quality and fish diversity of Pauna river, Maharashtra. *J. Aqua. Biol.* 21 (1) : 68-75.
- Nandan, S.N. and Jain, D.S., 2005. Study of algae tolerating organic pollution of Sonvad dam and Devbhane dam of Maharashtra. In : Advances in Limnology (Ed. S.R. Mishra), Daya Publishing House, Delhi, pp. 214-222.
- Niture, S.D. and Chavan, S.P., 2010. Fisheries management of Yeldari reservoir, Maharashtra, In : Advances in Aquatic ecology (Vol. 4), Edited by V.B. Sakhare, Daya Publishing House, Delhi, pp. 152-172.
- Patil, P.V., Kulkarni, M.Y., Kulkarni, A.N. and Walanikar, A.V., 2010.A note on some physico-chemical characteristic of Vanjarwadi reservoir, Beed. *The Ecoscan.* 4 (1) : 81-82.
- Pawar, Ashwini, P. and Mushan, 2012. Studies on diversity of zooplankton in four lakes of Solapur, Maharashtra. *Fishing Chimes.* 32 (3) : 47-49.
- Pawar, Sandhya and Pawar Sourabh.2012. Ichthyofaunal Diversity of Kanher Dam from Satara district of Maharashtra state of India. *International Journal of Fisheries and Aquaculture Sciences.* 1 : 15-19, paper available on http://www.irphouse.com.

- Penak, R.W., 1953. Freshwater invertebrates of United States. 2nd edition. John Willey and Sons Inc. 763.
- Pulle, J.S., Khan, M.A. and Pawar, S.K., 2003. Study of thermal regime of Isapur dam water, Yeotmal district (M.S.), India. *J. Aqua. Biol.* 18 (2) : 87-91.
- Pawar, S.K., Mane, A.M. and Pulle, J.S., 2006. The fish faua of Pethwadas dam taluka Kandhar in Nanded district, Maharashtra, India. *J. Aqua. Biol.* 21(2) : 55-58.
- Pawar, S.M. and Sonawane, S.R., 2011.Water quality index of Kanher dam of satara district (M.S.) India. The Ecoscan, 5(3 and 4) : 199-202.
- Rathod, S.R. and Khedkar, G.D., 2011. Impact of elevation, latitude and longitude on fish diversity in Godavari river. *Journal of Research in Biology*. 4 : 269-275.
- Sakhare, V.B., 2003. Studies on some aspects of fisheries management of Yeldari reservoir, Maharashtra. Ph.D. Thesis. Swami Ramanand Teerth Marathwada University, Nanded.
- Sakhare, V.B., 2005. Ecology and fisheries of Manjara reservoir in Maharashtra. *Fishing Chimes*. Vol.25 (3), pp. 42-45.
- Sakhare, V.B., 2005. Water quality of Hingni (Pangon) reservoir and its significance to fisheries. In : Advances in Limnology (Ed.S.R.Mishra), pp. 231-235, Daya Publishing House, Delhi.
- Sakhare, V.B., 2007. Reservoir Fisheries and Limnology, Narendra Publishing House, Delhi.
- Sakhare, V.B., 2012. Water quality of Mamdapur reservoir in relation to fisheries. *Ecol. Env. & Cons.* 18(3) : 547-549.
- Sakhare, V.B. and Joshi, P.K., 2004. Present status of reservoir fisheries in Maharashtra. *Fishing Chimes*. 24 (8) : 56-60.
- Sakhare, V. B.and Joshi, P.K., 2010. Physico-chemical limnology of Yeldari reservoir (Maharashtra) in relation to fisheries. In : Aquatic Ecosystem and its management (eds. K. Vijaykumar and B. Vasanthkumar), pp. 110-121, Daya Publsinhg House, New Delhi.
- Salaskar, P.B. and Yeragi, S.G. 2004. The fish fauna of Powai Lake, Mumbai, Maharashtra. *J. Aqua. Biol.* 19 (1) : 69-72.
- Salasakr, P.B. and Yeragi, S.G., 2004. Primary productivity of Powai Lake, Mumbai, Maharashtra. J.*Aqua.Biol.* 19 (1) : 19-22.
- Salasakr, P.B. and Yeragi, S.G., Gordon Rodriks, 2008. Initiative by Maharashtra State Anglers Association (MSAA) to conserve Powai Lake. In : Souvenir of ILEC-IAAB International Workshop on Integrated Lake Basin Management (ILBM), Hyderabad, India, 28-29th August 2008, pp.31-36.
- Sarwade, J.P. and Khillare, J.K., 2010. Fish diversity of Ujani wetland, Maharashtra, India. *The Bioscan*. 1 : 173-179.
- Sayyed Juned A. and 2Bhosle Arjun. B. 2012. Barium and Fluoride Evaluation of Sudha Dam Water Near Bhoker. *Advances in Environmental Biology*, 6(1) : 215-220.
- Sehgal, K.L. 1983. Planktonic copepods of freshwater ecosystem. Envirn. Sci.series. Interprint, New Delhi, pp. 69.
- Sehgal, K., Phadke, G.G., Chakraborty, S.K. and Vijay Kumar Reddy, S., 2013. Studies on Zooplankton Diversity in Dimbhe Reservoir, Maharashtra, India, *Advances in Applied Science Research*, 4(1) : 417-420.
- Shastri, Yogesh and Bhogaonkar, P.Y., 2006. Water quality of Talwade reservoir of Nashik district, Maharashtra. In : Ecology of Lakes and Reservoirs (Ed.V.B.Sakhare), pp 138-142, Daya Publishing House, Delhi.

- Shastri, Yogesh and Pendse, D.C., 2001. Hydrobiological study of Dahikhuta reservoir. *Journal of Environmental Biology*. 22(1): 67-70.
- Singh, K.M.P.1995. Final Project Report of Studies of Hydrobiology and Fisheries of Powai Lake, Mumbai.
- Sone, A.A. ad Malu, R.A., 2000. Fish diversity in relation to aquaculture in Ekburjii reservoir, Washim, Maharashtra. *J.Aqua.Biol.*15 (1 and 2) : 44-46.
- Surve, P.R., Ambore, N.E. and Pulle, J.S., 2004. Population dynamics of microzootic fauna of Khandar dam water, district Nanded (Maharashtra), India. *J.Aqua.Biol.* 19 (2) : 17-21.
- Ubharhande, S.B. and Sonawane, S.R., 2012. Study of freshwater fish fauna and water quality at Paintakli dam from Buldhana district, (M.S.) India. *Journal of Experimental Sciences*, 3(7) : 4-8.
- Valsangkar, S.V., 1993. Mahseer Fisheries of Koyana (Shivajisagar) in Maharashtra: Scrap to Bonaza. *Fishing Chimes*. 12 (10) : 15-19.
- Yardi, Dilip, Patil, S.S., Bandela, N.N. and Auti, R.G., 2008. Conservation of birds of Jaikwadi dam-A proposed ramsar site, Aurangabad. In : Proceedings of TAAL 2007. The 12th World Lake Conference (Eds. Sengupta, M. and Dalwani, R), pp. 547-553.
- Waghmare, V.N. and Mali, R.P., 2007. The study on phytoplankton of Kalamnuri minor irrigation dam, Kalamnuri, District Hingoli, Maharashtra. *J. Aqua. Biol.* 22 (2) : 59-62.

12

North-Eastern States

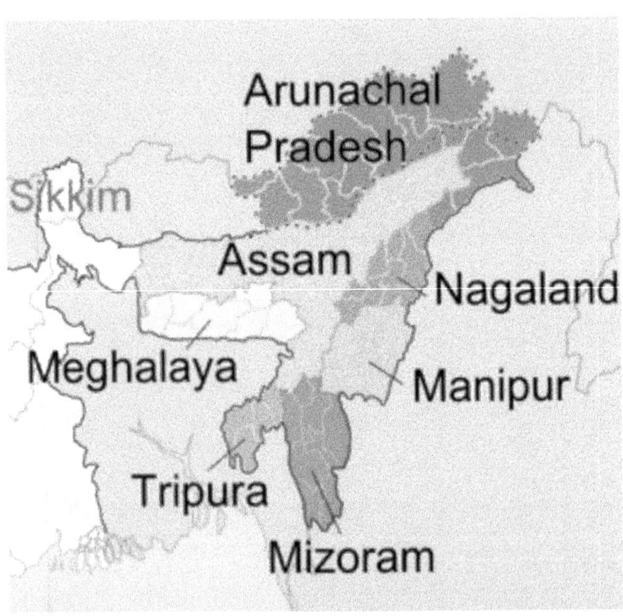

Fig. 12.1. Map of North-Eastern States (Not to scale)

The North-Eatsern region, comprising the seven states of Arunachal Pradesh, Manipur, Assam, Mizoram, Meghalaya, Nagaland and Tripura (22° – 29°30′N latitude and 89°47′ – 97°27′E is beast with a rigged hilly terrain and the Barak river basins constitute the main drainages and together account for over 50% of the hydel potential of the country. To overcome the flood problems few large storage reservoirs are constructed. These reservoirs besides boosting the irrigation and hydropower generation capacity of the region are also providing valuable resources for fish production. The details of some major reservoirs and their fisheries is as under:

Barapani Reservoir (Meghalaya): It is located 10 km from Shillong under the jurisdiction of the East Khasi Hills District Council. It is the oldest reservoir in the region. The reservoir is constructed on river Umium for hydroelectric purpose. The water-spread area of reservoir is about 500 ha. Except for stocking *Cyprinus carpio* in the early seventies, no fishery development activity has been undertaken in this reservoir.

Jyrdemkulai Reservoir (Meghalaya): The reservoir is located downstream of Barapani in East Khasi Hills District and created under stage III of Umtru-Umiam hydroelectric project. The water-spread area of the reservoir is about 90 ha. It is under ownership of Meghalaya State Electricity Board. Negotiations for leasing it out for fishery purpose to state fisheries department have now been successful.

Khandong Reservoir (Meghalaya and Assam): It was constructed in 1986 on the river Kopili hydroelectric project. The water-spread area of the reservoir is 12959 ha. The reservoir falls within the districts of North Cachar Hills (Assam) and Jaintia Hills (Meghalaya). Fishing of the reservoir is not developed on scientific basis.

Umrong Reservoir (Assam): It is constructed under the Kopili Hydro-electirc project satge –I in the year 1986. The water spread area of reservoir is about 5550 ha. The reservoir is located in North Cachar Hill district of Assam. North Eastern Electric Power Corporation is the owner of this reservoir. No fishery development activity has been initiated in this reservoir till now though efforts are being made to inttate the same in near future.

Gumti Reservoir (Tripura): It is located in the south Tripura district. Its construction was completed in the year 1976 under Gumti Hydroelectric project. The water-spread area of the reservoir is of 4500 hecatres, and comprises Gumti, Sarma and Raima river basins.

The reservoir is having some autostocking with fish breeding observed in certain areas. The resident fish species of economic importance *are Catla catla, Labeo calbasu, Labeo rohita, Cirrhinus mrigala, Heteropneustes fossilis, Clarias batrachus,* murrels and few species of weed fishes and minor carps.

The fishery in small quantities is supported by *Tor tor, Tor putitora, Anguilla bicolor, Wallago attu, Mystus aor, Mystus seenghala, Anguilla bicolor* and *Macrobrachium rosenbergii.*

Advanced fingerlings of Indian major carps (*Catla catla, Labeo rohita* and *Cirrhinus mrigala*) and exotic carps (*Hypophthalmichthys molitrix* and *Ctenopharyngodon idella*) are being stocked every year since 1978-79. Depending on their availability fingerlings @ 3.3 to 156.4/ha have been stocked during different years. The breeding of *H. molitrix* and *C. idella* has been observed in the reservoir.

The fish production has shown gradual improvement from 37.5 kg/ha in 1978-79 to 71.55 kg/ha in 1987-88. Indian major carps and exotic carps constitute about 15% of the total catch with minor carps (45%) and weed fishes (33%) forming the bulk. The reservoir contributed 2.3% of state's fish production during 1987-88.

Main fishing gears used in reservoir are gill-net (5 to 20 cm mesh size), baskets, long lines and dragnets. Licensed fishermen of eight co-operative socities do fishing. Nearly 1000 fishermen families of the surrounding villages are subsisting on the fishery of this reservoir. The reservoir has the inherent problem of submerged three stumps.

Rudrasagar Reservoir (Tripura): Rudrasagar Lake is situated at Melaghar in West Tripura district, about 40 km far from Agartala.Rudrasagar Lake was built by the ancient king of Tripura, Maharaja Bir Bikram Kishore Manikya in year 1930. The lake is very famous for boating as well as for fishing.

The Rudrasagar Lake is declared as National Lake No. 13 and recently, on 8th Novemer 2005, it was declared as an International Lake numbered 1572 as a Ramsar Site. The lake gets water from three different sources. These are Noyacharra, Kemtalicharra and Durlav Narayancharra. At the outlet, earlier, lake was liked with Gomati river but nowadays sluice gate is formed there. The lake is looked after by a fishermen Co-operative Society namely 'Rudrasagar Udbastu Fishermen Co-operative Society' which was established during 1951 with 2000 members which continues, as at present.

Fisheries of Lake: Roy and Mandal (2011) documented fisheries of Rudrasgar Lake. A major problem in respect of Rudrasagar Lake is siltation and increase in agricultural land area. Due to these, the lake is losing its depth, leading to reduction in fish production and fish biodiversity. Earlier, all the commercially important fish species such as catla, rohu, mrigal, calbasu, common carp, grass carp, pabda, chitala, tengra, rani fish, singhi, Esomus, chapila, colisa spp., chanda spp. were available in the lake and they used to breed there naturally. But, presently due to siltation, decrease in water area and pollution, these species are rarely found and their natural breeding has almost ceased. Earlier days about 600-800 active fishermen were fully dependent upon fishing in the lake. But now their number came down to 300-500. The fishermen are not satisfied with the catch they are getting and for this reason, many of them have started agriculture and some other allied activities. The society is also not getting good returns from the fishery activities, their main source of income being ferry boating. Recently the society has started stocking fingerlings of commercially important fish species. During year 2011-12, they have stocked around 9 lakhs of advanced fingerlings. Presently, the society has started giving lease of a particular area to individual fishermen for 6 to 7 months *i.e.,* June to December. During that time the fishermen remain engaged with their fishing activity only. During summer season the same area gets dried up and the fishermen convert it into agricultural land. During year 2010-11, the income of the society from fishery lease was around Rs.19, 39,310/- and during 2011-12, the income has reached to 19, 57,182/-. Seed stocking is done during April-May.

The fish production in the lake is decreasing almost every year (Table 12.1). The main reason for reduction of fish production in lake are as under: (1) Increase in agricultural land area around the lake, which led to reduction in water area for fisheries (2) Siltation is a major ad very severe problem in lake which badly affects fisheries of lake (3) pollutants and fertilizer remnants from nearby areas or agriculture land areas are finally coming into the lake and polluting the water which is harmful for flora and fauna (4) High water temperature and lesser depth during summer season is creating stressful conditions for fishes which is also another reason for the decline in fish production in Rudrasagar Lake (5) During late winter and summer time, the maximum part of the lake remains fully covered by Eichornia and other weeds, which hinder fish movement. These weeds prevent sunlight penetration into the lake water ad affect primary productivity. Operation of nets also becomes very difficult. Eichornia and other aquatic weeds should be cleaned ad removed in order to restore the past conditions of the lake. Eichornia can be used for producing bio-gas, vermin-compost and for other uses. The water body will be free and and there will be sufficient light penetration into the bottom and primary productivity of the lake will increase, and as a result, fish production will also increase.

Table 12.1. Fish production in Rudrasagar Lake

Sr.No.	Year	Fish Production (In Tonnes)
1.	2006-07	30.275
2.	2007-08	48.919
3.	2008-09	30.366
4.	2009-10	28.593
5.	2010-11	29.847

Crafts ad Gears used: The gears commonly used for fishing are drag net, faloon et, chapila net, latte net, atta net and aanta (trap). Among crafts mainly dingi is used in the lake.

Nongmahir Reservoir (Meghalaya): Nongmahir reservoir was constructed in the year 1979. The main earthen dam is 150 m long with a height of 50 m above the bed level of 638.72 m above MSL. The reservoir is more or less circular with a number of islands. The reservoir has a total waterspread area of 70 ha with a live storage capacity of 5.8×10^6 cm at full storage level and 2.16×10^6 cum at the dead storage level. The small catchment comprises hillrocks, bearing loose laterite soil. The morphometric features of Nongmahir reservoir are depicted in Table 12.2.

The reservoir is situated in the low slopes of east Khasi Hills where climate varies from warm sub-tropical to cool and bracing. The reservoir area receives precipitation ranging from 2.19 to 642.8 mm per month.

Table 12.2. Morphometric features of Naongmahir reservoir

1. Elevation at Full reservoir level : 672.07 m (Above MSL)
2. Area of the lake : 70 ha
3. Maximum water level : 673.78m (above MSL)
4. Minimum water level : 669.20 m (above MSL)
5. Dead storage capacity : 5.8×10^6 cum
6. Live storage capacity : 2.16×10^6 cum
7. Mean depth (volume/area) : 5.28 m
8. Length of the dam : 150m
9. Spillway level : 672.07 m (above MSL)
10. Tunnel intake level : 661.60 m (above MSL)
11. Power generation capacity : 30 MW

The basin sediments are marked by dominant sand and clay fraction with organic carbon percentage varying from 1.74 to 2.13. The soil invariably acidic in reaction (pH 4.6 –6.0), contained poor to moderate available nitrogen (29.91 to 44.59 mg/100 g) and phosphorus contents (0.38 to 1.12 mg/100 gm).The specific conductance was of low order (107.3 to 367.3 µmhos/cm) and organic matter ranged within 2.3 to 3.61%.

Water Quality: The water quality of Nongmahir reservoir is depicted in Table 12.3.

Table 12.3. Range of physico-chemical parameters in Nongmahir reservoir

1. Water temperature (°C)	25 – 31
2. Transparency (cm)	1 – 2.6
3. pH	6.6 – 7.40
4. Dissolved oxygen (mg/l)	6.9 – 7.28
5. Free carbon-dioxide (mg/l)	0.16 – 5.0
6. Total alkalinity (mg/l)	12 – 28
7. Total hardness (mg/l)	16.84 – 25.52
8. Chloride (mg/l)	6.66 – 22.56
9. Iron (mg/l)	0.09 – 0.3
10. Ammonical nitrogen (mg/l)	0.14 – 0.24
11. Nitrate nitrogen (mg/l)	0.02 – 0.82
12. Silicates (mg/l)	1.00 – 6.60
13. Phosphate (mg/l)	0.02 – 0.08
14. Calcium (mg/l)	13.9 – 20.88
15. Magnesium (mg/l)	2.56 – 9.10
16. Total dissolved solids (mg/l)	14 – 26
17. Specific conductivity (µ mhos/cm)	69.60 – 150.21

(Source: Sugunam and Yadava, 1991)

Plankton: Standing crop of net plankton in quantitative terms, indicated a poor community structure. The dry weight of total plankton, estimated from various sectors covering the four seasons, varied from 0.20 mg/l to 0.45 mg/l with an average value of 0.32 mg/l. The average plankton density in numerical terms was estimated at 5,440 units/l. Chlorophyceae, represented mainly by *Staurastrum* sp., *Chaetophora* sp., and *Dinobryon* sp., accounted for the bulk of plankton in the reservoir. The phytoplankton constituted more than 96% of the total plankton in the main reservoir. A higher density of the blue-greens, indicating eutrophication processes in the shallow enriched zones, characterized the plankton. The bays with littoral formations represent the trophogenic zones of the reservoir. (Sugunan and Yadavam, 1991).

The planktonic diversity recorded in the Nongmahir reservoir are depicted in Table 12.4.

Table 12.4. Plankton diversity in Nongmahir reservoir

Myxophyceae: *Microcystis aeruginosa, Spirulina* sp., *Anabaena* sp., *Phormidium* sp.
Chlorophyceae: *Ulothrix* sp., *Dinobryon* sp., *Staurastrum* spp., *Chaetophora* sp.
Bacillariophyceae: *Amphora* sp., *Navicula* sp., *staurastrum* spp., *Chaetophora* sp.
Dinophyceae: *Ceratium hirudinella.*
Rotifera: *Keratella* sp., *Brachionus* sp., *Polyarthra* sp., *Trichocerca* sp.
Cladocera: *Daphnia* sp., *Chydorus* sp.
Copepoda: *Cyclops* sp., *Diaptomus* sp., *Nauplii.*

Fish and Fisheries: The commonly recorded fishes in the *Catla catla, Labeo rohita, Cirrhinus mrigala, Hypophthalmichthys molitrix, Cyprinus carpio, Acrossocheilus hexagonolepis, Tor putitora, Puntius sophore, Danio rerio, Danio acquipinnatus, Danio dangila, Clarias batrachus,* and *Heteropneustes fossilis.*

Use of angle and bamboo traps is very common in fishing. Generally two to six people angle every day. The catch is very little and mostly comprises small specimens of common carp. Traps, lures and snares from the basic capture paraphernalia. Most common is the use of bamboo split mats, erected in the marginal areas of the reservoir. The pen forming a semi-circle extends its arms beyond the water phase, completely barricading the littoral zone. The pen facilitates entry of fish from the reservoir through a trap door, but prevents their return. Normally, the fishes trapped in the pen during the night are harvested in the morning, using hand nets, cast nets etc. Weed/trash fishes and *Heteropneustes fossilis* are commonly caught in these pens. Common carp is caught in the pens during monsoon period.

A single gill net owned by a local villager is occasionally used. The two country boats in the reservoir are more frequently used for collecting forest produce and for transporting the anglers.

Umaim Reservoir (Meghalaya): The Umiam reservoir lies in the Khasi Hills of Meghalaya (25°39′30″N and 91°43′51″E) at an altitude of about 900 m above the mean sea level.The reservoir was constructed in 1965. It falls under the jurisdiction of the Khasi Hills Autonomous District Council. Damming the river Umiam as stage 1 of the Umtrum-Umiam Hydroelectirc Project formed the reservoir

The morphometric features of reservoir are presented in Table 12.5.

Table 12.5. Morphmetric features of Umiam reservoir

Total water spread area	500 ha
Altitude	About 900 m above MSL
Maximum reservoir level (elevation)	3220 ft
Minimum reservoir level (elevation)	3150 ft
Gross storage (elevation)	3220 ft
Dead storage (elevation)	3150 ft
Total storage of dam	640 ft
Maximum height of dam	238 ft

The Umiam reservoir water is continuously drawn through a tunnel for hydroelectric power generation. Besides, a jackwell pumping station of army draws water regularly for drinking and other purposes. The continuous drawn down of water from reservoir and the variation in the inflow from the catchment areas diurnally and seasonally results in wide fluctuations in water level in reservoir. The water level fluctuations have been found to affect the fish catch considerably. The reservoir level increases from the month of June, with the onset of monsoon and reaches the highest level of 3220 ft in November. From December, the water level decreases and a lowest level of 3175.14 ft was recorded in May.

Table 12.6. Water quality of Umiam reservoir

Parameters	Pre-monsoon	Monsoon	Post monsoon	Mean
Temperature (°C)	21.27	26.43	18.79	22.16
Transparency (cm)	268.16	209.53	220.94	232.88
pH	7.62	9.11	7.12	7.95
Dissolved oxygen (mg/l)	7.60	9.08	6.98	7.89
Free CO_2 (mg/l)	1.74	0.15	3.39	1.76
Phenolphthalein alkalinity (mg/l)	0.33	3.79	0.0125	1.38
Methyl orange alkalinity (mg/l)	30.02	20.74	20.86	23.87
Hardness (mg/l)	25.99	17.23	17.73	20.32
Chloride (mg/l)	5.60	1.85	1.28	2.91
Nitrite (mg/l)	0.047	0.10	0.062	0.0697
Phosphate (mg/l)	0.06	0.095	0.05	0.682
Total dissolved solids (mg/l)	37.07	28.89	23.40	29.79

Fish diversity: Of the 29 species, 21 are native fish species, which are originally found at the time Umiam reservoir was formed, and 5 were introduced native fish species. The species *Cyprinus carpio* (var.communis and var.koi) and *Clarias gariepinus* were the exotic fish species introduced in Umiam reservoir.

Table 12.7. Check list of fishes recorded in Umiam reservoir

Species
Catla catla
Labeo rohita
Labeo calbasu
Labeo gonius
Cirrhinus mrigala
Cyprinus carpio var.communis
Cyprinus carpio var.koi
Clarias gariepinus
Badis badis
Brachydanio rerio
Chanda nama
Channa oreintalis
Channa punctatus
Clarias batrachus
Danio dangila
Garra gotyla
Garra lissorhynchus
Garra mccllendi
Glossogobius giuris
Glyptothorax spp
Heteropneustes fossilis
Lepiodocephalus guntea
Neolissocheilus hexagonolepis
Neolissocheilus hexastichus
Puntius sophore
Puntius shalynius
Schistura sikmaeinsis

Fishing Areas: Commercial fishing is not allowed in reservoir, although it is done on a small scale at three locales namely Mawdun, Dongrylla, and Umniuh. Cast nets and gill nets are commonly used in fishing. About 50 fishermen are involved in commercial fishing and the catches are collected by three women traders, who in turn sell them to the vendors at Bara Bazar fish market at Shillong and Lad Umsawo. A total of about 15 vendors are involved in the sale of fishes caught from Umiam reservoir. Besides commercial fishery, subsistence fishing takes place particularly in the bay areas of the reservoir and gears like cast nets, dragnets and traps are commonly used. The fishes caught are generally utilized for household consumption.

Sport Fishing: The reservoir attracts about 250 to 300 sport enthusiasts for angling on most of the days. The number of people who come for angling are more during holidays. They generally come early in the morning and continue to fish till late in the night and the fishes caught by angling are not sold.

Fishery: The common carp introduced in Umiam reservoir in early 1970s forms the mainstay of fishery and has been sustaining the livelihood of many people living in the hamlets around the reservoir. The other varieties commonly occur in reservoir are *Channa punctatus, Clarias batrachus, Clarias gariepinus, Neolisscheilus hexagonolepis* and *Chanda nama.* The small sized fishes like *Danio dangilla, Garra gotyla, Brachydanio rerio, Schistura sikmaeinsis, Badis badis* and *Lepidocephalus guntea* caught in the subsistence fishery. These fishes are also used as food. Due to their small size these species are also used as ornamental fishes.

During 2000-01, the stocked Indian major carp contributed substantially to the total fish catch of Umiam reservoir, which gave a fish yield of 97.87 kg/ha/yr (Vinod *et al*, 2003 b). During year 2005, average fish production of Umiam reservoir was 65.52 kg/ha/yr, which includes catches from subsistence and sport fishing, in addition to the commercial catches.

Common Carp Fishery: Common carp invariably formed the bulk of the catch during all months and seasons. The contribution of common carp to the total reservoir catch during different months is presented in Table 12.8 and 12.9.During June 2003 to May 2004, common carp contributed 30.54 to 92.86% to the total catch. In April and May, catches of common carp were comparatively less, wherein they contributed 40.19% and 30.54% respectively. Further, the overall

Table 12.8. Contribution of common carp catch to the mean daily total catch for June 2003 to May 2004

Months	Mean total daily catch (kg)	Mean daily catch of common carp (kg)	% contribution of common carp to the total catch
June 2003	151.06	114.07	75.51
July	48.37	45.51	94.09
August	77.81	67.7	86.20
September	76.75	65.12	84.85
October	74.73	55.16	73.81
November	130.00	118.51	91.16
December	148.19	135.82	91.65
January 2004	86.84	80.64	92.86
February	80.95	72.10	89.07
March	54.04	45.27	83.77
April	52.48	21.09	40.19
May	42.96	13.12	30.54
Mean	85.35	69.46	81.38

fish catch was also poor during April and May, amounting to only 52.48 kg and 42.96 kg respectively. During June 2004 to May 2005, common carp catches ranged from 14.27 to 98.92% of the total reservoir fish catch; the highest catch of common carp was recorded in July, while the lowest catch was observed in April. During June 2004 to May 2005 too, the contribution of common carp was very low and they accounted to 14.27 and 15.87% respectively in April and May.

Table 12.9. Contribution of common carp to the mean daily total catch for June 2004 to May 2005

Months	Mean total daily catch (kg)	Mean daily catch of common carp (kg)	% contribution of common carp to the total catch
June 2004	119.62	79.41	66.39
July	137.33	135.84	98.92
August	123.25	120.64	97.88
September	104.79	102.90	98.20
October	28.80	26.62	92.43
November	47.92	44.76	93.41
December	67.39	66.55	98.75
January 2005	53.92	52.65	97.64
February	40.48	31.34	77.42
March	49.96	28.01	56.06
April	45.19	6.45	14.27
May	90.14	14.31	15.87
Mean	75.73	59.12	78.07

Decline in Chocolate Mahseer Catch: *Neolissocheilus hexagonolepis* (chocolate mahseer) is endemic and highly preferred and good price fetching fish species of the reservoir. Studies have indicated that there has been a drastic decline in their catches in the last 15 years (Mahapatra *et al.*, 2004 a, Vinod *et al.*, 2007). Most of the chocolate mahseer caught in the commercial catches were very small. Earlier studies of Mahapatra *et al.* (2004a) also revealed that the adults (above 100 gm) contributed a mere 1.67% while the juvenile (below 100 gm) contributed 98.33% by number thereby indicating the rampant harvest of juvenile chocolate mahseer from Umiam reservoir. In subsistence fishing in bays also, often fry and fingerlings of chocolate mahseer are harvested in large quantities possibly leading to a drastic decline in their population.

Ward's Reservoir (Meghalaya): Ward's reservoir is situated in the heart of Shillong (Meghalaya, North East India) at an altitude of 1460 m (Lat. 25°34′N) and Long. 91°552′E). The reservoir has many names like Hopkinston's tank, Elliot's lake, Wards' lake, Nan Palok and Laath saab Ko Talao. The origin of the lake though the date isnot known was the initiation of digging by a covict, when water was struck that the district authorities took over ad extended the digging to the present lake. The idea of using this lake for recreation with a small park all around it goes to the credit of the the Chief Commissioner, William E. Ward 1893 after who the lake is presently known.

The lake has a maximum length of 333 meters ad maximum breadth 75 meters with a shore line of 1284 meters. The total water spread area of the lake is 23,800 square meters which encloses 80,920 cubic meters volume of water. The maximum depth is 6 meters with a mean depth of 3.4 meters.

Alfred and Thapa (1996) documented limnological investigation on Ward's reservoir. The average maximum temperature raged between 13.4°C and 25°C. The minimum temperature was recorded in January while the maximum in August. The maximum rainfall at reservoir site was recorded at 460.1 mm in the month of June, while it was nil in the months of January and December. Relative humidity was maximum (87%) in July and minimum (40%) in the month of December. The range of physico-chemical factors observed in Ward's reservoir is depicted in Table 12.10.

Table 12.10. Water quality of Ward's Reservoir

Sr.No	Parameters	Range
1.	Air temperature (°C)	8.3 – 25
2.	Water temperature (°C)	10.9 – 23.6
3.	Transparency (cm)	30 – 144
4.	pH	5.7 – 6.9
5.	Conductivity (μ mhos/cm)	49.4 – 240
6.	Dissolved oxygen (mg/l)	Nil – 9.8
7.	Free carbon-dioxide (mg/l)	2.2 – 39.9
8.	Alkalinity (mg/l)	25 – 66
9.	Phosphate (mg/l)	1.9 – 14
10.	Nitrate (mg/l)	0.2 – 1.6
11.	Silicate (mg/l)	0.1 – 0.9
12.	Calcium (mg/l)	1.0 – 5.0
13.	Magesium (mg /l)	22 – 293

Table 12.11 provides list of phytoplankton recorded in the reservoir. Phytoplankton were represented by classes such as chlorophyceae, euglenophyceae, bacillariophyceae, dinophyceae and myxophyceae.

Table 12.11. Phytoplankton diversity in Ward's Reservoir

Class : Cholorophyceae: *Chlamydomonas, Eudorina, Pandorina, Gonium, Gloeocystis, Dictysphaerium, Pediastrm, Coelastrum, Spirogyra, Hormidium, Ulothrix, Actinostrum, Sceedesums, Ankistrodesmus, Closteriopsis, Kircheriella, Oocystis, Seleastrum, Westella, Closterium, Cosmarium, Desmidium, Gymozyga, Micrasterias, Staurastrum.*
Class : Eugelnophyceae: *Eiglens, Phacus, Gleobotrys, Botryococcus, Chrysococcus, Syura, Tribonema, Dinobryon.*
Class : Bacillariophyceae: *Melosira, Tabellaria, Meridium, Syedra, Diatoma, itzschia, Cymbella, Amphipleura, Navicula, Pinnularia.*
Class : Dinophyceae: *Gymnodinium, Peridinium, Ceratium.*
Class : Myxophyceae: *Coelosphaerium, Microcystic, Glococapsa, Merismopedia, Chroococcus, Anacystis, Aphanocapsa, Oscillatoria, Spirulina.*

Zooplankton recorded in the reservoir were represented by five major groups-protozoa, rotifers, cyclopoda, cladocera and ostracoda (Table 12.12).

Table 12.12. Zooplankton diversity in Ward's Reservoir

PROTOZOA: *Centropyxis, Difflugia.*
ROTIFERA: *Keratella, Brachionus, Lecane, Epiphanes, Asplanchna, Trichocera, Testudiell, Proales, Cephalodella, Polyarthra, Lepadella.*
CYCLOPODA: *Cyclops, Diaptomus.*
CLADOCERA: *Diaphanosoma, Daphnia, Chydorus, Bosmina.*
OSTRACODA: *Cypris.*

A net primary productivity and gross primary productivity values at different stations were see to be higher in the euphotic zone with a reduction in their levels as one goes from the top to the bottom column layers of the water. The actual values recorded were early 900 mgC/m^3/day for net primary productivity and early 2000 mgC/m^3/day for gross primary productivity. These values definitely reveal the oligotrophic nature of the reservoir.

Loktak Lake (Manipur): Loktak lake is the largest freshwater wetland in North-Eastern India.It is also known as the only floating lake in the world due to the floating phumdis on it.Loktak is a shallow and somewhat acidic freshwater lake. It is the largest wetland in the North-Eastern region of India and has been referred as the lifeline of the people of Manipur due to its importance in the socioeconomic and cultural life. The lake is situated 38 kms south of Imphal city of Manipur state. It covers an area of about 286 sq.kms at the elevation of 768.5 m located between longitudes 93° 46′ and 93° 55′ E and latitudes 24° 25′ and 24° 42′ N. The depth of water during dry season ranges between 0.5 m to 1.5 m. The total water spread area of about 490 sq.kms. was recorded in 1966. Loktak lake plays an important role in the ecological and economic security of the region. The Lake has been the source of water for generation of hydroelectric power, irrigation and water supply. A large population living around the lake depends upon the lake resources for their sustenance. The lake is included on

Montreux record in 1993 as a result of ecological problems such as deforestation in the catchment area, infestation of water hyacinth, and pollution.

A number of streams flow directly into Loktak lake. These includes Nambul, Nambol, Thongjarok, Awang kharok, Awang khujairok, Oinan, Keinou and Irulok contribute maximum silt load to the lake. The indirect catchment area covers catchments of 5 important rivers *i.e.,* Imphal, Iril, Thoubal, Sekmai and Khuga and is spread over an area of 7157 sq.kms. It is rich in biodiversity and has been designated as a wetland of international importance under Ramsar convention in 1990.

The biodiversity of lake comprises 233 species of aquatic macrophytes of emergent, sumergent, free floating and rooted floating leaf types. The lake also provides refuge to thousands of birds, which belong to at least 116 species.

The lake is known for its unusual infestation with weeds, both dead and live, and also heavy siltation. This infestation piled up over years into various floating shapes, oblong, circular etc., with a depth of about 5 to 6 feet. These are known as 'phums'. They cover about half of the surface of the lake. Phums vary in size and can be as large as one ha. In a way, they have turned out to be a sort of fish aggregation devices that facilitate fish harvesting. Over years, the fishermen dependent on the lake fisheries improvised a certain alteration in the structure of several of the phums. Leaving a rim all round, they have removed the rest of the phum. The truncated area is used as a feeding place for fish which are offered fresh weeds, rice bran and ground nut oil cake powder as feed to attract their aggregation.

The lake is being stocked with fingerlings of carps since 1966. October to March is the season for phum fishing. Fishermen encircle a phum with a net to collect the fish. It is stated that phums yield between 15-40 t of fish annually, besides catches taken through use of hook and lines etc.

Phums get dispersed sometimes moving over long distances within the lake. When this happens, fishermen improvise fresh phums close to their habitations. Another feature one notices is that huts are built on phums and are mostly used as lodges for visitors or for fishermen's use.

In order to utilize the potential of the lake for beneficial purposes, the state government has set up Loktak Development Authority (LDA). The main purpose of the LDA was to set up a hydroelectric project, but beside this LDA is also charged with the responsibility of development of fisheries of the lake. The authority has taken up the work of cleaning the lake from phums and scattered weeds. To facilitate stocking, the authority has set up two major carp hatcheries.

Fish Fauna: The fish fauna of lake comprises 64 species. Loktak lake serves as the breeding ground for several species of migratory fishes such as *Labeo dero, L. angra, L. bata, Cirrhinus reba* and *Osteobrama belangeri*. These riverine species migrate from the Chindwin Irrwaddy river system in Burma to the upstream areas of Manipur river and breed in various shallowlakes in the valley. These fishes disappeared from the lake since the construction of Ithai barrage which has blocked their migratory route. The fisheries of lake is also supported by *Hypophthalmichthys molitrix, Ctenopharyngodon idella, Cyprinus carpio, Cirrhinus* *mrigala, Labeo rohita, L.calbasu, L.gonius, Catla catla, Puntius* spp., *Esomus danricus, Heteropneustes fossilis, Ambassis* sp., *Osteobrama cotio, Channa* spp., *Glossogobius* spp., *Notopterus spp, Amblypharyngodon mola, Anabus testudineus, Lepidocephalus* spp., *Clarias batrachus* and *Tilapia* sp.

Lokatak lake serves as the breeding ground for several species of migratory fishes such as *Labeo dero, Labeo angra, Labeo bata, Cirrhinus reba* and *Osteobrama belangiri.*

The fish catch is dominated by exotic carps (33%), followed by Indian major carps (21%), minnows (14%) and rest by other forms. The fish production of the lake is 78 kg/ha/yr.

The hooks used are mainly of no. 12 and 11. In each line there are about 5 to 7 hooks tied to different snoods of length 9 to 11 cm. The lines are of synthetic monofilament with size varying from 0.45 to 0.60 mm diameter. The hooks are baited and hurled to a distance of about 15 m from the shore. Such hooks and lines are set in series numbering about 2000 along the banks at a gap of about 0.5 m in between. Each line is wound around an empty can of 10 cm diameter and height of 14 cm for retrieving the hooks and the cans are weighted with stones parallel to the shore at a height of about 40 to 45 cm from the water surface so that they may not be easily swept. This innovation is to prevent the retrieving cans from being carried away by the fishes. As the lines rest on the supporting rope, when the hooked fish drags the retrieving can, it gets entangled in the supporting rope by gravity. After the bait gets disintegrated, the lines are hauled up. The lines which hook catches are pulled slowly and when the fish drags on the opposite direction the line is released slightly to reduce tension in the line. When the fish reaches the shore, it is scooped out of water using a scoop net having a mouth diameter of 65 cm and depth of 40 to 45 cm made with yarn of diameter 1.8 mm.

Another technique of fishing in Loktak lake is the use of pole and line with bait. In this method lightweight bamboo poles of length 2 to 3 m are used with nylon lines of length 5 to 6 m. A no. 18 hook is tied at the end of the line and after a gap of 4 to 5 cm; another hook of same size is tied. In between the two hooks, a snider made of lead is loaded and a float of dried stick is tied at a height of 1 to 1.5 m from the bottom hook. A platform is constructed with wooden planks and the four legs are fixed at the bottom of the canal and pole and line is operated from this platform. The bait is in the form of wheat flour mixed with water forming a dough and baited in small pieces of about 3 mm diameter. The fishes caught are kept in a container which is made of bamboo chips and is kept partly immersed in the water. The catch ranges from 0.5 to 3 kg per day per hook and the intensity of catch varies from season to season having a bimodal peak during March-April and August-September and a lean season during January-February.

In ripping method of fishing No. 9 or 11 hooks are used. The hooks are tied by means of rolling hitch knot with snood length of 9 to 11 cm and is tied to the end of the line which is about 4 m long, synthetic monofilament twine, with 0.66 mm diameter of which the other end is tied to a pole of length 2 to 3 m. These are operated when fishes exhibit surfacing during morning hours on some cloudy days, which may be due to low concentration of oxygen. At this time fishermen lash the fish quickly with a bunch of hooks holding of the end of the pole. When the hooks rip the fish, the catch is pulled towards the shore and finally scooped out of water with a scoop net.

Dip net of size ranging from 2×3 m^2 to 1.5×2.5 m^2 are also used at few points. The catch ranges from 1 to 5 kg per net per day with a peak during August to January.

Basudha and Vishwanath (1999) studied food and feeding habits of *Osteobrama belangeri* from Loktok lake and adjoining lakes.

In *O.belangeri* ranging 6-10 cm TL, the composition of food in the gut varied such that 51.91% was made up of zooplanktons like cladocerans (Daphnia), Copepods (Cyclops), rotifers (Branchionus), and prozoans (Arcella, Diffulgia). 16.33% was constituted by algae of which 9.61% and 6.72% were unicellur green algae and blue green algae respectively including Spirogyra, Nostoc and Anabaena. Diatoms, insects and worms formed 12.92, 4.26 and 4.86% of the gut contents respectively. Inorganic debris, mud, sand granules and small portion of aquatic plants formed upto 9.76%.

In fish ranging 10-20 cm (TL) indigested and chitinous matter of insects along with their nymphs and larvae, totally or partially intact, constituted 20.19% of the gut. The plant matter comprising of leaves, stems and roots of aquatic plants formed 40.64%. Unicellular algae, zooplanktons, worms and miscellaneous items formed 12.24, 15.56, 3.78 and 4.86% respectively. The insect diet comprised mainly of mayfly nymphs, larvae of addies fly, insect eggs and larvae and adults of chironemous. Zooplankton was represented by copepods and cladocerans, worms belongs to oligochaeta like *Branchiura* sp., *Tubifex* sp., and *Limnodrilus* sp, mud sand and inorganic debries constituted the miscellaneous part.

In fish ranging 20-26 cm TL, plant matter dominated over the food items (66.33 %) followed by algae (19.00 %), zooplankton (2.27%), insects (1.77), worms (0.33 %) and miscellaneous items (4.07).

Monthly observations revealed that stomach content of fishes varied in different seasons. A well pronounced feeding activity was noticed during October to April. It was extremely low from June to August.

Animal matters including zooplanktons, insects and worms are preferred by juveniles *O. belangeri* (40-60%). However, as the fish increased in its size it has higher preference for the plant food items. Adult fish prefers plant matters which includes large amount of leaves, stem and roots of aquatic plants. Other food items such as insects, worms, zooplanktons and phytoplanktons are also found in fairly large amount in the gut. Therefore, it may be concluded that *O. belangeri* is an omnivorous fish feeding mainly on zooplanktons and algae in the juvenile state and macrovegetation in the adult stage.

A dimenisting trend of gastro-somatic index during June-August period would suggest its low feeding intensity. High feeding intensity observed during September-February, appears to have correlation with post monsoon availability of sample amount of food in fish environment.

Threats and their Impacts

1. Over-exploitation, indiscriminate methods of fishing, extensive growth of phums are responsible for decrease in fish production. Construction of Ithai barrage has interfered with the fish migration from Chindwin-Irrawady river system of Myanmar and consequently brought changes in the catch.
2. Extensive deforestation and unscientific land use practices in the catchment area are responsible for deposition of silt in the lake.
3. The proliferation of phums and aquatic weeds have led to the reduced water holding capacity, deterioration of water quality, interference in navigation, and overall aesthetic value of the lake.
4. Inflow of organo-chlorine pesticides and chemical fertilizers used in the agricultural practices around the lake, muncipal wastes brought by Manbul river that runs through Imphal, soil nutrients from the denuded catchment area and domestic sewage from settlements in and around the lake are responsible for deterioration of water quality.

DOYANG RESERVOIR (NAGALAND)

The Doyang Reservoir, situated in Wokha District of Nagaland, is one of the largest reservoirs in the north-east region of the country. It is situated between the coordinates 26° 13′ 10″ N and 94° 17′ 90″ E. The impounded Doyang Reservoir with a catchment area of 2,606 ha falls under the category of "medium reservoirs"

Table 12.13. Morphometric and hydrographic details of Doyang Reservoir

1. Location (latitude and longitude)	26°13′10″ N and 94°17′90″ E
2. Length of the dam	525 m
3. Type of the dam	Rock-fill with impervious core
4. Height of dam	97 m
5. Reservoir area	3,407.73 ha
6. Catchment area	2,606 ha
7. Maximum depth at dam site (Bridge point)	50.1 m
8. Minimum depth at dam site (Bridge point)	43.2 m
9. Average depth at dam site (Bridge point)	46.39 m
10. Full reservoir level (FRL)	333 m
11. Storage capacity at FRL	565 m cum
12. Minimum drawdown level (MDDL)	306 m
13. Inflowing rivers	Doyang, Chumeya, Djupvu, Tzuza and Chubi
14. Outflowing rivers	Doyang
15. Average annual rainfall of Wokha	2,000 mm to 2,500 mm

and is mainly fed by the rivers Doyang, Chumeya, Djupvu, Tzuza and Chubi. It is important in view of its ichthyofaunal diversity and has been a source of livelihood and employment generation for many fishing communities surrounding it.

The fish fauna of Doyang Reservoir is comprised of 90 species belonging to 19 families. Important ichthyofauna of the reservoir based on their abundance is presented in Table 12.14. Among the families encountered, Cyprinidae dominated the fishery. Among the catfishes. The family Sisoridae, Bagiridae and Siluridae were predominant.The other families which contributed to the ichthyofaunal diversity were Balitoridae, Channidae, Mastacembelidae and Psylorhynchidae. The abundant fish groups included *Catla catla, Cyprinus carpio* var. *communis, Labeo bata, Labeo rohita* and *Mastacembelus armatus.* Other important ichthyofauna comprised of *Barilius* spp., *Cirrihinus mrigala, Devario devario, Hypophthalmichthys molitrix, Labeo calbasu, Neolissocheilus hexagonolepsis, Puntius* spp., *Tor* spp., *Channa* spp., *Crossocheilus latius, Ctenopharyngodon idella, Cyprinus carpio* var. *specularis,Garra* spp., *Labeo dyocheilus, Clupisoma garua, Sperata aor, Sperata seenghala, Bagarius yarelli, Glyptothorax* spp., *Psilorhynchus homaloptera, Clarias* spp., *Botia dario, Badis badis, Anguilla bengalensis bengalensis,Amblypharyngodon mola, Chagunius chagunio, Cyprinion semiplotum, Labeo dero, Labeo gonius, Salmostoma bacaila, Ompok* spp., *Nemacheilus botia, Schistura manipurensis.*

Table 12.14. Fish Fauna of Doynag Reservoir, Nagaland

Catla catla, Cyprinus carpio var. *communis, Labeo bata, Labeo rohita* and *Mastacembelus armatus,Barilius barna, Cirrihinus mrigala, Devario devario, Devario aequipinnatus, Hypophthalmichtys molitrix, Labeo calbasu, Neolissocheilus hexagonolepsis, Puntius chola, Tor putitora, Tor tor, Channa straita, Xenentodon cancila,Barilius bendelisis, Barilius tileo, Crossocheilus latius, Ctenopharyngodon idella, Cyprinus carpio* var. *specularis,Garra lissorhynchus, Labeo dyocheilus, Puntius sophore, Puntius ticto, Tor progeneius, Clupisoma garua, Sperata aor, S. seenghala, Channa punctata, Bagarius yarelli, Glyptothorax telchitta, Macrognathus pancalus, Psilorhynchus homaloptera, Clarias batrachus, Clarias gariepinus, Anabas testudineus, Botia Dario, Badis badis, Anguilla bengalensis bengalensis, Oreochromis mossammbicus mossambicus, Colisa fasciata,Amblypharyngodon mola, Barilius barila, Chagunius chagunio, Cyprinion semiplotum, Garra naganensis, Labeo dero, Notopterus notopterus, Labeo gonius, Salmostoma bacaila, Mystus* spp., *Ompok* spp., *Glossogobius giuris,Nemacheilus botia, Schistura manipurensis* and *Balitora burmanica.*

Stocking is the main reason for the dominance of Indian major carps like *Catla catla* in Doyang Reservoir. On the usability and fishery importance, it was found that majority (51%) of the fishes of Doyang reservoir are potential ornamental species and 35% are commercially important food fishes. Hence, there is a vast scope to develop ornamental fisheries as well, apart from stocking only food fishes. National Fisheries Development Board, Hyderabad extended financial assistance of Rs. 11.29 lakhs towards the fingerling stocking in Doyang reservoir of Nagaland for third consecutive year for the year 2011-12.

The estimated total fish production of Doyang reservoir from May '09 to April '10 was 207,256 kg. The catch was observed to reach its peak in the month of July (26,114 kg). The catch was seen to be low in the months of February to May, reaching its minimum in March (9,225.5 kg). The present productivity of the reservoir was worked out to be 79.53 kg ha^{-1} yr^{-1}. The Department of Fisheries, Nagaland reported 80.0 kg ha^{-1} yr^{-1} as average productivity of the reservoir.

The crafts and gears used in Doyang are mostly locally made crafts and traditional gears (except the nets, which are usually imported from outside). The various gears observed include wounding spears, pole and line, gorge, hooks, traps, nets (gill net and cast net). Groping and stranding which are very primitive techniques of fishing are also used. Though plank-built boats are the most commonly used crafts in the reservoirs. Harvesting of fishes in Doyang was mostly done using plank-built boats contributing more than 80% to the total catch. Gillnet was the most widely used major gear accounting for 80% of the fish catch from the reservoir followed by hooks and line (9%), traps (8%) and cast nets (3%).

Most of the fish caught from the reservoir are transported to other districts of Nagaland and nearby districts of Assam. There are five landing centres,namely – Salio, Janbemo, Luxio, Jentsu Station and Jonthungo. The latest addition was Jonthungo's landing centre which was started in September 2009. The annual catch data from different landing centres indicated maximum catches from Luxio landing centre with 56,301.5 kg and the lowest landing in Jonthungo's landing centre (13,895 kg). Variation in landings at different landing centres may be due to the choices of the fishermen, to whom they sell their landings.

The fish catch is sold off to retailers and dealers at landing centres, where it is packed in thermocol boxes and disposed to different places for further marketing. One of the major problems is the absence of ice plants or cold storage facility in the vicinity of the reservoir. At present, the ice blocks are procured from Assam. Bulk of the fish catch (55-70%) is marketed through Wokha (Dimapur,Kohima, Wokha, Mokokchung and other districts of Nagaland). The remaining catch is marketed either from Assam (30-35%; Dimapur, Dibrugarh, Jorhat, Golaghat, Tinsukia and Guwahati) or utilised for local consumption (10-15%) by the village communities in Doyang and nearby villages. The Doyang Reservoir has been a source of livelihood for many fishing communities in the villages surrounding it. As many as 250 fisher families (which often rises to 500 in the peak fishing season) belonging to 19 villages depend on the fisheries of Doyang Reservoir. There are, no registered fishermen co-operative societies in Doyang, though 7-8 unregistered societies or associations formed by the affected villages are manning and channelising the fish catch.

REFERENCES

- Alfred, J.R.B. ad Thapa, M.P.,1996. Limnological investigation on Ward's Lake-A wetland in Shillong, Meghalaya, N.E. India. Rec. Zool. Surv. India. Occ. Paper No. 169 : 1-125 (Published by the Director, Zool.Surv.India, Kolkata).
- Basudha, and Vishwanath, W., 1999. Food and feeding habits of an endemic carp, *Osteobrama belangeri* (Val.) in Manipur. *Indian J. Fish.* 46(1) : 71-77.
- Dixitulu, J.V.H., 2002. On Loktak Lake Fishery Development (Editorial), *Fishing Chimes*, 22 (4) : 4-5.

- Jayadev, W and Vishwanath, W., 2002. Fishing practices at water inlet area of Loktak Hydel Project. *Fishing Chimes*, 22(4) : 57-58.
- Mahapatra, B.K., Vinod, K. and Mandal, B.K., 2004a. Studies on chocolate mahseer, *Neolissocheilus hexagonolepis* (Mc Clelland) fishery and the cause of its decline in Umiam reservoir, Meghalaya. *J.Natcon.*, 16(1) : 199-205.
- Roy, Deepayan and Mandal, S.C. 2011. Rudrasagar Lake (Ramsar Site) in Tripura State Present Status of fish Production and Suggestions for future Development. *Fishing Chimes*. 31 (7) : 38-40.
- Sinha, M., 1990. Reservoir fisheries –its present status and future potentials in North eastern region. P.57-64. In : Jhingran, Arun G. and V.K.Unnithan (eds). Reservoir Fisheries in India. Proceedings of the National Workshop on Reservoir Fisheires, 3-4 January, 1990. Special Publication 3, Asian Fisheries Society. Indian Branch, Mangalore, India.
- Subenthung Odyuo and Nagesh, T.S., 2012. Fisheries and management status of Doyang Reservoir, Nagaland, north-east India. *Indian J. Fish.*, 59(2) : 1-6.
- Sugunan, V.V. and Yadava, Y.S., 1991. Feasibility studies for fisheries development in Nongmahir reservoir, Meghalaya, Central Inland Fisheries Research Institute, Barrackpore.
- Vinod, K., Mahapatra, B.K. and Mandal, B.K., 2003b. Studies on the growth of stocked Indian major carps and their impact on fisheries in Umiam reservoir, Meghalaya, *Aquacult*, 4(2) : 247-252.
- Vinod, K.Mahapatra, B.K. and Mandal, B.K., 2007. Umiam reservoir fisheries of Meghalaya (Eastern Himalayas) strategies for yield optimization. *Fishing Chimes*. 26 (10) : 8-15.

13

ORISSA

Fig. 13.1. Map of Orissa State (Not to Scale)

Orissa is one of the important maritime states of India. It is bestowed with ample inland resources in the form of tanks, ponds, reservoirs, rivers and also a vast stretch of land suitable for coastal aquaculture due to its long coastline of 480 km. The freshwater aquaculture resources of Orissa comprise 1,21,010 ha of ponds and tanks out of which 50,310 ha are panchyat tanks, 35,933 ha are revenue tanks, 34,767 ha are under private ownership, with corresponding average pond size being 0.79 ha, 2.69 ha and 0.23 ha respectively.

Table 13.1. Details of Freshwater Aquaculture Resources in Orissa of Freshwater Aquaculture

Sr.No.	Resources	Water Area (in lakh ha.)	Present production in 2009-10 (in lakh MT)
1.	Culture Fisheries (Tanks and Ponds)	1.21	1.90
2.	Culture based capture fisheries (Reservoirs)	2.00	0.12
3.	Capture fisheries (Lakes/swamps/bheels)	1.80	0.02
4.	Rivers and canals	1.71	0.11
5.	**Total**	**6.72**	**2.15**

HIRAKUD RESERVOIR

This was the first post-independence major multipurpose river valley project in India and also the largest man made reservoir in Asia with the longest dam in the world and is situated in undivided Sambalpur district (Sambalpur, Jharsugudu and Bargarh). Construction of dam was completed in the year 1951 and the reservoir was filled in the year 1957. It was formed by damming the confluence of rivers Mahanadi and lb. The reservoir is situated in between the geographical ordinates of 21°30′N to 21°50′N Latitude ad 83°30′E to 84°05′E Longitude. The reservoir has a water spread area of 74592 hectares at FRL. The catchment area of the reservoir is 83,395 km^2 with a shoreline of 643.6 km. It is a multipurpose project with 307.5 MW capacity of power generation through its two power houses .It irrigates about 264038 ha of land. While the main objectives for the construction of the dam are flood control, irrigation and power generation, the fisheries in the reservoir have also become important as they form bread and butter directly to more than 4000 fisher families living around the reservoir.

The bed of the reservoir mainly consists of stones, boulders, and tree stumps. The forest around the reservoir consists of a wide variety of vegetation along with deciduous plants like Sal, Shishu, Bija, Mohua, Haldi, Bamboo etc.

The fishing rights in the river system which were earlier open to the people were withdrawn in the year 1960 and were made over by the state government to its Department of Revenue since there was a need for the development of fisheries in the reservoir the government *i.e.*, Revenue Department transferred the rights of leasing out of fishery of the reservoir to the fisheries department. The fisheries department with the cooperation of the dam authority reclaimed 24 sq.miles of area at 590′ – 630′ contour by cutting jungles on the right dyke area in 1956-57 to facilitate fishing in shallow areas.

Table 13.2. Salient Features of Hirakud Reservoir

1. Geographical ordinates	21°30′N to 21°50′NL and 83°30′E to 84°05′E Long
2. River	Mahanadi and Ib
3. Maximum water level	196.92 mtr
4. Full reservoir level	192.02 mtr
5. Dead storage level	179.83 mtr
6. Water spread area at full water level	73,328 ha
7. Water spread area at full reservoir level	71,963 ha
8. Water spread area at dead storage level	36190 ha
9. Maximum length	85 kms
10. Maximum width	17 kms
11. Mean width	8 kms
12. Maximum depth	64 mtrs

Fish Fauna: The checklist of fish fauna of the Hirakud reservoir is depicted in Table 13.3. Before construction of the dam, the fish fauna of the Mahanadi river system was surveyed by many workers. They reported that the fish fauna of economically important species like *Notopterus notopterus, Barbus sarana,Catla catla, Labeo boggut, Labeo fimbriatus, Labeo calbasu* and *Mystus aor*. After the formation of the reservoir in the year 1956, the fish fauna of the Mahanadi river underwent a significant change. Most of the species have dwindled due to change of eco-environmental conditions and lack of conditions congenial for migration.

Table 13.3. Fish fauna of Hirakud reservoir

CARPS: *Catla catla, Labeo rohita, Labeo calbasu, Labeo finbriatus, Labeo bata, Labeo gonius, Tor mosal, Cirrhinus mrigala, Cirrhinus reba, Hypophthalmichthys molitrix.*
CATFISHES: *Wallago attu, Ompak bimaculatus, Ompak pabda, Mystus aor, Mystus tengra, Mystus cavassius, Rita chrysea, Bagarius bagarius, Silonia silondia, Pangasius pangasius, Eutropiichthys vacha.*
FEATHERBACKS: *Notopterus notopterus, Notopterus chitala.*
LIVE FISHES: *Chana gachua, Channa punctatus, Channa striatus, Clarias batrachus, Heteropneustes fossilis.*
OTHER FISHES: *Puntius ticto, Puntius sarana, Mastacembelus aramatus, Nandus nandus, Glossogobius giuris, Gudusia chapra, Esomus danricus, Rasbora daiconius, Amblypharyngodon mola, Rohtee cotio, Rhinomugil corsula, Chelabacaila, Sciaeid* spp.

The availability of 100 mm size mahseer in Hirakud reservoir indicates that the species has adapted itself to the ecological conditions of the reservoir ad has its breeding grounds in the reservoir. The fish is caught throughout the year but maximum landings take place during the summer season.

Fish Catch: Since 1958-59, the highest catch was recorded in the year 1980-81 (483 mt), with an average catch of 15.6 kg/ha (Table 13.4).

Table 13.4. Year wise fish catch from Hirakud Reservoir

Year	Catch (mt)	Year	Catch (mt)
1958-59	2.976	1980-81	843.080
1959-60	5.911	1981-82	302.018
1960-61	32.736	1982-83	287.639
1961-62	34.535	1983-84	321.160
1962-63	32.400	1984-85	207.757
1963-64	14.401	1985-86	330.134
1964-65	15.092	1986-87	408.681
1965-66	4.315	1987-88	183.478
1966-67	15.791	1988-89	368.386
1967-68	54.700	1989-90	263.510
1968-69	63.906	1990-91	332.837
1969-70	65.676	1991-92	380.238
1970-71	220.245	1992-93	336.940
1971-72	214.370	1993-94	426.387
1972-73	181.148	1994-95	400.289
1973-74	197.856	1995-96	358.600
1974-75	148.934	1996-97	267.301
1975-76	137.100	1997-98	260.174
1976-77	178.882	1998-99	274.781
1977-78	576.621	1999-2000	259.236
1978-79	690.900	2000-01	234.931
1979-80	779.300		

Fish catches from the reservoir shows wide fluctuations in different years. The actual fish catch in the reservoir is difficult to know as a good quantity of fish is illegally fished and smuggled out. The production figures, however, shows decreasing trend year to year. It is due to indiscriminate fishing (irrespective of species and size), using small meshed shore seines (drag nets). Sometimes, there is an upward fluctuation which indicates over exploitation through increase in effort and by way of operating small meshed nets. There was an increasing trend in the annual fish catches since 1958-59 reaching the highest in 1980-81. There has been a sharp fall in the following years. The highest fish yield was in 1986-87 (408.681 t) in the last 20 years, while the minimum yield was recorded in

1958-59 (2.976 mt). The reason for the increase in production from 1958-59 to 1980-81 can be attributed to the gradual introduction of improved fishing gears and increasing fishing effort. After 1980-81, over-exploitation started resulting in gradual drop in production. This was the result of indiscriminate exploitation of brooders and juveniles of almost all species of fish. From the trend of fish catches from the reservoir it is seen that *Labeo rohita* has a dominant presence in the reservoir among Indian major carps and *Mystus aor* is dominant among the catfishes.

The most dominant in the catches are catfishes (48.69%) followed by major carps (32.21%) and rest 19.11% are weed fishes. The trend towards exploitation of catfish is due to good consumer preference and their higher value outside the state.

The fish production of Hirakud reservoir was estimated to be only 15 kg/ ha/yr (Khan *et al.* 1992).

Fishing Tackles: Gill nets, drag nets, cast nets, stake nets and long lines comprise the major fishing tackles operated in Hirakud. Gill nets, mainly of 3types are operated within 3-10 water depth. Drag nets of 4-5 types, simple to big shore seines and hooks and lines are used round the year. Gear with varying mesh sizes (0.5″ to 8.0″) are regularly used, though the regulation prohibits the use of gill net of <4″ mesh size (Sugunan and Yadava, 1992). Commercial fishing is carried out exclusively at night with set gill nets which are hauled next morning. The maximum fish catch is recorded in August and September and the minimum fish catch is recorded in the period of April and July. During monsoon period, a large number of juveniles of Indian major carps are caught by the use of shore-seines. Hence it is necessary to impose seasonal restrictions on the use offshore seines.

The crafts used in Hirakud reservoir are mostly wooden boats with sizes ranging from 18 – 20″. Larger boats are used for transportation.

Trawling Operation: The high opening trawl operated in reservoir landed a total of 2705 kg of fish with an average catch rate of 48.35 kg/ha. The occurrence of bottom and off bottom fishes like scienids, *Notopterus notopterus, Rita chrysea, Mystus* spp, *Rohtee cotio, Wallago attu* etc., indicate that bottom trawling is effective for the removal of these fishes (Dawson, 2002). Trawling fishing has been successfully conducted in Hirakud reservoir. The results obtained in bottom trawling provide scope for introducing this fishing technique in other Indian reservoir where trawlable grounds are available.

Stocking: Stocking of Hirakud reservoir with advanced carp fingerlings started in 1956-57. There are three fish farms in the vicinity of reservoir. The reservoir is stocked with advanced fingerlings raised at these farms. The details of seed stocking in reservoir are depicted in Table 13.5.

Revenue from the Reservoir: Department of Fisheries ears revenue from the fisheries of the Hirakud reservoir by lease of water body, fishing licenses, departmental fishing, and 1/3 share of the fish catches laded by the CIFT during the experimental fishing in the reservoir conducted by it. Department of fisheries issues licenses to local fishermen for fishing in sector 3 of reservoir. License fees are collected, on annual basis, at varying rates related to different crafts and ears used. The highest collection of revenue at Rs. 39,027 from issue of fishing license was recorded in the year 2000-01.

Table 13.5. Seed stocking in Hirakud reservoir

Year	Stocking (nos.)
1956-57	32750
1957-58	2,45,750
1958-59	39,903
1961-62	8004
1962-63	7940
1963-64	12000
1965-66	2,92,750
1988-89	6,74,110
1989-90	14,12,050
1990-91	23,75,000
1991-92	13,35,000

The total waterspread area of the reservoir is divided into six sectors, out of which five sectors are given on annual lease to five separate primary fishermen's cooperative societies on a lease value of Rs.100/sq.mile/year. The remaining one sector (sector 3) with an extent of 120 sq.miles is retained by the state fisheries department for its experimental fishing. The remaining 168 sq.miles of water spread area, given o lease to co-operatives, generates a total revenue of Rs. 16,800 only per annum.

Fisheries Co-operatives: Of the six sectors, five have been leased out to five primary fishermen's co-operative societies. The details of the co-operative societies are furnished in Table 13.6.

Table 13.6. Details of Primary fishermen's Cooperative Societies of Hirakud Reservoir

Co-operatives	Fishing Area (ha)	Members				Total shares		Average catch (kg)	Yield (kg/ha /yr)
		SC	ST	Others	Total	Members	Govt.		
Tamedi	6734	403	44	108	555	3333	2000	29510	4.38
Lachhipali	4144	195	118	66	379	21830	23200	29832	7.20
Thebra	11137	233	324	100	657	48010	207600	122682	11.02
Mahamadpur	11655	248	472	102	822	12870	42200	30691	2.63
Ib	9842	452	107	58	617	6170	9500	30201	3.07

(Data pulled from Sugunan and Yadava, 1992)

Experimental Fishing: Department of fisheries conducts experimental fishing in the reservoir and the catch is sold out at the rates prescribed by the government.

The CIFT's research centre at Burla (a place close to the reservoir) also conducts experimental fishing in the reservoir. It deposits 1/3 of the sale proceeds of fish catch to the account of the fisheries department. The total revenue eared from year 1996-97 to 2000-01 is shown in Table 13.7. The rates of license and lease are unchanged for the last several years.

Table 13.7. Revenue collected from Hirakud Reservoir

Year	License (Rs.)	Lease (Rs.)	Departmental Fishing (Rs.)	CIFT's 1/3rd share (Rs.)
1996-97	14,729	16800	789	5768
1997-98	18068	16800	275	Nil
1998-99	25137	16800	390	3335
1999-2000	30437	16800	Nil	3952
2000-01	39027	16800	Nil	2961

Socio-Economic Variables of Fishermen: Balasubramaniam and Bihari (2002) accounted socio-economic variables of fishermen in two fishing villages such as Larbhanga and Sanutikira in Hirakud reservoir area near the CIFT Research Centre Burla. The results revealed that the respondents were employed for about 100 days in a year including the days spent on subsidiary occupation. Though 38.63% of the respondents in Larbhaga and 13.63% in Sanutikira were involved in cultivation oriented activities and other avocations, the average annual income of fishermen in Sanutikira centre (mea Rs. 52,900) was found to be significantly higher than the respondents in Larbhanga centre (mean Rs. 34,500).

The average investment on fishing craft (Rs. 8772.770) and also on fishing nets (Rs. 14818.2) were found to be significantly higher in Sanutikira than in the Larbhanga village (Rs. 5727.3 and Rs. 10811.4 respectively). Though in both villages, the respondets were engaged in reservoir fishing on a small scale level, the higher investment made in Sanutikira on fishing craft and gear could be a possible reason for their higher income.

The results showed that more than 40% of the respondents did not have any formal education and another 45-52% of them had only primary education and nost of them (96.6%) beloged to the scheduled caste-scheduled tribe communities. The study also revealed that about 27% of respondents in Larbhaga and 11% in Sautikira did not own any craft, though they had owned fishery gear. The fishermen had used plank-built canoes and the dimensions of these canoes ranged from 5.6 to 8.2 m in length and 0.43 to 1 m in breadth. The dragnets were used in the shallow marginal areas of the reservoir while the gill nets were operated in the reservoir as surface column and bottom set nets. Gill nets and shore seines were used by more number of respondents in Sanutikira than in Larbhanga. It was found that though George *et al.* (1982) had introduced trawling in Hirakud reservoir long back,none of the fishermen had used trawls, as it involved larger investment. About 59% of respondents in Larbhanga and 86% in Sanutikira had owned radio sets for information.

Table 13.8. Physico-chemical characteristics of Hirkud Reservoir

Parameter	Range	Mean
Atmospheric temperature (°C)	21 – 34	28.1
Surface temperature (°C)	20.5 – 33.8	27.4
Electrical conductivity	117.5 – 165.38	147.73
pH	7.64 – 7.9	7.52
Total dissolved solids	93.75 – 177.5	128.69
Turbidity	5.1 – 11.5	7.63
Dissolved oxygen	5.9 – 7.7	6.87
Total hardness	51.84 – 95.5	69.2
Phenolphthalein alkalinity	3.64 – 21.15	13.19
Total alkalinity	119.75 – 163.07	142.53
NH_3	0.131 – 0.265	0.197
Mg^{+2}	4.19 – 8.56	6.16
Ca^{+2}	11.63 – 25.9	17.58

Water Quality: The colour, odour and taste of the reservoir water vary in different seasons. The water is mostly light green to deep green colour, algae odour and swampy taste during the moths of October to middle of the June, and muddy brown to light brown colour, soil odour and muddy taste during the rest of the year (Samal *et al.*, 2005). The pH values varying from 7.64 to 7.9. It indicates that the reservoir water is slightly alkaline in nature. The electrical conductivity is maximum during summer and minimum during monsoon. The dissolved oxygen of reservoir is maximum during the winter season and minimum during monsoon. The total dissolved solids, turbidity, total alkalinity, hardness and NH_3 were gradually increased from winter onwards ,ad the maximum values were noted during monsoon.

Age and Growth study of carps: Mathew and Zacharia (1982) studied age ad growth of Indian major carps (*Catla catla, Labeo rohita* and *Cirrhius mrigala*) from Hirakud reservoir. It was found that in these three fishes rings were laid down almost during the same period *i.e.,* May to August. The result obtained from scale study is compared with Petersen Method. A close agreement can be seen in their values obtained. Von Bertalanffy's growth equation was used for fitting the growth data obtained by scale study and it adequately describes the actual growth of those fishes. The length at age and the rate of growth of these fishes were compared and it was found that rate of increase in length was faster during the initial periods of life and decrease when they become old.

REFERENCES

- Balasubramaniam, S. and Bihari Banki, 2002. Socio-economic variables of fishermen in Hirakud reservoir and the technical adoption (Boopendranath, M.R.Meeakumari, B., Joseph, J., Sankar, T.V., Pravin, P., and Edwin, L. Eds.), pp. 426-432, Society of Fisheries Technologists (India), Cochin.
- Dawson, Percy. 2002. Trawling in Indian reservoirs. In : Riverine ad Reservoir Fisheries of India (Boopendraath, M.R. Meeakumari, B., Joseph, J., Sankar, T.V., Pravin, P., and Edwin, L. Eds.), pp.264-267, Society of Fisheries Technologists (India), Cochin.
- Khan, A.A., Dawson, P., Rao, J.S.R., and Varghese, M.D., 1992. Fishing in impounded waters-A case study in Hirakud reservoir, Orissa, CIFT, Cochin.
- Mahapatra, Dipti Kumar., 2003. Preset status of fisheries of Hirakud reservoir, Orissa. *Fishing Chimes*. 22 (10 and 11) : 76-79.
- Mathew, Kuruvila and Zacharia, P.U., 1982. On the age and growth of three Indian majora carps from Hirakud reservoir. *Bull. Dept. Mar. Sci. Univ. Cochin, 1982, XIII, 81-95.*
- Samal, S.K., Pradhan, B. and Tiwari, T.N., 2005. Water quality status of Hirakud reservoir and its suitability for irrigation. In : Advances in Limnology (Ed. S.R. Mishra), pp. 91-97, Daya Publishing House, Delhi.
- Sugunan, V.V. and Yadava, Y.S., 1992. Hirakud reservoir-strategies for fisheries development. Bulletin No. 66, April 1992. CIFRI, Barrackpore.

14

PUNJAB

Fig. 14.1. Map of Punjab State (Not to scale)

Punjab is a state in northwest India. The state has made great strides in aquaculture during the last three decades. The areas under fish farming and also capture fish production have gone up by about 30 times. Around 70% of the total fish production of the state comes from aquaculture and 30% from capture sector. There is a vast network of natural water resources including 17543 km of rivers and canals and 23000 ha of reservoirs/lakes/wetlands. These water resources have immense potential for fishery development. The reservoir productivity (50 kg/ha/yr) though higher than the national productivity (20 kg/ha/yr), the actual potential is much higher, being 100 kg/ha/yr (Dhawan, 2010).

HAIRKE WETLAND

Harike wetland (31°10′15″N, 74°57′E) covering 4100 ha area is included in the list of Ramsar sites in 1990. The wetland has assumed international importance as it is a breeding ground and habitat for a large variety of migratory as well as domiciled birds.Harike wetland located at the confluence of two major rivers Stulej and Beas.It is one of the largest wetlands in North India and is very important source of water supply for the state of Punjab. The wetland area is spread over in about 41 sq.km. Of this, 2269 acres are upland ad 6904 acres are under water. Out of upland area, 853 acres area is marshy and 1416 acres are covered with a variety of plants. Harike wetland spreads in four districts of Amritsar, Ferozpur, Kapurthala and Jalandhar. There are 31 villages in the wetland zone.

Soil texture of wetland was mainly sandy loam. The entire wetland bed soil has alkaline reaction. Organic carbon and available phosphate and nitrogen all had high values indicating sufficient dissolved solids and productive soil (Moza and Mishra, 2008).

Water within wetland was full of suspended material except during winter especially along Beas inlet side. The average values of water parameters of Harike for year 2004-05 are depicted in Table 14.1.

Harike wetland is a very important abode for the birds migrating from across the international frontiers. It supports more than 400 species of birds (Tiwana *et al.* 2008). The Wigeon, Common Teal, Pintail, Shoveller and Brahminy ducks are commonly see durig winter season. Other birds such as Scape duck, falcated teal, white headed stifftailed duck, Greylag goose, Barheaded goose and common crane are regular winter visitors. The common mammals fount at Harike wetland include the Smooth Indian Otter, the jungle cat, jackal Indian wild boar and common mongoose. The reptiles fount in lake include 7 species of turtle belonging to 6 genera. Water hyacinth (*Eichhornia crassipes*) is the main weed which has invaded Harike wetland. The area under the plant doubles every 6.2 to 15 days depending on the nutrient status of water.

The wetland is getting silted up as vast areas along the right side of river Beas falling under villages Chambakallan, Kamboh Dhaiwala, Kirrian and Harike. Vast area is seriously degraded with deep cuts due to formation of ravines over the years. Erosion in this area transfers heavy load of silt into Harike wetland. Consequently the storage capacity of the Harike lake has drastically reduced.Eutrophication in the lake has resulted infestation of water hyacinth in heavy amount ad resulting in the choking of water ways. It is reported that there are 60 species of fish in Harike wetland.

Table 14.1. Average values of water parameters of Harike for year 2004-05

Parameters	Average values
1. Water temperature (°C)	24.08
2. Transparency (cms)	26.41
3. pH	6.82
4. Dissolved oxygen (mg/l)	5.26
5. Total alkalinity (mg/l)	75.66
6. Total hardness (mg/l)	152.08
7. Specific conductivity (μmhos/cm)	249.5
8. Calcium (mg/l)	31.16
9. Magnesium (mg/l)	17.73
10. Silicate (mg/l)	3.28
11. Chloride (mg/l)	25.5
12. Available phosphate (mg/l)	0.37

Dua and Prakash (2009) reported 61 species of fishes belonging to 17 families (Table 14.2). Maximum number of species (27) recorded were of family cyprinidae followed by families bagridae (7 species), siluridae, schilbeidae, channidae, mastacembelidae (3 species each), gobiidae, notopteridae, sisoridae, ambassidae and belontiidae (2 species each), clupeidae, clariidae, heteropneustidae, synbranchidae, belonidae and nandidae (1 species each). Two species, most abundantly found in the Harike wetland were *Cirrhinus mrigala* and *Cyprinus carpio communis* and these constitute 23% of the total fish catch.

Table 14.2. Fish fauna of Harike wetland

Fish fauna
Family : Notopteridae
1. *Notopterus notopterus* (Pallas)
2. *Notopterus chitala* (Hamilton-Buchanan)
Family : Clupeidae
3. *Gudusia chapra* (Hamilton-Buchaan)
Family : Cyprinidae
4. *Catla catla* (Hamilton-Buchanan)
5. *Cirrhinus mrigala* (Hamilton-Buchanan)
6. *Cirrhinus reba* (Hamilton-Buchanan)
7. *Cyprinus carpio communis* (Linnaeus)
8. *Cyprinus carpio specularis* (Linnaeus)

Contd...

Table 14.2: Contd...

9. *Labeo angra* (Hamilton-Buchanan)
10. *Labei bata* (Hamilton-Buchanan)
11. *Labeo caeruleus* (Hamilton-Buchanan)
12. *Labeo calbasu* (Hamilton-Buchanan)
13. *Labeo dero* (Hamilton-Buchanan)
14. *Labeo gonius* (Hamilton-Buchanan)
15. *Labeo lippus* (Fowler)
16. *Labeo rohita* (Hamilton-Buchanan)
17. *Osteobrama cotio cotio* (Hamilton-Buchanan)
18. *Puntius jerdoni* (Day)
19. *Puntius sarana sarana* (Hamilton-Buchanan)
20. *Puntius sophore* (Hamilton-Buchanan)
21. *Puntius terio* (Hamilton-Buchanan)
22. *Puntius ticto* (Hamilton-Buchanan)
23. *Salmostoma bacaila* (Hamilton-Buchanan)
24. *Salmostoma horai* (Silas)
25. *Salmostoma phulo* (Hamilton-Buchanan)
26. *Amblypharyngodon mola* (Hamilton-Buchanan)
27. *Esomus danricus (Hamilton-Buchanan)*
28. *Parluciosoma daniconius* (Hamilton-Buchanan)
29. *Schizothorax richardsonii* (Gray)
30. *Gara gotyla gotyla* (Gray)
Family : Bagridae
31. *Aorichthys aor* (Hamilton-Buchanan)
32. *Aorichthys seenghala* (Sykes)
33. *Mystus bleekeri* (Day)
34. *Mystus cavasius* (Hamilton-Buchanan)
35. *Mystus horai* (Jayaram)
36. *Mystus vittatus* (Bloch)
37. *Rita rita* (Hamilton-Buchanan)
Family : Siluridae
38. *Ompak bimaculatus* (Bloch)
39. *Ompak pabda* (Hamilton-Buchanan)
40. *Wallago attu* (Schneider)
Family : Schibeidae
41. *Clupisoma garua* (Hamilton-Buchanan)
42. *Eutropiichthys murius* (Hamilton-Buchanan)
43. *Eutropiichthys vacha* (Hamilton-Buchanan)
Family : Sisoridae
44. *Bagarius bagarius* (Sykes)
45. *Glyptothorax punjabensis* (Mirza and Kashmiri)
Family : Clariidae
46. *Clarias batrachus* (Linnaeus)

Contd...

Table 14.2: Contd...

Family : Heteropneustidae
47. *Heteropneustes fossilis* (Bloch)
Family : Belonidae
48. *Xenentodon cancila* (Hamilton-Buchanan)
Family : Synbranchidae
49. *Monopterus cuchia* (Hamilton-Buchanan)
Family : Ambassidae
50. *Chanda nama* (Hamilton-Buchanan)
51. *Pseudembassis ranga* (Hamilton-Buchanan)
Family : Nandidae
52. *Nandus nandus* (Hamilton-Buchanan)
Family : Gobiidae
53. *Glossogobius giuris* (Hamilton-Buchanan)
Family : Belontiidae
54. *Colisa fasciatus* (Schneider)
55. *Colisa lalia* (Hamilton-Buchanan)
Family : Channidae
56. *Channa marulius* (Hamilton-Buchanan)
57. *Channa punctatus* (Bloch)
58. *Channa striatus* (Bloch)
Family : Mastacemblidae
59. *Macrognathus aral* (Bloch and Schneider)
60. *Macrognathus pancalus* (Hamilton-Buchanan)
61. *Mastacembelus armatus* (Lacepede)

Moza and Mishra (2008) studied ecology and fishery of Harike wetland. The fish catch was all time high, 286.7 t/yr during 1999-2000 with a average of 28.67 t/m, when the sanctuary area too was used for fishing activity to some extent. The catch decreased sharply thereafter, culminating in an average decrease of 57%. The details of the average fish catch and composition are given in Table 13.3. Fish composition of Harike indicate that the area holds maximum commercial fishery where in Indian major carps is dominant, present in the range of 26.63 to 51.48%, followed by common carp 8.96 to 33.54% and large size catfishes, 4.32 to 23.65%. Existence of *Clarias gariepinus* at Harike since year 2002 is a disturbing trend and emphasize the polluted nature of water body.

Dua and Kumar (2006) studied age and growth patterns in *Channa marulius* from Harike wetland during 1998-99. Linear relationship with a high degree of correlation was observed between total fish length and the lateral scale radius. Age determination studies revealed 5 age groups. The harvestable size falls just below the 2nd year. Various growth parameters indicate a hardy nature of the fish and the suitability of habitat ecology for its optimum growth.

Table 14.3. Average fish catch and composition at Harike between 1999-2005

Year	Total Landing t/yr	PERCENTAGE COMPOSITION					
		IMC	Minor Carps	*C.carpio*	Large size catfish	Asso. groups	*C.gariepinus*
1999-00	286.7	12.80	32.91	22.78	11.91	19.50	–
2000-01	104.0	26.63	11.63	8.96	23.65	29.13	–
2001-02	129.5	37.92	6.33	18.07	7.28	30.42	–
2002-03	26.0	32.69	2.30	20.0	10.38	34.24	0.28
2003-04	46.40	43.66	7.10	33.54	6.02	9.69	0.38
2004-05	117.90	51.48	8.33	18.41	4.32	16.28	0.18

ROPAR WETLAND

It is situated at 31°01'N 076°30'E covering 1365 ha area. Wetland came into formation in the year 1952 with the construction of head works. The area surrounding Ropar wetland is hilly in the north west and plain in the south and south east. Small streams and Choes that empty into Ropar wetland are very important from ecological point of view. Since this wetland is located right by the side of badly damaged and absolutely eroded Shivalik foothills, a large amount of silt and nutrients gets transported into the wetland every year. The excessive siltation is reducing the water holding capacity of the lake. 55 species of fish and 318 species of birds have been reported from this wetland (Tiwana *et al.* 2008). Scaly anteater, Sambar and Hog deer are found in area. Migratory birds such as Rudy Shelduck, Northern Pintail, Common Teal, Mallard, Gadwall, Eurasian Wigeon, Northern Shoveler, Pochards use the lake as their winter home.

REFERENCES

- Dhawan, Asha, 2010. Punjab to go ahead in aquaculture. *Fishing Chimes.* 30 (1) : 166-168.
- Dua, Anish and Chander Prakash, 2009. Distribution and abundance of fish populations in Harike wetland—a Ramsar site in India. *J. Environ. Biol.* 30 (2) : 247-251.
- Dua, Anish and Kumar, Kanwaljit, 2006. Age and growth patterns in *Channa marulius* from Harike wetland (A Ramsar Site), Punjab, India. *J. Environ. Biol.* 27 (2) : 377-380.
- Moza, Usha and Mishra, D.N., 2008. Current status of Harike wetland vis a vis its ecology and fishery. In Proceedings of TAAL -2007: The 12th World Lake Conference (Sengupta M. and Dalwani, R.Eds.), pp. 1470-1476.
- Tiwana, N.S., Jerath, Neelima, Sexena, S.K. and Sharma Vivek, 2008. Conservation of Ramsar sites in Punjab. In : Proceedings of TAAL -2007. The 12th World Lake Conference (Sengupta M. and Dalwani, R.Eds.), pp. 1463-1469.

15

RAJASTHAN

Fig. 15.1. Map of Rajasthan (Not to scale)

Rajasthan has been the house of royalty in India. It is located in the northwest of India. It encompasses most of the area of the large, inhospitable Great Indian Desert, which has an edge paralleling the Sutlej-Indus river valley along its border with Pakistan. The state is bordered by Pakistan to the west, Gujarat to the southwest, Madhya Pradesh to the southeast, Uttar Pradesh and Haryana to the northeast and Punjab to the north. Rajasthan covers 10.4% of India, an area of 342,269 square kilometers.

PUSHKAR LAKE

Pushkar Lake (26°29′14″N 74°33′15″E) is located in the town of Pushkar in Ajmer district.It has mythological significance and is a touristic delight, a holy place where people come from all over the world. Pushkar Lake is a sacred lake of the Hindus.Pushkar Lake is surrounded by 52 bathing ghats, where pilgrims throng in large numbers to take a sacred bath, especially around Kartik Poornima when the Pushkar fair is held. Over 500 Hindu temples are situated around the lake precincts. Tourism and deforestation in the surroundings have taken a heavy toll on the lake, adversely affecting its water quality, reducing the water levels and destroying the fish population.

Haemoglobin parameters in the blood of *Clarias batrachus* netted from polluted Gau Ghat and Jaipur Ghat area of Pushkar lake were studied and compared with control fish caught from Bisalpur reservoir to study the possible toxic effects of effluent pollutants. The parameters included haemoglobin (Hb), white blood corpuscles (WBC), red blood corpuscles (RBC), packed cell volume (PCV), mean corpuscular volume (MCV), mean corpuscular hemoglobin (MCH) and mean corpuscular haemoglobin concentration (MCHC). These parameters showed a drastic change in the treated samples when compared with control, revealing the adverse effect of pollution on hematological parameters of fish inhabiting Pushkar lake (Summarwar, 2012).

HEMAWAS RESERVOIR

Hemawas dam is the second largest reservoir of Pali district of Rajasthan, connected with a extensive network of seasonal rivers namely Khari, Bandi, Jajari, Sumer nadi etc. Sharma and Mohan (2010) recorded fourteen fish species belonging to 4orders and 5 families during December 20006 to June 2009. The most dominating order recorded was cypriniformes which was represented by 10 species. The ext dominant order siluriformes was represented by two species, while osteoglossiformes and perciformes were represented by one species each. Six species were identified as Lower Risk-near threatened (LRnt), six species as Lower-Risk-least concern and two species as vulnerable species. Several external factors seem to be responsible for reduction in fish faunal diversity of the reservoir. The notable ones among them are erratic rainfall ad increased anthropogenic activities.

Table 15.1. Fish faua of Hemawas reservoir, Rajasthan

1. *Notopterus notopterus* (Native)
2. *Catla catla* (Native)
3. *Cirrhius mrigala* (Native)
4. *Cyprinus carpio* (Exotic)
5. *Ctenopharyngodon idella* (Exotic)
6. *Labeo bata* (Native)
7. *Labeo calbasu* (Native)
8. *Labeo rohita* (Native)
9. *Puntius sophore* (Native)
10. *Puntius ticto* (Native)
11. *Hypophthalmichthys molitrix* (Exotic)
12. *Mystus vittatus* (Native)
13. *Wallago attu* (Native)
14. *Channa punctatus* (Native)

BISALPUR RESERVOIR

Bisalpur reservoir is constructed on the river Banas, which originates from Gongunda area of Southern Rajasthan. The river flows through Eastern Arawati ranges passing through Tonk and Sawaimadhopur districts. It ultimately merges in the river Chambal.It is one of the largest reservoirs in state covering catchment area of 8655 km^2. The reservoir lies between altitude 26°55′20″N and 75°27′30″E. The reservoir is situated near village Bisalpur in Devli tehsil of Tonk district.

Mathur and Summarwar (2007) conducted preliminary survey of fisheries of Bisalpur reservoir. During their survey period 44 species of fishes were recorded (Table 15.2). They also reported fastest growth rate of *Catla catla* (1.5 to 2 kg/ year) followed by *Labeo rohita* (about 800-900 gms/year) and *Cirrhinus mrigala* (600-700 gm/year). The exotic fishes such as *Cyprinus carpio* var.communis and *Ctenopharyngodon idella* are stocked in reservoir with seed of catla, rohu and mrigal.The seed is stocked in the form of fry stage. The details of seed stocking programme in Bisalpur reservoir for the period of 10 years (1994-95 to 2003-04) are provided in Table 15.3.

Table 15.2. List of fishes recorded in Bisalpur reservoir

1. *Catla catla*
2. *Labeo rohita*
3. *Labeo calbasu*
4. *Labeo fimbriatus*
5. *Labeo bata*
6. *Labeo gonius*
7. *Cirrhinus mrigala*
8. *Cyprinus carpio*
9. *Ctenopharyngodon idella*
10. *Tot tot*

Contd...

Table 15.2: Contd...

11. *Puntius sarana*
12. *Puntius sophore*
13. *Puntius ticto*
14. *Osteobrama cotio*
15. *Gadusia chapra*
16. *Oxygaster bacaila*
17. *Oxygaster clupeorides*
18. *Rasbora daniconius*
19. *Esomus daniconius*
20. *Garra gotyala*
21. *Barilius bola*
22. *Danio devario*
23. *Glossogobius giuris*
24. *Channa gachua*
25. *Channa marulius*
26. *Channa punctata*
27. *Shanna striatus*
28. *Xenentedon cancila*
29. *Mastacembelus armatus*
30. *Notopterus notopterus*
31. *Notopterus chitala*
32. *Mystus cavasius*
33. *Mystus aor*
34. *Mystus seenghala*
35. *Mystus vittatus*
36. *Rita rita*
37. *Eutrophchthys vacha*
38. *Silonia silondia*
39. *Heteropneustes fossilis*
40. *Clarias batrachus*
41. *Bagarius bagarius*
42. *Wallago attu*
43. *Ompak pabda*
44. *Ompak bimaculatus*

Table 15.3. Seed stocking in Bisalpur reservoir (from 1994-95 to 2003-04)

Sr.No.	Year	Catla	Rohu	Mrigal	Total
1.	1994-95	4250	7300	86750	2,00,000
2.	1995-96	—	—	—	—
3.	1996-97	—	1,20,000	1,30,000	2,50,000
4.	1997-98	—	—	—	—
5.	1998-99	—	—	—	—

Contd...

Table 15.3: Contd...

6.	1999-00	26,66,750	89,900	26900	27,83,550
7.	2000-01	19,03,510	2,67,835	11,42,695	33,18,040
8.	2001-02	34,73,870	12,59,910	33,805	47,71,585
9.	2002-03	31,87,500	7,46,430	5,65,110	44,99,040
10.	2003-04	37,11,015	9,97,500	5,84,500	52,92,015

The fish production in Bisalpur reservoir for the period of 1994-95 to 2003-04 varied from 2455 kg (1994-95) to 1,77,619 (1999-00). The details on carp and catfish production is depicted in Table 15.4.

Table 15.4. Fish production in Bisalpur reservoir, Rajasthan

Sr.No.	Year	Carp production (kg)	Catfish production (kg)	Total fish production (kg)
1.	1994-95	1949	506	2455
2.	1995-96	18613	4934	23547
3.	1996-97	22985	4053	27038
4.	1997-98	15402	13183	28585
5.	1998-99	65936	17040	82976
6.	1999-00	147755	29864	177619
7.	2000-01	253369	100576	353945
8.	2001-02	273861	284132	557993
9.	2002-03	531853	304631	836484
10.	2003-04	352733	91162	443895

The predatory and air breathing fishes fetch good market when compared to Indian major carps.

The lease amount of reservoir was Rs. 5800/- in year 1994-95,which reached up to Rs. 53, 18,747 during year 2003-04 (Fig. 15.2).

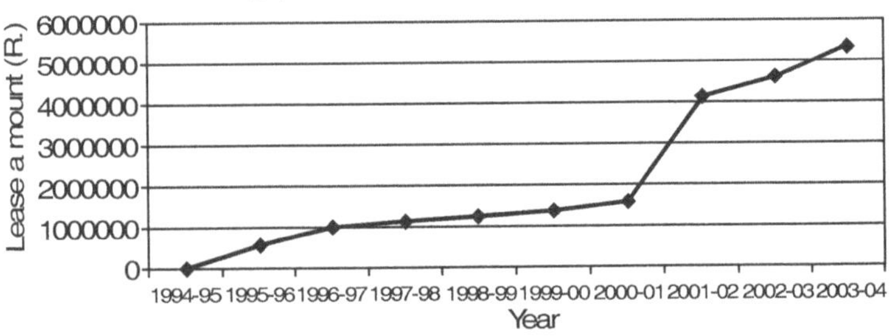

Fig. 15.2. Lease money of Bisalpur (pair by contractor from 1999-2000)

GUDHA RESERVOIR

The Gudha reservoir is built on River Mej in Bundi district in south-eastern part of Rajasthan. The catchment area of the reservoir is about 7.53 sq.km and is fed by river Mej and some local rivulets. The annual storage is 70 mc.ft. and the water spread area is about 1928 ha.

Sharma and Bharadhwaj (1980) reported fisheries of Gudha reservoir. They recorded a total of 45 fish species belonging to 7 orders, 12 families and 28 genera. The total fish harvested during 1975-76 was 78,511 kg, which went down to 77,986 kg in 1976-77. From the percentage composition, *Catla catla* was noted to be dominated (56.11%) in 1975-76, but sharply declined to 36% in 1976-77. On the other hand *Cirrhinus mrigala* rose from 2.26% (1975-76) to 17% in 1976-77. Whereas, the catch of *Labeo rohita* (2.24 to 3%), *Labeo calbasu* (0.86 to 3.5%) and Tor tor (1.65 to 2.5%) remained very poor in both the years. The ext species in importance after major carps include *Mystus seenghala, Wallago attu* and *Silonia silondia*. These contribute about 28.14%. The minor carps on the other hand constitute 2.75% in the total catch.

The fishing is done by gill-nets, drag nets, cast nets and hook and lines. Although the gill nets are of varied types, commonly they are about 30 m in length, 2 to 4 m in depth and 4 to 20 cm in bar mesh.

To protect the breeding of prospective brooders, it might be useful to prohibit fishing operations in immediate areas of spawning grounds, particularly at the mouths of rivulets of river Mej during monsoon.

GADISAR LAKE

It was the only source of water for the Jaisalmer city in the olden days. It is an artificial lake in Jaisalmer. It was constructed by Raja Rawal Jaisal, the first ruler of Jaisalmer. Later on Maharaja Garisisar Singh rebuilt and revamped the reservoir. Artistically carved Chattris, Temples, Shrines and Ghats surround the banks of Gadisar Lake Jaisalmer. Located towards the south of Jaisalmer city the entrance to the Gadisar Lake or Garsisar Tank is through a magnificent and artistically carved yellow sandstone archway that is known as the Tilon-Ki-Pol. One can see rare migratory birds that stop for a drink at the lustrous Gadisar Lake that reflects the mellow yellow of the picturesque sandstone banks; enjoy a boat ride on the lake or just sit on any of the Ghats and see the yellow sandstone banks change hues of ochre with the setting sun, a delightful panorama that you would not miss for the world. The serene Gadisar Lake springs to life during the annual Gangaur celebrations. The view of the Jaisalmer Fort from the Gadisar Lake is breathtaking.

PICHHOLA LAKE

Pichhola is an old lake believed to be constructed by a Banjara at the end of 14th century which later renovated in 1560 A.D. by Maharana Udai Singh. The river Sisarma, a tributary of the river Kotra is the main source of water for the lake. Lake pichhola is situated in udaipur district of Rajasthan at Latitude 24°34′N and Longitude 73°40′E. The water spread area of the lake is 6.96 km^2. The length of the lake is 3.6 Km and the maximum depth of the lake is 8 m. towards the central western part. The maximum and mean width of the Lake is 2.61 and 1.93 km. respectively. Lake Pichhola commands a total catchment area of about 12,700 hr.

Sharma *et al.* (2011) studied limnolgical characteristics, plankton diversity and fishes in Pichhola Lake of Udaipur and reported 15 species of fishes belonging to 6 family and 13 genera from Pichhola lake namely *Notopterus notopterus Catla catla, Cirrhinus cirrhinus, Ctenopharygodon idella, Labeo gonius, Labeo rohita, Puntius sarana sarana, Puntius ticto, Chela cachius, Garra gotyla gotyla, Aorichthys seenghala, Mystus cavasius, Heteropneustes fossilis, Xenentodon cancila* and *Gambusia affinis.*

The average value of air and water temperature was 31.28°C and 27.3°C respectively. The water remained moderately alkaline (pH 7.5) while electrical conductance (0.3958 mS/cm), TDS (237.5 mg/l), chloride (176 mg/l), hardness (174.33 mg/l) and alkalinity (207.16 mg/l) showed low mean values. Average dissolved oxygen levels were at 5.75 mg/l while average nitrate and phosphate levels were 3.70 mg/l and 2.79 mg/l respectively. On the basis of water quality parameters in general, lake Pichhola was found to be eutrophic. A high rate of primary production (302.085 mgc/m^2/hr), diversity of phytoplankton (58 forms), zooplankton (104 forms) and fish (15 species) were also observed by Sharma *et al.* (2011).

Total 58 forms of phytoplankton were identified and out of these 28 belonged to chlorophyceae, 11 to bacillariophyceae, 9 to myxophyceae, 4 to dinophyceae, 3 to desmidiaceae and 3 to xanthophyceae in the water of lake Pichhola. The most prominent phytoplankters during the study were *Oedogonium* sp., *Spirogyra* sp., *Volvox* sp., *Ulothrix* sp. and *Pediastrum* sp. from group *Chlorophyceae.While Microsystis* sp. *and Coccochlaris* sp. dominated myxophyceae.

Table 15.5. Phytoplankton diversity in Pichhola Lake, Rajasthan

Volvox sp., *Eudorina* sp., *Pandorina* sp., *Scenedesmus* sp., *Chlorella* sp., *Ankistrodesmus* sp., *Coelastrum* sp., *Spirogyra* sp., *Oedogonium* sp., *Ulothrix* sp., *Cladophora* sp., *Chlamydomonas* sp. *Mougeotia* sp., *Pediastrum* sp., *Oocystis* sp., *Zygnema* sp., *Hydrodictyon* sp., *Microspora* sp., *Spaerocystis* sp., *Asterococcus* sp., *Closteriopsis* sp., *Schizomeris* sp., *Oedocladium* sp., *Actinastrum* sp., *Kirchneriella* sp. *Nephrocytium* sp., *Zygnemopsis* sp., *Pleodorina* sp., *Cosmarium* sp., *Desmidium* sp., *Sphaerozosma* sp., *Chlorobotrys* sp., *Botrydiopsis* sp., *Botryococcus* sp., *Microcystis* sp., *Agmenellum* sp., *Anabaena* sp., *Oscillatoria* sp., *Nostoc* sp., *Spirulina* sp., *Coccochlaris* sp., *Gomphosphaeria* sp., *Lyngbya* sp., *Glenidium* sp., *Peridinium* sp., *Ceratium* sp., *sphaerodinium* sp., *Cyclotella* sp., *Synedra* sp. *Fragillaria* sp., *Navicula* sp., *Pinnularia* sp., *Nitzschia* sp., *Asterionella* sp., *Amphora* sp., *Gomphonema* sp., *Cymbella* sp., *Bacillaria* sp.

Lake Pichhola harbour diverse taxonomic groups of zooplankton representing Protozoa, Rotifera, Cladocera, Copepoda and Ostracoda. During present investigation, 9 forms of protozoans belonging to 8 genera and 9 species were reported. Rotifers were represented by 20 genera and 40 species. Along with these, 29 species of Cladocerans belonging to 12 genera, and 10 genera and 11 species belonging to Copepoda are enlisted. Besides these, 5 species of Ostracoda were also recorded. After including occasional zooplankters like insects and their larvae, crustacean larvae, spiders and mites total 104 forms of zooplankters were recorded.According to the Menhinick's index of diversity, Rotifers indicated highest diversity throughout the study period followed by Cladocerans, Protozoans, Copepods,Ostracods, Insects and others.

Table 15.6. Zooplankton diversity in Pichhola Lake,Rajasthan

Volvox, Euglena acur, Euglena sp., *Amoeba* sp., *Arcella discoida, Difflugia* sp., *Paramecium* sp., *Peridinium* sp., *Phacus* sp., *Brachionus angularis, Brachionus angularis bidens, Brachionus calyciflorus, Brachionus calyciflorus, Brachionus calyciflorus, Brachionus diversicornis, Brachionus diversicornis, Brachionus quadridentatus, Brachionus quadridentatus, Brachionus falcatus, Brachionus falcatus, Brachionus forficula, Brachionus caudatus, Brachionus bidentata, Keratella tropica, Keratella tropica asymmetrica, Keratella tropica heterospina, Keratella tropica, Keratella vulga, Keratella cochleris, Lopocharis salpina, Mytilina ventralis, Anuraeopsis fissa, Trichotria tetractis, Trichotria similis, Lecane luna, Lecane depressa, Cephalodella mucronata, Cephalodella exigua, Monostyla bulla, Monostyla quradridenta, Lepadella ovalis, Lepadella patella, Tricocerca cylindrico, Tricocerca longiseta, Platyias quadricornis, Asplanchna herricki, Asplanchna brightwelli, Asplanchna priodonta, Asplanchnopsis, Polyarthra vulgaris, Filinia longiseta, Filinia terminalis, Filinia tetramatris, Testudinella patina, Horella mira, Hexarthra mira,* Bdelloid, *Philodina, Diphonosoma leuchtenbergianum, Diphonosoma brachyurum, Ceriodaphnia rigaudi, Ceriodaphnia laticaudata, Ceriodaphnia lacustris, Ceriodaphnia acanthine, Daphnia lumholtzi, Daphnia ambigua, Daphnia dubia, Simocephalus vetulus, Scapholeberis kingi, Moina micrura, Moina macrocopa, Moina rosea, Bosminopsis deitersi, Bosmina longirostris, Bosmina coregoni, Macrothrix rosea, Chydorus globosus, Chydorus gibbus, Chydorus sphaericus, Chydorus ovalis, Chydorus faviformis, Leydigia, Alona macrocopa, Alona karau, Alonella nana, Alonella globosa, Alonella dentifera, Allodiaptomus raoi, Heliodiaptomus viddus, Phyllodiaptomus, Rhinediaptomus, Neodiaptomus, Cyclops leuckarti, Mesocylops hyalinus, Paracyclops affinis, Microcyclops bicolor, Mesocyclops leuckartii, Nauplii, Heterocypris, Cyclocypris, Stenocypris, Eucypris, Centroypris, Insects, Insects larva,* Water mites, Arachnids water spiders.

Table 15.7. Fish fauna of Pichhola Lake, Rajasthan

Ctenopharyngodon Idella
Catla catla
Aorichthys seenghala
Mystus Cavasius
Cirrhinus cirrhinus
Labeo gonius
Heteropneustes fossilis
Notopterus notopterus
Labeo rohita
Puntius sarana sarana
Xenentodon cancila
Puntius ticto
Chela Cachius
Garra gotyla gotyla
Gambusia affinis

DAYA RESERVOIR

The Daya reservoir of Udaipur was constructed across river Daya tidi. Height above the lowest foundation is 32 meters with cultivable command area of about 1840 hectares. The length of dam is 140 meters and irrigation potential of the reservoir is 1040 hectares. The designed spillway capacity is 630 m^3/s.

Kumar *et al.* (2007) studied the food and feeding habits of *Catla catla* from Daya reservoir. Crustaceans (cladocerans and copepods) formed the main item of gut contents forming 28.19 per cent by volume and 41.89 per cent by occurrence. The major genera of cladocerans and copepods in the diet of the species were Daphnia, Moina, Macrothrix, Bosmina, Cyclops, Diaptomus, Nauplius larvae and Canthocampus. The rotifers were next in the order of dominance forming 18.20 per cent by volume and 14.81 per cent by occurrence in the gut contents of *Catla catla*. This group was mainly respresented by Keratella, Brachionus, Hexarthra, Trichocerca and Filina. Bacillariophyceae (diatoms) formed the next important item in the gut contents forming 15.45 per cent by volume and 7.40 per cent by occurrence. Amongst the Bacillariophyceae,the abundant genera were Fragilaria, Synedra, Pinnularia, Diatoma and Navicula. Aquatic insects and their instars formed 13.74 per cent by volume and 11.11 per cent by occurrence. Chlorophyceae (green algae and desmids) group also formed a part of gut content constituting only 9.30 per cent by volume and 9.26 per cent by occurrence. The major genera of Chlorophyceae in the diet of the species were Pediastrum, Cosmarium, Gonatozygon, Hydrodictyon, Protococcus and Coelastrum. Percentage of Myxophyceae (blue green algae) in the gut contents of species were 8.12 per cent by volume and 5.56 per cent by occurrence. In Myxophyceae, the major genera were Nostoc, Polycystis, Anabaena, and Aphanizomenon. Remnants of macro-vegetation were next in the order of occurrence and represented 5.67 per cent by volume and 6.19 per cent by occurrence. These plant remnantsinclude the fragments of Typha. Decayed and semi-decayed organic matter constituted only 1.33 per cent by volume and 3.78 per cent by occurrence.

The basic food of catla in Daya reservoir mainly comprised crustaceans (62.52%) and rotifers (14.27%) followed by insects (8.082%), Bacillariophyceae (6.053%), Chlorophyceae (4.559%), Myxophyceae (2.390%) and plant material (1.858%). The decayed organic matter formed negligible (0.266%) amount of the gut content.

The catla is planktophagus and feeds primarily on zooplankton. Studies on gastrosomatic index (GaSI) of *Catla catla* revealed that the feeding intensity remained high during winter months (non-spawning period) and reduced during summer months (pre-spawning period).

JAISAMAND LAKE

Jaismand Lake is the Asia's first manmade lake in Udaipur district. It is located at 24°14′N Latitude and 75°57′E Longitude. It is more than 250 years old, impounded on river Gomti in the year 1730 AD. The total water spread area of the reservoir is 7304 ha at FRL with a catchment area of 1127 sq.km. The maximum and mean depth of the reservoir are 32 m and 15 m respectively.

The fishery of the lake has been managed by Rajasthan Tribal Area Development Co-operative Federation Limited from 1978-79 onwards. The federation has been constituted with few societies of local tribal which train them in fishing activities and provides subsidy for purchasing crafts and gears. The fish marketing is also arranged by the federation through contract lift system and through their own retail outlets. Recently the federation has acquired a few hatchery units from the state fisheries department to produce required fish seed and as such the dependence for fish seed on other states has almost been reduced to half.

A total of 44 species of fish have been recorded from the lake (Anon, 1984). The important fishes were *Catla catla, Labeo rohita, Labeo calbasu, Labeo fimbriatus, Labeo gonius, Labeo boggut, Labeo bata, Cirrhinus mrigala, Cirrhinus reba, Tor tor, Puntius sarana, Wallago attu. Mystus cavasius, Mystus seenghala, Mystus aor, Heteropneustes fossilis, Ompak pabda, channa marulius, Channa striatus, Channa punctatus, Notopterus notopterus, Mastacembelus armatus* etc. The major gears used were surface gill nets, tapa nets, hook and lines and drag nets.

Tilapia (*Oreochromis mossambicus*) appeared in the fish catch during 1990-91 for the first time. The low occurrence in the beginning (1-2 specimens occasionally) has registered a sharp (5% to 10% in 1992-94) to very sharp (60%) increase in the total catch of the reservoir (Jain, 1997).

In earlier years catch composition of fishes was dominated by catfish followed by minor carps and Indian major carps. Indian major carps were contributing 24% to 30% to the total fish catch and was dominated by *Catla catla*. However, after the appearance to tilapia the percentage of minor carps increased to 60% and above. In the year 1995-96, the total fish landing was 163.64 tonnes only, in which minor carps including tilapia contributing 83.3% while the contribution of major carps was only 16.7% (Jain, 1997). Tilapia started appearing in the year 1990-91 with one or two occasional specimens of about 300-500 g, which increased to 10-25 kg per day in 1993-94, with few specimens weighing up to 1.5 kg, which has however, declined in recent year and now averaging 200 gm only (Jain, 1997).

JAWAHAR SAGAR RESERVOIR

Sarang and Sharma (2010) studied length-weight relationship and harvestable size of *Labeo calbasu* from Jawahar Sagar dam (24°82.7922 N latitude and 76°52.072 2 E longitude) of southern Rajasthan was studied using key scales. The length-weight relationship of this fish was observed to conform to the formula Log W = – 3.34124 + 1.91260 Log TL. The coefficient of correlation (r) 0.959 showed a high degree of correlation between total length and weight. Maximum three annual rings were found in the cycloid scales of fish to assess growth data in the samples representing 1 to 4 year classes. On the basis of growth parameter the harvestable size of 60 cm at the age of 3+ year class has been calculated for Labeo calbasu of Jawahar Sagar dam.

Sarang *et al.* (2009) determined age, growth and harvestable size of *Cirrhinus mrigala* from Jawahar Sagar Dam. Maximum three annual rings were found and used to assess various growth characteristics. These growth parameters indicated higher value of index of population weight growth intensity (QCW = 7.43 g) and index of species average size (Qh = 15.83 cm). The calculated weight of fish (Wg) was found to increase upto the third year of life. On the basis of growth parameters studied, a harvestable size of 66 cm has been assigned for *Cirrhinus mrigala* from the Jawahar Sagar Dam.

REFERENCES

- Anon, 1984. Studies on energy flow and productivity of Lake Jaisamand in relation to fisheries potential. Dept. Limn. Fish., RCA, Udaipur., Final Report of the ICAR Project : 24.
- Jain, Atul Kumar, 1997. Impact of *Oreochrmis mossambicus* on the fish yield ,catch composition and native fauna of Lake Jaisamand (Rajasthan), pp.139-142. In : K.K. Vass and M.Sinha (Eds.), Changing Perspectives in Inland Fisheries, Proceedings of the National Seminar, 16-17th March, 1997, Inland Fisheries Society of India, Barrackpore.
- Mathur, Nalini and Summarwar, Sudha, 2007. Survey of the status of fisheries in Bisalpur reservoir, Rajasthan, India. *Eco. Env. & Cons.* 13 (2) : 279-282.
- Raj Kumar, Sharma, B.K. and Sharma, L.L., 2007. Food and feeding habits of *Catla catla* (Hamilton-Buchanan) from Daya Reservoir, Udaipur, Rajasthan. *Indian J. Anim. Res.,* 41 (4) : 266-269.
- Sarang, N. and Sharma, L. L., 2010. Length weight relationship and harvestable size of *Labeo Calbasu* from Jawahar Sagar Dam of Southern Rajasthan, India. *Journal of Inland Fisheries Society of India.* 42(1) : 53-58.
- Sarang, N.; Sharma, L. L.; Saini; V. P., 2009. Age, growth and harvestable size of *Cirrhinus mrigala* from the Jawahar Sagar Dam, Rajasthan, India. *Indian Journal of Fisheries.* 56(3) : 215-218.
- Sharma, Archana and Mohan Devedra, 2010. Fish faunal diversity of Hemawas dam, Pali, Rajasthan. *J. Aqua. Biol.* 25(2) : 37-40.
- Sharma, K.P. and Bharadwaj, R.S., 1980. Fisheries of Gudha Reservoir, Rajasthan. *India Today & Tomorrow.* 8(4) : 175-171.
- Sharma, Riddhi, Sharma Vipul, Sharma Madhu Sudan, Verma Bhoopendra Kumar, Modi Rachana and Gaur Kuldeep Singh, 2011. Studies on Limnological Characteristic, Planktonic Diversity and Fishes (Species) in Lake Pichhola, Udaipur, Rajasthan (India). *Universal Journal of Environmental Research and Technology.* Available Online at: www.environmentaljournal.org. 1(3) : 274-285.
- Summarwar, Sudha, 2012. Compartive haematolgical studies of *Clarius batrachus* in Bisalpur reservoir and Pushkar Lake. *Indian Journal of Fundamental and Applied Life Sciences,* 2(2) : 230-233.

16

TAMIL NADU

Fig. 16.1. Map of Tamil Nadu (Not to scale)

Tamil Nadu is gifted with rich fishery potential. Fishing is one of the oldest occupations of human kind, which provides a rich and easily available source of protein, and it plays a vital role in improving the dietary standards. Tamil Nadu is one of the important Maritime States with rich inland and marine resources. The inland fisheries sector has about 3.7 lakhs ha of water spread area comprising reservoirs, major irrigation and long seasonal tanks, short seasonal tanks and ponds, estuaries, backwaters etc. which are suitable for both capture and culture fisheries. About 5,000 ha of water-spread area is being utilised presently for fresh water aquaculture under the programmes of the Fish Farmers Development Agencies.

Table 16.1. List of reservoirs under the control of Fisheries Department, Tamil Nadu

Sr.No.	Name of the Reservoir	Water Spread Area (ha)
1	Kolavai	700
2	Poondi	3269
3	Vidur	773
4	Manimuktha	778
5	Gomuki	363
6	Veeranam	6100
7	Wellington	834
8	Mettur	15346
9	Krishnagiri	1298
10	Barur	245
11	Kelavarampalli	430
12	Pambar	220
13	Pillur	718
14	Sholayar	520
15	Vaigai	2590
16	Kuduaganaru	496
17	Vembakottai	480
18	Kullurchandai	316.4
19	Manimuthar	940
20	Pechiparai	1515
21	Perunchai	962
22	Citttar I&II	707
23	Chinnar	170

Contd...

Table 16.1: Contd...

24	Anaimduvu	107
25	Thumbalahalli	192.7
26	Nagavathy	117.6
27	Vaniyar	109.3
28	Kesarikulihalli	120
29	Thoppiar	110
30	Manjalar	197
31	Nangangiyar	169
32	Vadakku pechaiyar	184
33	Nambiyaru	153

BHAVANISAGAR RESERVOIR

Bhavanisagar reservoir has an area of 7,720 ha at full reservoir level and 3,695 ha at mean reservoir level while the mean depth is 11.10 m. This reservoir is characterized by a strong oxycline, kilinograde distribution of oxygen, accumulation of carbon dioxide and low pH in the bottom layers and chemical (bicarbonate) stratification, which support high photosynthetic and tropholytic activities (Sreenivasan, 1964). The organic carbon and nitrogen level is fairly high (Anon, 1981). In spite of high phytoplankton productivity, *Catla catla* has fairly establish itself into a viable stock, and therefore, it is necessary to stock this reservoir with the fingerlings (8 to 10 cm long) of this species at a rate of about 750 fingerlings per hectare every year and reduce predatory fishes like *Wallago attu* by selective fishing using hook and lines (Devaraj *et al.*, 1990).

The annual catch ranged from about 70,090 kg for an effort of 43,000 gill net days in 1983-84 to 340,560 kg for 258,000 gillnet days in 1981-82, while the average annual catch was 193,488 kg for the average annual efforts of 123,333 gillnet days. The extremely high catch in 1981-82 (340,560 kg) was due to very low level of water in the reservoir. The average annual catch for 1977-78 to 1983-84 period for which species composition data is available was 212,900 kg.

National Fisheries Development Board, Hyderabad sanctioned Rs. 8.4 lakhs and released 50% as first installment *i.e.,* Rs. 4.2 lakh to Tamil Nadu Fisheries Development Corporation Limited for stocking in Bhavanisagar reservoirs under 4th consecutive year stocking programme for the year 2011-12. The second installment of Rs. 4.2 lakhs was also sanctioned by NFDB.

The list of common fish fauna recorded in Bhavanisagar reservoir is depicted in Table 16.2.

The gross photosynthetic productivity in reservoir has been found to be $(8.162\ cal/m^2/year) \times 10^6$.

Table 16.2. Fish fauna of Bhavanisagar reservoir in Tamil Nadu

Catla catla, Hypopthalmichthys molitrix, Puntius carnaticus, Labeo rohita, Labeo fimbriatus, Labeo calbasu, Puntius Puntius, Labeo bata, Cirrhinus mrigala, Cirrhinus cirrhosa, Cirrhinus reba, Puntius hexagonolepis, Puntius sarana, Cyprinus carpio, Tilapia mossambica, Mystus aor, Mystus seenghala, Silonia silonida, Pangasius pangasius, Ompak bimaculatus, Puntius dubius, Etroplus suratensis, Tor tor, Channa marulius, Wallago attu, Anguilla bicolor, and *Notopterus notopteus.*

Food and feeding habits of *Notopterus notopterus* from Bhavanisagar reservoir were studied by Jesu Arockia Raj *et al* (2004). The total length and body weight of N.notopterus ranged between 17 and 29 cm and 60 and 174 of respectively. The gut contents of fish consisted of algae, copepoda, rotifer, insect larvae, cladocaera, crustacean, detritus, annelid worms, small fish, fish scale, sand granules, prawn juveniles and insects. Aquatic insect larvae (23.14%), algae (17.18%), fish scale (16.25%), and annelid worms (15.5%) showed that highest value of index of preponderance. Presence of *N. notopterus* youngones in the gut confirmed that the fish is cannibalistic in nature. The zooplankton and phytoplankton, insect larvae, annelid worms, fish scale are mostly preferred food items of *N. notopterus*. However, as the fish increased in size, it has higher preferable food items of their own youngones, insect larvae, aquatic insects, prawn juveniles and annelid worms. The remaining food items *viz.*, copepoda, cladocera, crustacean, rotifera, and detritus matter are also found in small sized fish at very low quantity in the gut. The anatomical features of the alimentary canal and their preferable food items are also concluded that *N. notopterus* is an omnivorous fish.

METTUR RESERVOIR

Mettur reservoir (Stanley Reservoir) is the only large reservoir in the state of Tamil Nadu.It is located at 11°50′N and 77°50′E. The reservoir across the river Cauvery was built between the hills of Seethamalai and Palamalai during 1920 to 1924. The detail morphomery of the reservoir is depicted in Table 16.3. The reservoir water is used for irrigation, navigation and drinking purpose by large number of peoples in Salem district.

Table 16.3. Morphometry of Mettur Reservoir

1. River	Cauvery
2. Year of construction	1920-1924
3. Water spread area (at FRL)	14690 ha
4. Water storage capacity	2546 million m^3
5. Maximum length of dam	53 kms
6. Width of dam	8.85 kms
7. Maximum depth	45.7 m
8. Mean depth (m)	18.3 m
9. Shoreline (km)	293

The dam was sealed in 1934.Fish was stocked in the headwaters ahead of closure of the dam. *Catla catla* was introduced in 1928, while the other major carps like *Labeo rohita, Cirrhinus mrigala* and *Labeo calbasu* were introduced from 1948 onwards. *Etroplus suratensis* and *Chanos chanos* were also stocked (Sreenivasan, 1998).

Mathivaran *et al.* (2005) reported status of water quality from output and input of Mettur dam. The investigation revealed that the water is polluted. Range of physico-chemical parameters in Mettur reservoir is depicted in Table 16.4.

Table 16.4. Physico-chemical parameters of Mettur Dam

Parameters	Range
Turbidity (NTU)	0 – 87
Total Dissolved Solids (ppm)	290 – 623
Electr.Conductivity (μ.mhoscm^{-1})	414 – 890
pH	7.8 – 8.69
Total Hardness (mg/l)	122 – 232
Dissolved oxygen (mg/l)	4.8 – 6.9

Stocking is done regularly with seed of Indian Major Carps as well as with *Labeo fimbriatus, Labeo kontius, Cyrpinus carpio* etc. The size at stocking is not regulated.

During the period 1963-64 to 1983-84, the annual catch ranged from 159,100 kg in 1977-78 for 430 coracle units to 579,540 kg in 1979-80 for 743 coracle units. The average annual catch was 351,374 kg and the average annual effort 494 coracle units. The number of licenses issued remained at 305 to320 from 1963-64 to 1972-73 except in 1968-70 (408 to 428), but nearly doubled to 539 to 750 from 1974-75 onwards except in 1975-76 (463) and 1977-78 (430).

Fishing in Mettur reservoir has almost reached the Maximum Sustainable Yield (MSY) level of about 500 tonnes from 1979 to 1982, for an annual effort of about 750 coracle units.

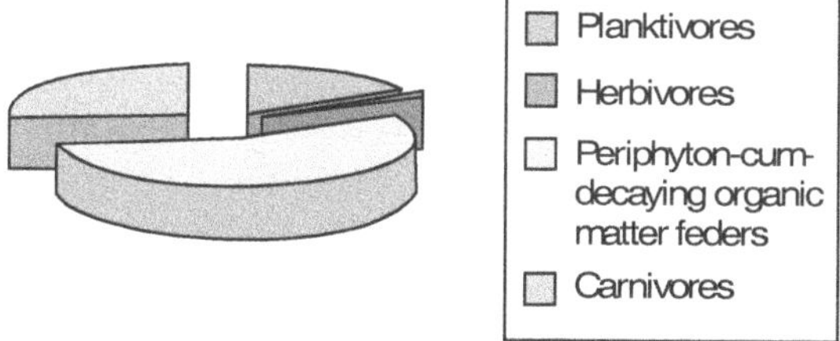

Fig. 16.2. Species composition of fish from Mettur reservoir, Tamil Nadu

A critical analysis of post impoundment fisheries data available for various periods from 1943 to 2003 in respect of Mettur reservoir indicates that a major change in species composition of reservoir fish landings has taken place and that a few native species such as *Cirrhinus cirrosa, Labeo kontius, Puntius dubius, Silonia silondia* etc have disappeared, giving way for the stocked species of Indian Major

Carps which dominate in the current fishery of the reservoir (Palaniswamy and Manoharan, 2010).

Rangnathan and Natarajan (1969) and Sreenivasan (1995) have reported that many fishes such as *Cirrhinus cirrhosa, Silonia silonida, Puntius dubius, Pangassius pangassius* and *Labeo fimbriatus* have disappeared due to introduction of exotic fishes. Tilapia less than 100 gm are very common in reservoir. However, for some reason, these were not procured by the co-operative society and the fishermen had to sell them to merchants. Many small sized fishes such as *Cirrhinus reba, Mystus* spp., were not procured by the society. The state Fisheries Department of Tamil Nadu furnished the following recent data on seed stocking and fish catch from Mettur reservoir.

Table 16.5. Seed stocking and fish catch from Mettur Reservoir

Sr.No.	Year	Seed stocking (in lakhs)	Fish production (tonnes)
1.	2007-08	25	254.010
2.	2008-09	25	166.31
3.	2009-10	42.86	132.72
4.	2010-11	44.30	289.67
5.	2011-12	23.02	671.422

The fishery is exploited by licensed fishers whose numbers are regulated by the Department of Fisheries. The main fishing gear is gill nets operated from a coracle. Rod and line, cast nets, shore nets etc. are also in use. All the licensees are members of the Fishermen's Co-operative Marketing Society and their catches are handed over to the co-operative at prices fixed each year. They are collected from different landing centers in a boat carrying ice and brought to the assembly centre. After making local sales, the major part of the catches are handed over to the middlemen who is the highest bidder. The fish are gutted, iced, packed in baskets and transported to the railhead for onward dispatch to Kolkata by train.Enroute re-icing is done at a couple of stations.

SATHNUR RESERVOIR

Located at an altitude of 230 m above mean sea level, the reservoir has an area of 1,635 ha at FRL. The reservoir was constructed about 11 km from Sathnur village.The reservoir was constructed for irrigation purpose and is located in the Ponnaiyar reserve forest. There was no human displacement, as the affected area had no human habitation.The 40 m high Sathnur dam presents an imposing example of engineering skills, both formiodable and admirable. The spillways over which the water cascades down present a hypnotic sight. The hills surrounding the dam, resplendent with the greenery of the flora of Ponnaiyar reserve forest, give a majestic look to the place. The 7.5 MW hydro-electricity unit of Tamil Nadu Electricity Board is built near Sathnur dam. It was commissioned in September 1999. Though favourable morphometric edaphic and thermal conditions prevail, primary productivity was found to be low owing to high turbidity (Sreenivasan, 1968).

The fish fauna of the reservoir during the period of impoundment consisted of *Cyprinus carpio* and few catfishes. In the year 1959, the Tamil Nadu Fisheries Department took charge of reservoir fishery. Seeds of fast growing species of fishes such as *Catla catla, Labeo rohita, Cirrhinus mrigala* etc. were stocked in the reservoir. Later on the fishery rights of the reservoir were transferred to the Tamil Nadu Fisheries Development Corporation Limited (TNFDC).

A self-sustaining population of Catla was established and this continued for decades. About 90% of the catches were carps, of which over 80% were catla. From 1977-78, a population balance was established by mesh modification and stocking to obtain catches of rohu, mrigal, *Labeo calbasu, Labeo fimbriatus* etc.

The fishery of reservoir is mainly supported by the fishes such as *Catla catla, Labeo rohita, Labeo calbasu, Labeo fimbriatus, Cirrhinus mrigala, Cirrhinus cirrhosa, Cyprinus carpio, Tilapia mossambica, Mystus aor, Mystus seenghala, Heteropneustes fossilis, Clarias batrachus* and *Wallago attu.*

The annual fish catch ranged from 45000 kg in 1966-67 for an effort of 6 coracle unit per day for 365 days a year to 184,000 kg in 1977-78 for 19 coracle units per day. The fishing effort progressively increased from 6 coracles units per day in 1966-67 and 1967-68 though 8 units in 1968-69, 12 units during 1969-70 to 1971-72, 14 units in 1972-73,15 units both in 1973-74 and 1974-75, 19 units in 1977-78 and 21 units in 1978-79 to 39 to 40 units since 1979-80.

Species composition for the period of 1977-78 to 1983-84, when the average annual catch was 128,687 kg, indicates that the fishery was constituted by the planktivores (66.78%-exclusively of *Catla catla*).

National Fisheries Development Board, Hyderabad sanctioned Rs. 5 lakhs and released 50% as first installment *i.e.,* Rs. 2.5 lakh to Tamil Nadu Fisheries Development Corporation Limited for stocking in Sathnur reservoir under 4th consecutive year stocking programme for the year 2011-12. The second installment of Rs. 2.5 lakhs was also sanctioned by NFDB.

The fish catches from the reservoir are sold to the public through various retail outlets at the rates fixed by the TNFDC.The functioning outlets are at the places like Sathnur dam, Thiruvannamalai, Polur, Chengam, Authur, Polur, Cheyyar, Thirupatthur, Sankarapuram, Arakkalulur and Vadavasi.

The details of the fish seed stocking,fish catch and fish catch per hectare in Sathnur reservoir have been presented in Tables 16.6, 16.7 and 16.8 respectively.

Table 16.6. Seed stocking in Sathnur Reservoir

Species	Seed stocked	
	1999-2000	2001-02
Catla catla	5.50	2.655
Labeo rohita	2.53	2.06
Cirrhinus mrigala	1.54	1.21

Table 16.7. Details of total fish catch in Sathnur reservoir

Year	Fish catch (tonnes)
1992-93	165.05
1993-94	140.48
1994-95	80.06
1995-96	142.64
1996-97	133.82
1997-98	113.76
1998-99	Not Available
1999-00	130.21
2000-01	375
2001-02	127.13

Table 16.8. Fish catch per hectare of Sathnur reservoir

Year	Fish catch (kg/ha)
1991	58.7
1992	58.7
1993	91.7
1994	78
1995	44.5
1996	79.2
1997	74.3
1998	84

Each fishing units consists of two fishermen, one coracle and 20 gill nets. The units operate under the close supervision of the Deputy Manager. The nets used by the TNFDC are called Rangoon nets. These are basically gill nets of mesh size 15 c without ottom rope and sinkers. The stretched length and depth of each net are 30 and 16 m respectively. The fishermen are not permitted to use nets of smaller mesh size to avoid capture of fingerlings. Net operation in the upper reaches of the reservoir during the breeding season is prohibited. The range of water quality parameters is depicted in Table 16.9.

Table 16.9. Water quality of Sathnur Reservoir in year 1998

Sr.No.	Parameters	Range
1.	Total solids (mg/l)	346 – 628
2.	Suspended solids (mg/l)	10 – 267

Contd...

Table 16.9: Contd...

3.	Total dissolved solids (mg/l)	333 – 410
4.	Total hardness (mg/l)	175 – 215
5.	Calcium hardness (mg/l)	22 to 34
6.	Chloride (mg/l)	63.8 – 85.1
7.	Magnesium (mg/l)	21.6 – 36
8.	Sulphate (mg/l)	13.4 – 22.1
9.	Total alkalinity (mg/l)	230 – 280
10.	pH	7.8 – 8.9
11.	Electrical conductivity (µ mho cm^{-1})	616 – 760

PERUMCHANI RESERVOIR

The Perumchani reservoir is located in the South Western Ghats in Kanyakumari district of Tamil Nadu at 89° 22′ latitude and 77° 22′ east longitudes. It is a straight gravity masonry dam of 373.1 m long consisting of 275.28 m of bulkhead section. It is built on Paraliyar river.The water from this dam is taken to Nagarcoil town and nearby areas mainly for irrigation but during water scarcity in other dams, this water is also diverted for domestic uses including drinking source. The catchment area has a few rubber plantations.

Raj and Lawrence (2012) reported water quality of Perumchani Reservoir. The annual average for water transparency ranged from 0.93 m to1.97 m with moderately high levels of water transparency during hotter months, and still higher levels during colder months. Monsoons season recorded the lowest transparency levels. After the monsoon there was an increase in the transparency level in water. The overall low transparency level in Perumchani reservoir may be one of the reason for very low productivity thereby fish fauna here showing the negative correlation between transparency and productivity. Summer months exhibited slight high levels of water transparency. This may be due to evaporation of water, rich in nutrients, increase in temptation and development of plankton flora. The annual average of pH ranged from 5.86 to 6.38. The summer months and post-monsoon season also recorded slightly higher values of pH. The annual average of alkalinity fluctuated between 15.30 mg/l to 24.80 mg/l. The low alkalinity recorded in the water of this reservoir is thereby affecting the productivity of the water body. Higher BOD levels were also observed during months before the rains. The higher BOD level during summer months is due to increased bacterial activity affecting the BOD levels. The high phosphate level during summer months is due to high wind speed, decrease of water level,increased evaporation and release of nutrients at increased temperature from decomposition of flora.

Raj (2010) studied sediment profile in Perumchani reservoir and found higher silt accumulation was during summer and winter months at the reservoir outlet. The month wise analysis of sediments showed 2.25% – 47.05% clay, 9.08 – 43.18

% silt, and 8.81 – 87.89 % sand. The annual average of Phosphorus ranged from 0.080 mg/*l* to 0.156 mg/*l* with higher level recorded during monsoon season. High level of Sediment Organic Carbon was noticed during premonsoon. The summer months had high nitrogen content in the sediments and lowest during winter months and month after rains. The annual average of Carbon Nitrogen ratio ranged from 6.15 – 58.67. Rate of siltation is fast while all other parameters are within the permissible limit.

The fishery of reservoir is predominantly of Tilapia (Oreochromis mossambicus). Numerous tilapia-breeding pits of about one-meter diameters are seen along the margin of the reservoir. Other fishes found in reservoir are catla, mrigal, common carp and *Mystus* spp.

ALIYAR RESERVOIR

Present fish yield in Aliyar reservoir is one of the highest in the country. The State Fisheries Department launched a stocking programme in the reservoir immediately after its formation in 1962. The stocking density was erratic in the earlier years with a high preponderance of medium and minor carps including *Puntius carnaticus, Puntius dubius, Labeo kontius, Labeo fimbriatus, Labeo calbasu, Cirrhinus reba, Cirrhinus cirrhosa, Acrossocheilus hexagonolepis* and *Tor khudree malabaricus.*

The fishing in Aliyar reservoir was done both by Department of Fisheries and private fishing units, on a royalty basis, without any restriction on meshsize. This resulted in indiscriminate exploitation of stocked fishes much below the desirable size. The cumulative effect of a number of irrational reservoir management practices resulted in a very low yields, ranging from 2.67 to 54.7 kg/ha/yr, and an average of 26.21 kg/ha/yr. The reservoir was subjected to scientific investigations since 1983 and a number of measures were initiated to develop the fisheries. The results of scientific management started reflecting in the yield rate from 1985. The management options used for enhancing fish yield were:

1. Stocking was limited to Indian Major Carps
2. Size of stocking was increased to advanced fingerlings (>100 mm).
3. Stocking density was reduced to 236 to 312 nos/ha.
4. The stocking schedule was spread over different months of the year
5. A strict mesh regulation banning nets <50 mm in mesh bar
6. Ban on catching Indian Major Carps less than 1 kg in size

A direct outcome of the adoption of scientific management practices in Aliyar reservoir was a sustained increase in fish yield to the tune of 193.58 kg/ha/yr. The yield ranged during 1985-86 to 1989-90, from 77.75 to 193.58 kg/ha/yr, with major carps constituting 84.04 to 96.25% (Sugunan, 1995).

KRISHNAGIRI RESERVOIR

Krishnagiri reservoir is constructed across the Ponnaiyar river basin near Periyamuttur village about 10 Km from Krishnagiri town in Krishnagiri district,

Tamilnadu. It is located at a latitude of 12°28′ N and a longitude of 78° 11′ E. The Ponnaiyar river, takes its birth in the eastern slopes of Chennakesava hills, north west of Nandi Durg and flows towards the south east direction. The total drainage area of the river is 15,101 km^2. The river flows through Krishangiri, Dharmapuri, Vellore and Cuddalore and join the Bay of Bengal near Cuddalore town.The water spread area of Krishnagiri reservoir is 1280 ha at FRL and its water storage capacity is 2410 million cu.ft.The reservoir is influenced both by the south-west and North-east monsoons.Its maximum depth is 17.4 m and mean depth is 5.3m.The volume development is 0-94 and shore development 1.5.Its catchment area is 210 sq.miles (538 sq.km). The dam was constructed in 1957.

The range of physico-chemical parameters of reservoir in the surface zone are depicted in Table 16.10.

Table 16.10. Range of physico-chemical parameters of Krishnagiri reservoir in the surface zone

Parameter	Range
Temperature (°C)	23.5 – 31.4
Carbonate alkalinity (ppm)	8.0 – 50
Bicarbonate alkalinity (ppm)	146 – 373
pH	7.6 – 8.6
Dissolved oxygen (ppm)	4.4 – 8.4
Total hardness (ppm)	76 – 146
Conductivity (¼s/cm)	425 – 575
Total solids (ppm)	291 (av.)
Silicates (ppm)	1.9 (av.)

Phytoplankton was dominated by bluegreens, especially *Microcystis*. Diatoms were poorly represented.

The common fishes recorded in the catches are *Labeo kontius, Labeo fimbriatus, Puntius sarana, Cirrhinus reba, Cirrhinus mrigala, Notopterus notopterus, Ompak* sp, *Mystus aor, Wallago attu, Glossogobius giuris, Mastacembelus armatus, Chela* sp., *Cyprinus carpio, Tilapia mossambica,* and *Rhinomugil corsula.*

Data on stocking and harvesting are presented in Table 16.11 and 16.12 respectively. The total catch per year rose from 14,132 kg in 1965-66 to a peak of 90,003 kg in 1990-91 (70.3 kg/ha). During this period, the shre of each of the fishing units that were operating swelled upto 3750 kg/yr. This is a satisfactory share compared to many similar reservoirs. After 1991-92, however, the catches dwindled year by year reaching a 17 year low of 18,502 kg (14.45 kg/ha) in 1997-98.

A detailed year by year analysis of stocking and harvesting would indicate the deterioration in the management of this productive reservoir. From 1965-66 till 1979-80 the catches varied from 14132 kg/yr to 17400 kg/yr. The initial lower yields were due to higher turbitdity. Construction of dams on three tributaries

Vaniar,Pamban and Sulagiri Chinnar (all less than 200 ha in area) reduced the silt turbidity in Krishnagiri reservoir, resulting in higher fish yields. It is worth noting that fish catches from the above mentioned three small reservoirs (Sreenivasan, 1999) were a bonus. The improved catches in Krishnagiri from 1982-83 may be due to this factor.

The stocking rate in reservoir has been seemingly good. The rates per annum averaged as follows:

Duration	Fingerlings/ha/yr
1967-68 to 1971-72	920
1972-73 to 1976-77	517
1983-84 to 1987-88	1019
1988-89 to 1992-93	1462
1993-94 to 1997-98	387

These figures are not reconcilable with yields. Doubts are expressed about the authenticity of the figures. Though stocking figures of major carps are impressive the harvested quantities are disappointing as may be seen from the following figures:

Duration	% of major carps
1967-68 to 1971-72	23.14
1972-73 to 1976-77	6.36
1977-78 to 1982-83	4.24
1983-84 to 1987-88	8.72
1988-89 to 1992-93	5.58
1993-94 to 1997-98	4.87

A very peculiar feature of the reservoir is the alternating dominance of tilapia and the freshwater mullet (*Rhinomugil corsula*). From table, it can be seen that in the initial years, from 1967-68 to 1987-88, when mullets dominated, tilapia was low in percentage yield but from 1988-89, tilapia catches reached a peak with 77% and mullet dwindled to 13.1% and disappeared from 1993-94 (when tilapia contributed to 61%). These species seem to be competitively and mutually excessive.

Krishnagiri has a high MorphoEdaphic Index (MEI) of 67.3, ranking second among all Tamil Nadu reservoirs. The methyl orange alkalinity is also high. It can be seen that except for the period 1966-67 to 1971-72, when major carps accounted for 25.32% of the total yield, it was always less than 7% in the next 25 years. The major catch was that of low value fishes such as tilapia and mullets. Catfishes were conspicuous during some years, from 1981-82 to 1989-90, even accounting for 21.3% of the total catch in 1981-82. It was over 6.0% till 1989-90. *Notopterus notopterus* was also caught in noticeable quantities in some years (for ex., 13.7% in 1983-84).

Krishnagiri is an example of a productive but poorly managed reservoir. The potential for production is great but due to poor management and lack of planning, it has become a loss-making reservoir. It is necessary to intensively fish out the low value fish, stock real numbers of advanced fingerlings of major carps (no common carp), prevent 'leakage' and arrange for profitable marketing in nearby urban centers, like Dharmapuri, Hosur, Salem etc. A 100 m.t. harvest would fetch revenue of Rs. 20 lakhs at a modest price of Rs. 20/kg. Proper monitoring, assessment and management are imperative to make this reservoir profitable.

Table 16.11. Stocking in Krishnagiri reservoir of Tamil Nadu

Years	Stocking (x 1000)				Total
	Catla	Rohu	Mrigal	Common carp	
1993-94 to 97-98 Avg/yr	697139.4	714142.8	35070	30060	2465493
1988-89-93 to 92-93 Avg/yr	790263	524175	671224	870290	56141871
1983-84 to 1987-88 Avg/yr	118.6539.55	1414471	821274	149.849.9	39121304
1977-78 to 1982-83 Avg/yr	NA	NA	NA	NA	NA
1972-73 to 1976-77 Avg/yr	43.348.07	408.581.7	——	955191	3061662
1967-68 to 1971-72 Avg/yr	5.11.0	1640.7328.1	——	765.3133.1	58851176.6

Table 16.12. Fish harvesting and revenue from Krishnagiri reservoir of Tamil Nadu

Harvest (kg)						Revenue
Catla	Rohu	Common carp	Tilapia	Mullets	Total	(Rs.)
6719	4606	885	155354	—	254195	1209998
1344	925.2	177	31071	—	50839	241999.80
10672	5776	701	173022	38841	372221	954913
3557	1925	234	57674	9710	744444	191000
3221	3348	2993	29635	105642	230444	595585
644	670	599	5927	21108	46090	119117
1020	1220	856	5724	37138	96322	NA
510	610	428	2862	7427	19224	NA
123	660	659	6780	12895	66062	62814
25	132	132	1356	2579	13212	12563
9717	459	1342	1909	16128	61575	52343
1943	92	268	382	3224	12315	10468
4036	75	65	60	78	13534	10950

Renuga and Ramanibai (2010) studied zooplankton composition in Krishnagiri reservoir. 23 taxa were found within 4 groups, copepod nauplii did not considered as a group and therefore not included in this list. The organisms of the reservoir classified into four main groups rotifera, cladocera, copepoda and ostracoda. Rotifera represented by 8 species, cladocera represented by 5 species, copepoda by 6 species and ostracoda by 4 species.

Table 16.13. Stocking and harvesting of carps in Krishnagiri reservoir of Tamil Nadu

Years	Catla		Rohu		Mrigal		Common carp	
	S	H	S	H	S	H	S	H
1967-68 to 71-72	5100	9717	1640	459	700	4072	5885	1342
1972-73 to 76-77	40340	123	408500	660	9,54,900	3397	3061	659
1977-78 to 1982-83	NA	1021	NA	1220	NA	1926	NA	856
1983-84 to 87-88	1,18,650	3321	1414000	3348	8,21,000	15919	3912	2993
1988-89 to 92-93	7,90,000	10672	524000	5776	6,71,000	4327	5614	701
1993-94 to 97-98	7,97,000	6719	714000	4606	3,50,000	1057	2465	885

S = Numbers stocked., *H* = Kg of fish harvested

Table 16.14. Zooplankton diversity in Krishnagiri Reservoir

I. Rotifera 1. *Brachionus calyciflorus* 2. *Brachionus caudatus* 3. *Brachionus forficula* 4. *Brachionus falcatus* 5. *Brachionus longiseta* 6. *Filinia longiseta* 7. *Keratella tropica* 8. *Keratella valga*
II. Cladocera 1. *Ceriodaphnia cornuta* 2. *Dhypanosoma excisium* 3. *Macrothrix laticornis* 4. *Bosmina longirostris* 5. *Moina micrura*
III. Copepoda 1. *Mesocyclops aspericornis* 2. *Mesocyclops hyalinus* 3. *M. thermocylopoides* 4. *Heliodiaptomus viduvs* 5. *Eodiaptomus shihi* 6. *Neodiapotmus satanas*

Contd...

Table 16.14: Contd...

IV. Ostracoda
1. *Cypris protubera*
2. *Cypris subglobosa*
3. *Strandesia indica*
4. *Strandesia elongata*

AMARAVATHY RESERVOIR

Amaravathy reservoir is situated at the foot of Amaravathy hills, 20 km south of Udumalpet Taluk, Coimbatore District at a latitude of 10 29′ and longitude of 77 10′. The catchment area is 839 square kilometers with a rainfall of 125 mm to 320 mm. It is an important reservoir engaged with fisheries activities.

Thirumathal (2006) reported cladoceran diversity and physico-chemical environment of Amaravathy reservoir.Five species of cladocerans such as *Daphnia pulex, Ceriodaphinia cornutta, Moina cornutta, Bosmina longistris,* and *Chydorus parvus* were recorded.Generally the cladocerans were found more during winter months and less in months like June, July, February and April. The population was completely absent in summer months,which may be due to changes in water pH and temperature. The nutrients also act as one of the important parameter in regulating the growth of cladocerans. Monthly variations of pH, temperature and nutrients in the Amaravathy reservoir are shown in Table 16.15.

Table 16.15. Monthly variations of pH, temperature and nutrients in the Amaravathy reservoir, Udumalpet from June, 2002 to May, 2003

Months	pH	Temperature	Sulphates	Chlorides	Nitrates	Phosphates
June, 2002	7.3	24.6	9.8	30.6	9.5	3.8
July	7.6	26.5	7.5	30.2	9.2	3.5
Aug	7.5	25.8	8.1	28.7	8.6	4.2
Sep	7.8	26.3	8.5	27.9	9.2	3.1
Oct	7.6	27.6	7.9	29.4	8.6	4.6
Nov	7.5	24.8	8.5	30.6	8.1	4.5
Dec	7.4	25.6	7.6	28.6	8.7	3.8
Jan, 2003	7.6	23.8	9.2	29.5	9.2	3.6
Feb	7.9	25.7	8.7	26.8	9.6	4.2
Mar	7.9	26.6	9.6	30.4	9.6	3.9
Apr	7.9	27.9	9.1	29.8	9.8	4.8
May	7.8	28.3	8.8	30.6	9.5	4.6

(Temperature is given as °C and sulphates, chlorides, nitrates and phosphates are given as mg/l)

KODUMUDIARU RESERVOIR

The Kodumudiaru reservoir was constructed in year 2004. The reservoir was built across the river Kodumudiaru, which originates from Western Ghats in Tirunelveli District at a height of above 200 meters. The total length of the dam is 43.2 meters with a top level breadth of 5 meters. The dam water is used for an area of 5781 acres. The amout of water that comes out from the reservoir through the canal is 7944 cub/sec. The maximum amount of water is passed from the reservoir through the canal is 17084 cub/sec. People of more than 50 villages are using this water for drinking and agricultural purposes.

Santhi (2012) studied the water quality of Kodumudiaru reservoir and concluded that the concentration of all the water quality parameters are within the prescribed limit. The water is suitable for drinking and irrigation purposes. Total alkalinity ranged from 56.58 mg/l to 150.88 mg/l. The amount of total hardness ranged from 28 mg/l to 76 mg/l. In the water samples pH ranged from 6.67 to 8.13. The electrical conductance ranged between 70 µmho/cm and 115 µmho/cm. Chloride and salinity ranged from 14.683 mg/l to 25.361 mg/l and 26.943 mg/l to 46.537 mg/l.

THIRUMOORTHY RESERVOIR

Thirumoorthy Reservoir with a water spread area of 388 ha at full reservoir level and 234 ha at average storage level is located at 10° 28′ N and 77°09′ E in Coimbatore District. Earlier the fish yield was very low *i.e.*, 56.33 kg/ha/yr. Later on following the scientific management the reservoir was stocked (Selvaraj *et al.* 1997). The reservoir was stocked with pure seed of major carps of more than 100 mm in length from 1991-92 onwards at reduced stocking density from 215-406 nos/ha/yr with an average of 343 nos/ha/yr. The density and ratio of different species were determined by their growth rate under the existing ecological condition of the reservoir. There had been a conspicuous improvement in the yield right from 1991-92 (87.3 kg/ha/yr) and a record yield of 182.1 kg/ha/yr was achieved during 1995-96. The contribution by major carps also improved (88.5 – 96.4%) simultaneously, resulting in substantial increase in the revenue.

Palaniswamy *et al.* (2011) studied the population dynamics of major carps and common carps cultured in reservoir. A total of 14,315 numbers of fin clipped *Catla catla, Cirrihinus mrigala, Labeo rohita* and *Cyprinus carpio* var. Communis were artificially recruited into the reservoir. The assessment of population parameters revealed that the fishery of *Catla catla* in this reservoir is good but the asymptotic length has not reduced significantly thus showing scope for increasing the stocking density. In the case of *Cyprinus carpio* and *Labeo rohita,* the present stocking density appeared to be close to being optimum. However, recapture of rohu is very poor, discouraging additional stocking; rather reducing the stocking density would avoid incurring monetary loss. But for mrigal, the present stocking density is suboptimal but the minimum harvestable size takes more than one year. Hence increasing the stocking density of mrigal will further delay the minimum harvesting size.

VEERANAM LAKE

The Veeranam Lake is oldest and largest lake of Tamil Nadu. This is a very significant lake in south India, and nearly 45,000 acres of agricultural land areas are irrigated by this lake and in addition since- 2004 the lake supplies the large quantity of water for Chennai metropolitan city for drinking purpose. The Veeranam lake (Lat. 11°17′N); Long. 79°32′E) is situated in the Cuddalore district of Tamil Nadu state, India (Fig. 16.3). It was formed by the Prince Rajathithar son of Paranthaga Cholan during the 9th century and as on date its age is 1038 years. The lake is located in the western half of Chidambaram taluk and southern half of Kattumannarkoil taluk in the Cuddalore district. The lake is being used for multipurpose utility such as irrigation, fish catching washing and bathing. This lake is one of the major water resource supply to the Chennai metropolitan city corporation. The lake covers on water spread 15 sq. miles (38.85 sq. km) area of length 15 km from North-South, 5 km width from East-West, minimum depth 5.43 m, water storing capacity 45.50, ft = 990 mct (0.99 TMC) (mean sea level), = 990 mct (0.99 TMC) with 34 sluices irrigated land area approximately 45,000 acres.

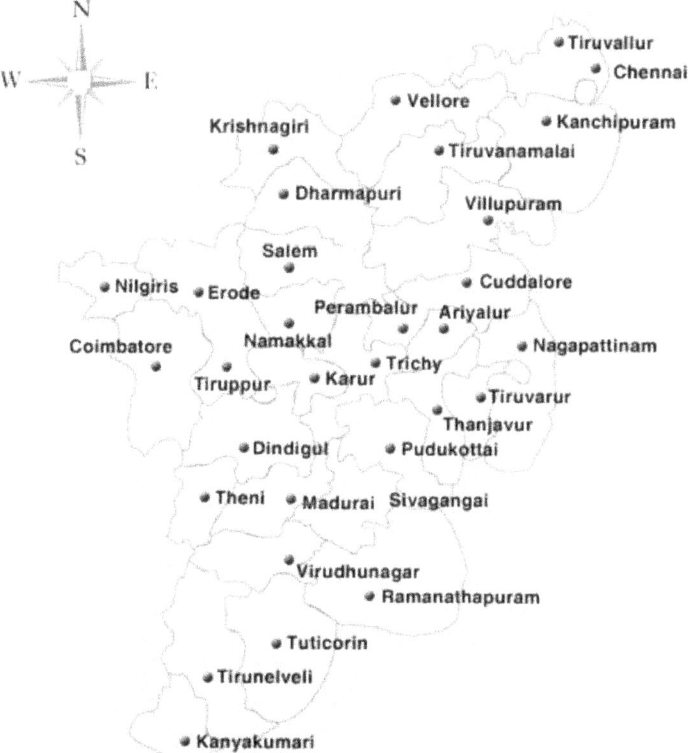

Fig. 16.3. Map of Veeranam Lake of Cuddalore district

Senthilkumar and Sivakumar (2008) studied phytoplankton diversity in response to abiotic factors in Veeranam lake in the Cuddalore district of Tamil Nadu. The physico-chemical parameters of water such as air-temperature, water temperature, pH, salinity, dissolved oxygen, electrical conductivity and total dissolved solids were observed and their ranges were: 30.1 – 36.5°C, 29.0 – 34.4°C, 7.9 – 8.4, 1.2 – 2.5 mgl^{-1}, 7.6 – 9.2 ¼ S and 2.5 – 5.2 mgl^{-1}. Totally 160 species of phytoplankton belonging to different taxonomic groups were identified. Among these 74 species to belong to bacillariophyceae, 43 species to chlorophyceae, 38 species to cyanophyceae and 5 species to euglenophyceae. The phytoplankton density was high (1705 cells l^{-1}) during the summer season and low (760 cells l^{-1}) during the winter season. Bacillariophyceae formed the dominant group.

PECHIPARAI RESERVOIR

The European Engineer Mr. Minchin Alexander constructed Pechiparai Reservoir during the period of 1897-1906. The catchment area of the reservoir is 207.19 km^2. Vidhya and Nair (2012) studied planktonic diversity and hydrobiological characteristics of Pechiparai Reservoir of Kanyakumari district for the period of March 2010 to August 2010. From this observation, five divisions of different phytoplankton (104 species) and four different groups of zooplankton (30 species) were identified. The phytoplankton was dominated by bacillariophyceae (65.16%) followed by cyanophyceae (10.44%), chlorophyceae (10.32%), euglenophyceae (9.62%) and dinophyceae (4.45%). Four groups of zooplankton viz., rotiferans were the most dominant forms (35.40%) followed by crustaceans (30.21%), protozoans (21.17%) and cnidaria (13.21%) were recorded. The phytoplankton were predominant members of this biotope.

ORATHAPALAYAM RESERVOIR

The reservoir is located on the Noyyal River between Chennimalai and Kangayam in Tirupur District. The dam is situated 16 km north of Kangayam and 26 km east of Tirupur.The dam was built in 1992. It has an ayacut of over 10,000 acres in Tirupur and Karur Districts. It was used by the farmers only for five years as it became a storage tank for textile effluents after that. The farmers, who depended on the dam and river for irrigation, stopped the farm activity in their land. The height of dam is 21.7 meters with length of 2290 meters. The capacity and surface area of dam is 17480 m^3 and 2.2 km^2 respectively.

Orthapalayam reservoir consists of a stone masonry dam and spillway of length 99.50 metres across Noyyal river with eastern dam on both flanks for a total length of 2.19 km. The floodwater stored in this reservoir (of capacity 617 mcft) can be discharged into Noyyal river through a sluice built in the masonry dam. This project were started in July 1984 and completed in 1991. Immediately after the completion of the reservoir, the Fisheries Department started fisheries activities at Orathapalayam reservoir in 1993. The fingerlings stock at Orthapalayam reservoir was 385,000 in 1993. But the stock was increased to 700,000 during 1993-94 and 801,000 in 1996-97. But in subsequent years no stocking took place in the reservoir. Towards the end of 1997, mass fish mortality occurred at Orathapalayam reservoir. In early December, dead fish were floating near the dam site, which caused serious health hazards. Subsequently the District Collector made arrangements to bury the dead fish, which came to several tonnes. The Government also ordered the Fisheries Department to make immediate arrangements for removing all the fish in the reservoir to avoid further fish mortality and health hazards. The PWD also requested the Fisheries Department to stop the fisheries activities in Orathapalayam. Subsequently the Director of Fisheries ordered the Regional Fisheries Office at Erode to stop the fisheries at Orathapalayam on 12/3/1998.

After realizing the pollution implication on Orthapalayam reservoir the Commissioner of Fisheries conducted a detailed study. The study was conducted by the Hydrology Research Station, Department of Fisheries, Chennai (1996). The salient findings were:

(*a*) River is unfit for aquatic organisms.

(*b*) Fish landing was observed downstream of Tiruppur and Orthapalayam reservoir – Tilapia is the main variety, but size is small.

(*c*) Plankton pollution was observed in huge quantity in reservoir.

(*d*) Bio-assay test conducted revealed that the fishes are bleached.

(*e*) Fish growth rate will be affected.

(*f*) Survival of Tilapia may not be a problem, but present condition is unsuitable for the growth of other carps.

(*g*) Fish caught from the reservoir get spoiled in a short period of two hours. Hence fish cannot be marketed to distant places.

(*h*) High possibility of frequent fish mortality in future. The fisheries activities at Orathapalayam reservoir (which was started by State Fisheries Department with great expectation in 1993) was completely stopped and continuation is also highly doubtful.

At present even though the river, tanks and reservoir are completely affected by textile pollution, some informal fishing is still taking place in the water bodies and the fish caught is sold in the local and neighbouring markets. Generally the inland fishing is extremely important in this area since the possibility of getting sea fish (Bay of Bengal and Arabian Sea are quite far) is low. Even the fish that survive in the Orthapalayam reservoir may be of doubtful quality. Azo dyes, which were used widely are banned recently due to harmful properties including carcinogenesis.It is found that many species of fish survive in the Orthapalayam reservoir where all the chemical wastes from Tiruppur get collected. It is quite possible that the fishes accumulate these chemicals and are transferred to human beings while consuming fishes."Based on the fish catch data, the estimated overall annual productivity loss in fisheries sector as Rs. 1.5 million (in which Rs.14,000 in Noyyal river, Rs. 258,000 in system tanks and Rs. 1.2 million in Orathapalayam reservoir). The capitalized value of the loss in fish catch was Rs. 12.5 million in current prices.

REFERENCES

- Anon, 1981. All India Coordinated Research Project on Ecology and Fisheries of Freshwater reservoir, Bhavanisagar, 1971-81.
- Appasamy, P.P. and Nelliyat, Prakash, 2007. Compensating the loss of ecosystem services due to pollution in Noyyal river basin,Tamil Nadu, Working Paper No. 14/2006, Madras School of Economics, Gandhi Mandapam Road, Chennai, p 50.

- Devaraj, M.Prabhavathy G., and Goswami, P.K., 1990. Assessment of optimum fish yields from reservoirs in Tamil Nadu. pp.118-130. In : Jhingran, Arun G. and Unnithan, V.K. (eds.). Reservoir Fisheries in India. Proceedings of the National Workshop on Reservoir Fisheries, 3-4 January, 1990. Special Publication 3, Asian Fisheries Society, Indian Branch, Mangalore, India.
- Jesu Arockia Raj. A., Seetharaman, S., Haniffa, M.A. and Singh, S.P., 2004. Food and feeding habits of a threatened featherback, *Notopterus notopterus.J, Aqua.biol.* 19 (1) : 115-118.
- Mathivanan V., Vijayan P., and Sabhanakam, Selvi., 2005. An assessment of water quality of river cauvery at Mettur, Salam district, Tamil Nadu in relation to pollution. In fundamentals of Limnology (ed. Prof. Arvind Kumar), APH Publishing Corporation, New Delhi. PP.36-43.
- Palaniswamy, Rani and Manoharan, S., 2010. Mettur Reservoir Fisheries : Past and Present Status. *Fishing Chimes.* 30(1) : 32-135.
- Raj, Deleep Packia. 2010. Sediment profile of Perumchani Reservoir of Kanyakumari district, Tamil Nadu, *Journal of Basic and Applied Biology,* 4(3) : 174-180.
- Raj, Deleep Packia and Beena Lawrence 2012. Ecological investigation of water of Perumchani Reservoir, Kanyakmari district, Tamil Nadu, *Applied Research and Development Institute Journal,* 5(3) : 23-30.
- Ranganathan, V. ad Natarajan, V., 1969. Fisheries of Mettur Reservoir, an artificial impoundment on Cauvery. Proc. of the Seminar on the Ecology ad Fisheries of Freshwater Reservoirs, Central Inland Fisheries Research Institute.
- Renuga, K. and R. Ramanibai, R., 2010. Zooplankton composition present in Krishnagiri reservoir, Tamilnadu, India. *Current Biotica.* 3 (4) : 519-525.
- Santhi, D., 2012. Assessmet of water quality parameters in Kodumudiaru dam at Nanguneri Taluk of Tirunelveli District. *Poll. Res.* 31(1) : 83-86.
- Selvaraj, C., Murugesan, V.K. and Manoharan, S., 1997. Record fish yiled from Thirumoorthy reservoir under scientific managaeet. p. 114-118. In : K.K.Vass and M. Sinha (eds.), Changing Perspectives in Inland Fisheries. Proceedings of the National Seminar, 16-17 March, 1997, Inland Fisheries Society of India, Barrackpore.
- Senthilkumar, R. and K. Sivakumar, K.,2008. Studies on phytoplankton diversity in response to abiotic factors in Veeranam lake in the Cuddalore district of Tamil Nadu. *Journal of Environmental Biology.* 29(5) : 747-752.
- Sreenivasan, A., 1964. A hydrological study of a tropical impoundment, Bhavanisagar, Madras State, India, for the years 1956-61. *Hydrobiol.* 24 : 514-592.
- Sreenivasan, A., 1995. Where have all these fish species gone?. *Fishing Chimes,* May 7-9.
- Sreenivasan, A., 1998. Integrated development of reservoir fisheries of India: Production to marketing. *Fishing Chimes.* 18 (1) : 60-63.
- Sreenivasan, A., 1998. Fifty years of reservoir fisheries in Mettur dam,India:Some lessions. Naga, The ICLARM quarterly, October-December 1998 : 4-7.
- Sreenivasan, A., 2000. Fish production from a medium–sized hardwater reservoir (Krishnagiri) in Tamil Nadu. *Fishing chimes.* 20 (5) : 32-34.
- Thirumathal, K., 2006. Cladocerans of Amaravathy reservoir, Udumalpet, Coimbatore District, Tamil Nadu, India. In : Ecology of Lakes and Reservoirs (Dr. V.B. Sakhare), Daya Publishing House, Delhi.
- Vidhya, V. and Radhakrishnan Nair, 2012. Planktonic diversity and hydrobiological characteristics in Pechiparai reservoir, Kanyakumari district, Tamil Nadu. India. *Poll. Res.* 31(2) : 315-318.

17

UTTARAKHAND

Fig. 17.1. Map of Uttarakhand State (Not to scale)

Uttarakhand became the 27th state of the Republic of India on 9th November 2000. The State came out of Uttar Pradesh. It occupies 17.3% of India's total land area with 51,125 sq kms. Uttarakhand is a place with great diversity of the region where snow-clad mountains, green hills, fertile valleys, flowing rivers and thriving lakes add to the natural beauty. It is a region with several natural lakes, which are the center of attraction of the town and good for sporting activities.

The important reservoirs in state are Sharadasagar (2958 ha), Nanaksagar (4089 ha), Baigul (2995 ha), Dhaura (1295 ha), Haripura (1144 ha), Baur (1295 ha), Tumaria (2377 ha) and Tehri (4200 ha). The Sharadasagar has about 43% area in India and remaining in Nepal.The total area of reservoirs in Uttarakhand is 20,353 ha (Kathia *et al.*, 2009).

DHAURA RESERVOIR

Dhaura reservoir is located in the Tarai region of Udham Singh Nagar of Uttarakhand.It was constructed in 1961 across the River Dhaura which receives a number of rivulets in the course of their flow. The details of morphometry of reservoir are depicted in Table 17.1. Mishra *et al.* (2010) recorded 30 species of phytoplankton belonging to Chlorophyceae (10 species), Bacillariophyceae (11 species), Cyanophyceae (6 species) and Dinophyceae (3 species) (Table 17.2). Deorari (1993) reported 33 species of phytoplankton from Dhaura reservoir.

Table 17.1. Morphometry of Dhaura Reservoir, Uttarakhand

1. Latitude	28°53′N
2. Longitude	79°34′E
3. Altitude (m)	200
4. District	Udham Singh Nagar
5. Feeder Rivers	Dhaura and Katna
6. Year of construction	1961
7. Gross storage capacity (m.c.ft)	4442
7. Area at Full Reservoir Level (ha)	1280
8. Type of dam	Earthen
9. Total length of bundh (km)	9.05
10. Top width of the dam (m)	4.0
11. Maximum height of reservoir (m)	14
12. Mean depth (m)	2.2
13. Spillway	Manual
14. Number of gates	3
15. Total catchment Area (km^2)	134.68
16. Average rainfall (mm)	1700

The commonly occurring green algae were *Pediastrum spp.*, *Scenedesmus quadricauda*, *Spirogyra* sp., *Zygnema* sp., *Volvox* sp., and *Mougeotia* sp. Among these, *Pediastrum* spp. were dominant and observed in all the seasons. *Spirogyra* sp. and *Zygnema* sp. Were dominant during winter. *Chlorella vulgaris*, *Eudorina elegans*,

Cosmarium sp. and *Ankistrodesmus falcatus* were observed occasionally. The density of *Pediastrum* spp. was found to be the maximum (40,000 cell/l) during June, 2006. *S. quadricauda* was in the highest density (10,000 cells/l) in lentic sector in June 2006. Generally, the members of chlorophyceae were in high numbers from the onset of summer to the beginning of autumn seasons. It may be due to the input of allochthonous. Organic matter with increased nitrate concentration in July favouring the growth of phytoplankton from August onwards.

Table 17.2. Phytoplankton diversity in Dhaura Reservoir

Chlorophyceae: *Volvox* sp., *Mougeotia* sp., *Pediastrum* spp., *Chlorella vulgaris, Scenedesmus quadricauda, Ankistrodesmus falcatus, Cosmarium* sp., *Eudorina elegas, Zygnema* sp., *Spirogyra* sp.
Cyanophyceae: *Oscillatoria* sp., *Microcystis aeruginosa, Anabaena spiroides, Spirulina* sp., *Nostoc* sp., *Aphanocaspa* sp.
Dinophyceae: *Peridinium* sp., *Gymnodinium* sp., *Ceratium* sp.
Bacillariophyceae: *Navicula viridula, Fragillaria crotonensis, Cymbella sp., Synedra acus, Gomphonema* sp., *Melosira granulate, Nitzschia palea, Amphora ovalis, Pinnularia* sp.

Table 17.3. Zooplankton Diversity in Dhaura Reservoir of Uttarakhand

Rotifera: *Lecane* sp., *Keratella* spp., *Filinia* sp., *Notholca* sp., *Brachionus* spp.
Copepoda: *Diaptomus* sp., *Mescyclops leuckarti, Eucyclops* sp., *Cyclops* sp., *Thermocyclops* sp., *Nauplius larvae.*
Cladocera: *Moina micrura, Chydorus* sp., *Bosmina longirostris, Daphnia* spp., *Bosminopsis.*

Diatoms were represented by *Synedra acus, Navicula viridula, Nitzschia palea, Melosira granualata. Fragillaria crotonensis, Gymphonema* sp., *Neidium* sp., *Cymbella* sp., and *Diatoma* sp. *S.acus* exhibited the highest density (10150 cells/l) in the lentic sector in July 2006 while it was absent in December 2006.

Among blue green algae, *Microcystis aeurginosa, Anabaena spiroides, Oscillatoria* sp., *Spirulina* sp., *Nostoc* sp., and *Aphanocapsa* sp. were recorded. *M. aeruginosa, A.spiroidae* and *Oscillatoria* sp. were observed throughout the study period in all the sectors with the maximum quantity in winter season. *M. aeruginosa* ranged from 37750 to 74780 cells/l during November, 2006 and Jauary, 2007 respectively. Deorari (1993) recorded *Microcystis* spp. As dominant member of phytoplankton in Dhaura during his study. According to Deorari (1993), cyanophyceae seemed to thrive well within a temperature range of 15.5 to 25°C during winter season.

Dinoflagellates represented by *Ceratium* sp., *Peridinium* sp. and *Gymnodinium* sp. were observed during summer and winter seasons. The density of *Ceratium* sp. varied from nil to 16,730 cells/l in Dhaura reservoir (Mishra *et al.* 2010).

Chlorophyceae was the dominant group in most of the months to a maximum of 52.24% during August 2006. The population density ranged from 40533 (November, 2006) to 86,133 cells/l (June, 2006). In September and December, 2006 and April-May, 2007, chlorophyceae was not observed in reservoir.

The population density of diatoms was between 30508 (March, 2007) and 53266 cells/l (November 2006). They contributed 54.82% (September, 2006). The blue green algae constituted a good proportion with 26.5% (February, 2007) to 71.02% (December, 2006). The density of this group varied from 27816 (September, 2006) to 91966 cells/l (November 2006). Dinophyceae was poorly represented. It was recorded maximum (28.97%) in December 2006.The members of this group were not observed in June-September 2006 and January-March 2007. The range of population density of dinophyceae was between 7833 (March 2007) and 19435 cell/l (February 2007).

Composition and Abundance of Zooplankton: A total of 15 species of zooplankton including 5 species each of Rotifera, Cladocera and Copepoda were observed by Mishra *et al.* (2010). Deorari (1993) recorded 8 species of rotifers,8 species of cladocerans and 6 species of copepods from the same reservoir.

The members of rotifer recorded were *Keratella* spp., *Filina longiseta, Lecane* sp., *Brachionus* spp. and *Notholca* sp. *Brachionus* spp. was observed in good numbers with density varying from 0 to 870 units/l. The highest density of Keratella spp. Was recorded at 1041 units/l in transition zone of Dhaura during January 2007. The members of this group were not observed in September and December 2006.

The members of cladocera were *Daphnia* spp., *Moina micrura, Chydorus* sp., *Bosmina longirostris* and *Bosmiopsis* sp. *Daphnia* spp. Was more abundant than the other cladocerans. Bosminopsis sp. was sparse in the reservoir. The population of Daphnia spp. Varied from nil to 700 units/l with highest in lentic sector during October 2006.

Copepods were represented by *Cyclops* sp., *Diaptomus* sp., *Mesocyclops leuckarti* and *Eucyclops* sp. while *Thermocyclops sp* and *Nauplius larvae* were rare. Peak abundance of Cyclops sp. was observed during October 2006 in lentic sector of Dhaura at 880 units/l. The population of *Diaptomus sp.* was the highest during July, 2007 in the lentic sector of Dhaura (1070 units/l). Copepods were not observed in August 2006.

The population density of rotifer ranged from nil (September 2006 and May 2007) to 2046 units/l (July 2006). The contribution of rotifers varied from nil to 56.58% in January 2007. Cladocera constituted a significant component of zooplankton community for most part of the study period. They fluctuated from nil to 53.26% in august 2006. Population density ranged from nil (June, July, September and December 2006; January and May 2007) to 2700 units/l in October 2006.The population density of copepod ranged from nil (August, 2006) to 3493 units/l (October 2006).Copepods contributed to the total zooplankton with percentage ranging from nil to 57.84% (July, 2006).

The percentage distribution of phytoplankton and zooplankton shows that the maximum contribution was by phytoplankton.

In Dhaura reservoir, weed fishes *viz., Puntius* spp., and *Chela labuca* constitute 75 per cent of the total fish yield. *Oxygaster bacaila, Amblypharyngodon mola, Esomus danricus, Chanda nama, Chanda ranga* and *Xenentedon cancila* are the other weed fishes commonly found in the Dhaura reservoir. Bhaumik *et al.* (2009)

estimated fish production potential of the reservoir at 290 kg /ha/yr. The reservoir was stocked @ 413 numbers /ha. The fish production during the study period was 106.6 kg/ha against pre-adoption production of 32 kg/ ha.

REFERENCES

- Bhaumik, Utpal; Singh, U.P.; Paria, T., 2009. Ecology and management of the Dhaura reservoir, Uttarakhand for enhancing fish production. *Indian Journal of Fisheries*. 56 (3) : 189-193.
- Deorari, B.P., 1993. Productive potential of a manmade reservoir of Tarai of Uttar Pradesh with particular reference to fish fauna. Ph.D. Thesis, Rohilkhand University, Bareiliy, pp. 291.
- Kathia, P.K., Vass, K.K., Das, A.K. and Shrivastava, N.P., 2009. Improved productivity of reservoir fisheries in India, Central Inland Fisheries Research Institute, Barrackpore.
- Mishra, Ashutosh, Chakraborthy, S.K., Jaiswar, A.K., Sharma, A.P., Deshmukhe, Geetanjali and Mohan Madan, 2010. Plankton diversity in Dhaura and Baigul reservoirs of Uttarakhand. *Indian J. Fish.* 57(3) : 19-27.

18

UTTAR PRADESH

Fig. 18.1. Map of Uttar Pradesh (Map not to scale)

Uttar Pradesh is endowed with natural wealth in abundance. This wealth lies hidden below a variety of rocks of different ages found in lofty mountain ranges of the Himalayas in the North and Vindhyan ranges in the South. The diversity of flora and fauna displayed here due to vast area, big and small rivers, varieties of climatic conditions, and different kinds of soil are hard to find elsewhere. It was created on 1st April 1937 as the United Provinces with the passing of the States Reorganisation Act and renamed Uttar Pradesh in 1950. The capital city is Lucknow. Lucknow is the second largest metropolitan area of north and central India after Delhi. Kanpur was the largest industrial town in seventies, but due to economic growth in the state, several other industrial towns like Bhadohi, Muradabad, Ghaziabad and Meerut have become important. On 9th November 2000, a new state, Uttarakhand, was carved from the mountainous Himalayan region of Uttar Pradesh.

NANAKSAGAR RESERVOIR

Nanaksagar reservoir is constructed across the rivers Deoha and Kamin. The reservoir is located within the geographical coordinates of 29°55′N and 79°40′. It covers an area of 4,662 ha and is located at a distance of 70 km from Pantnagar University. The reservoir was built in 1962 primarily to facilitate irrigation in the nearly 40,000 ha downstream areas, however, it has also made a substantial contribution to the state economy through fish production.

The littoral region of the lake basin is shallow with irregular periphery; the mean depth of the reservoir is 10 m and the maximum depth is 20 m. Generally, the reservoir gets filled to the full level in September. During January to June, the level of water recedes continuously due to draw down and evaporation. With an average fluctuation in water level between 0.01 and 3.8 m, areas under 3m of water depth vary between 10.5 to 20.6% of the entire reservoir. The gross storage capacity of the reservoir is $20 \times 10^7 m^3$. The shoreline index works out to be 1.65 and the volume index 1.2.

The bottom region of the reservoir is sandy clayey loam in texture with an organic reserve varying in percentage between 0.5 and 1.75. Annual mean and range of physico-chemical features of the reservoir water are depicted in Table 18.1.

Table 18.1. Annual mean and range of physico-chemical parameters of Nanaksagar reservoir

Parameter	Range	Mean
Temperature (°C)	10 – 33	27
Transparency (cm)	12.5 – 206	155
pH	7.4 – 8.3	7.8
Dissolved oxygen (mg/l)	7.2 – 15.2	9
Alkalinity	82 – 140	74
Phosphate phosphorus (ug)	1 – 7	5
Nitrate nitrogen (ug/l)	35 – 340	145
Silicate (ug/l)	200 – 900	650

Fish Community: The common fishes recorded in the reservoir are *Gadusia chapra, Labeo rohita, L. gonius, L. calbasu, Cirrhinus reba, Channa punctatus, Mastacembelus armatus, Puntius sarana, P. sophore, Mystsus seenghala, Mystus tyengra, Wallago attu, Notopterus chitala* and *Ompak bimaculatus.*

Macrophytic Productivity: The macrophytes in the littoral region of the reservoir include submerged species *viz., Potamogeton pectinatua, P. crispus, Hydrilla verticillata* and *Vallisneria spiralis.*

Phytoplankton Productivity: The gross primary production (GPP) in the euphotic zone of the reservoir ranged from 82.4 to 864.20 mg ($m^{-3}d^{-1}$).

The seasonal variation evinced typical example of a tropical aquatic ecosystem with lower values being found in the winter (82.4 to 125 mg ($m^{-3}d^{-1}$) and higher values during summer (207 to 864.2 mg ($m^{-3}d^{-1}$). The maximum GPP was observed in May. The net primary production (NPP) varied from 64.2 to 530.5 mg ($m^{-3}d^{-1}$) during the study period; the maximum value of NPP coincided with that of the GPP in May. The NPP comprised 44 to 82% of the GPP. Community respiration comprised 18 to 45% of the GPP with a mean contribution of 34% to the annual GPP.

Bisht *et al.* (2000) found *Fusarium moniliformes* and *F. udum* as natural pathogens of freshwater fish in Nanaksagar reservoir, causing mycosis and high mortality in *Wallago attu, P. sophore, C. punctata, L. rohita* and other economically important fishes. Fusarium species paralyze fish more commonly during summer through rainy season. A temperature above 25°C, coupled with relatively low pH (7.1 to 7.7) and dissolved oxygen (8.3 to 9.5 mg/l) encouraged association and infection of these fungi, whereas low temperature during winter (<20°C) adversely affected their colonization on the fish. Notably mycosis due to watermolds is prevalent during winter-spring, while extra-aquatic fungi dominate during summer through the rainy season, thus posing a continual threat to fish in the reservoirs. This necessitates an integrated approach to combat mycosis in reservoir. Besides, prophylatic measures to protect fingerlings before being introduced into the reservoir, intensive research on biological control of fish diseases caused by *Fusarium* and other fungal species is warranted, not only to increase production in reservoir but also to conserve several rare fish species.

The fungal species isolated from diseased fish of Nanaksagar reservoir are depicted in Table 18.2.

Joshi and Sharma (1980) studied the relative fecundity of *Labeo gonius* from Nanak sagar reservoir. The weight of the two ovaries together accounted for 14·71 to 21·87% of the total weight of the fish. Specimens about 1 kg in weight had ovaries weighing about 205·312 g on an average and this worked out to about 19·93% of the total weight of the fish. The average number of eggs per g weight of fish and ovary was 271 and 1424 respectively. The fecundity per kg body weight of fish was 286111.

Table 18.2. Fungal species isolated from diseased fish in Nanaksagar reservoir

Funal species	Fish species						
	Puntius sarana	*Channa punctatus*	*Labeo rohita*	*Mastacembelus armatus*	*Mystus tengra*	*Puntius sophore*	*Wallago attus*
Alternaria aternata	+	+	–	+	–	+	+
Aspergillus niger	+	+	–	–	–	+	+
Cladosporium cladosporioides	–	+	+	–	+	+	+
Curvwaria lunata	+	–	–	+	–	–	+
Drecshlera australiensis	–	+	–	+	–	+	–
Fuarium moniliforme	–	+	+	+	+	+	+
Fusarium oxysporum	–	+	–	–	+	–	–
Fusarium udum	+	–	+	+	+	–	+
Penisillium chrysogenum	–	+	+	–	–	+	–
Penicillium expansun	+	–	+	–	+	+	–
Trichoderma harzianum	+	–	+	–	–	+	+

KOHARGADDI RESERVOIR

Kohargaddi dam is one of the most important dams of district Balrampur at Indo-Nepal border. Dam was constructed to solve the water problem for irrigational purposes during pre-independence in the foothills of Himalayas. As we know that nearby foot hills of Himalayas water crises exist as a serious problem because bore-well and tube-wells are not success here. Therefore, the excess water from hills downstream is collected in the catchments area in form of Kohargaddi reservoir.

Kohargaddi reservoir is situated at 82° – 36′ east ad 27° – 36′ north altitude at Pachperwa town of Balrampur district in Uttar Pradesh. It is about 8 km in length having 120 meter level. The height of the dam is 36 m, which discharges about 24000 cumex minimum water every month. The reservoir water is utilized to irrigate 50,340 acres unirrigated lad of Balrampur district. It covers about 126 km catchments area. Local inhabitants *viz.* Tharus and their pet also use this water for drinking purposes.Mishra *et al.* (2008) investigated physico-chemical parameters of reservoir water. The study revealed that reservoir water is not suitable for human or cattle consumption (Table 18.3).

Table 18.3. Physcio-chemical properties of Kohargaddi reservoir

Sr.No.	Properties	Range
1.	Water temperature (°C)	22.3 to 30.5
2.	pH	7.3 – 8.1
3.	Dissolved oxygen (mg/l)	4.3 to 7.2
4.	Total dissolved solids (mg/l)	30.2 to 61.2
5.	Total hardness (mg/l)	32.0 to 52.1
6.	Calcium (mg/l)	17.5 to 29.2
7.	Magnesium (mg/l)	10.9 to 21.3
8.	Sulphate (mg/l)	11.3 to 18.0
9.	Chloride (mg/l)	15.3 to 21.4
10.	Nitrate (mg/l)	6 to 13

Singh *et al.* (2012) reported phytoplankton and zoobenthos diversity in freshwater bodies of Kohargaddi dam. The taxa observed in the water body were classified under four categories : (1) consistent-present for 9-12 months (*Aeruginosa, Diatoms, Microcystis* and *Synachocystis* sp) (2) subconsistent-present for 6-8 months (*Microcystis robusta* and *Synachococcus* sp.) (3) inconsistency-present for 2-5 months (*Anabaena* sp., *Anabaena circularis, Chiorococcum* sp., *Chlorella* sp., *Cosmarium botrytis, Eucapsis minuta, Raphidiopsis* sp. and *Scenedesmus bijugatus* and (4) sporadic –present for one month only. Among algae (6) taxa were pollution tolerant, (5) moderate to pollution and remaining clear water forms. Zoobenthos showed distinct quantitative variations. Number of zoobenthos increased in the summer months while decreased in the winter months. Different groups of zoobenthos exhibited their distinct peaks in different months of the year.

TARAI RESERVOIR

The reservoir is located in the intermediate zone between the plains and Bhabar of district Nainital of Kumaun region (29°55′N Lat., 79°40′E Long.). The reservoir was constructed in the year 1962 across the rivers Deoha and Kamin.The reservoir impounds an area of 46.62 km^2 and is situated at a distance of 70 km from the Pantnagar University campus. The reservoir is primarily used for irrigation and water supply pruposes.There are total seven spill way gates through which the water is drawn out.

The climate of the area is humid and is influenced by southwest monsoon. The mean minimum temperature varies from 6.7°C in January to 25°C in June, and the mean maximum temperature from 21.5°C in January to 39.6°C in June. Monthly rainfall varies from nil to 840 mm, with annual rainfall of 1707 mm.

JARI RESERVOIR

Jari reservoir near Allahabad of Uttar Pradesh is mainly a rainfed water body. The reservoir, in addition to rainwater receives water from Tons canal. The reservoir water is mainly used for irrigation purpose and is also used for carp culture.

Physico-chemical Environment: Water temperature was at the maximum (31.9°C) at 15.00 hrs during summer and minimum temperature (12.5°C) at 0.300 hrs. Transparency values ranged between 8.0 and 24.0 cm. The pH had a range of 7.7 to 9.6. Total alkalinity was maximum at 6.00 hrs (140 ppm) in summer and minimum (82.5 ppm) at 15.00 hrs in winter. Carbonates were in maximum quantity (18 ppm) at 3.0 p.m. in summer ad minimum (2.0 ppm) in winter and monsoon. The dissolved oxygen had a rage of 1.8 to 14.5 ppm. Both the values were recorded during summer at 3.00 pm and 6.00 am. respectively.

Phytoplankton Diversity: Diurnal variations of phytoplankton correspond to the general behaviour of plankton during 24 hour cycle, maximum value of 1,98,500 units/l was observed in summer. Monsoon samples indicated a significant fall in phytoplankton population. The values varied between 2450 units at 15.00 hrs and 350 unitsl at 24.00 hrs. *Anabaena* sp., *Oscillatoria* sp., *Synedra* sp., and *Melosira* sp., *Scenedesmus* sp., *Spirogyra* sp., *Closterium* sp., were dominant phytoplankters showing the diurnal distribution.

Zooplankton Diversity: Summer study showed low plankton density in 1800 units/l at 06.00 hrs, which increased to maximum 15,550 units/l at 9.00 hrs. Minimum number of 1250 units/l was recorded at 15.00 hrs. Between 18.00 to 24.00 hrs the desity gradually increased but registering another fall to 1600 units/l at 3.00 hrs. Monsoon study showed wide fluctuations in zooplankton count in 24 hours cycle. At 06.00 hrs minimum of 1050 units/l was recorded which increased to a maximum of 2250 units/l at 09.00 hrs. Subsequently between 12.00 hrs to 03.00 hrs there was a gradual fall in the population density. The diurnal variations in winter revealed much less count than the summer and monsoon. A minimum population of 350 units/l was recorded at 06.00 hrs. The population increase attained a peak (1550 units/l) at 12.00 hrs. Between 15.00 to 18.00 hrs density increased gradually and a sharp fall was registered from 21.00 hrs reaching to 650 units/l at 24.00 hrs. Again at 03.00 hrs a increase (100 units/l) in zooplankton population was observed. *Keratella* sp., *Asplanchna* sp., *Mesocyclops* sp., and *Nauplii* were the dominant zooplankters.

KEETHAM RESERVOIR

The Keetham reservoir in Agra was constructed in 1925 with a maximum water spread area of about 306 ha with a total storage capacity of 28,937 C.ft., and a maximum depth gauze of 7.2 m at L.S. 544.70. The average water depth varies from 2.5 to 3.5 m. The bottom reservoir is almost flat with old alluvium soil. The reservoir area receives average annual rainfall in the range of 500 to 750 mm with a wind velocity 2.5 to 7.0 km/ hour. Dwivedi and Clador (1980) studied the physico-chemical characteristics of Keetham Reservoir.

Table 18.4. Physico-chemical characteristics of Keetham Reservoir

1. Water level (m)	2.7 to 6.6
2. Water Temperature (°C)	17.4 to 36
3. Turbidity (ppm)	11 to 152
4. pH	7.2 to 9.2
5. Total Alkalinity (ppm)	152 to 176
6. Carbonates (ppm)	Traces to 21.7
7. Bicarbonates (ppm)	46 to 208.4
8. Dissolved Oxygen (ppm)	3.8 to 17.6

In the bottom loamy and sand texture with pH 9.2, Organic carbon 0.68%. The data indicates that the lake is medium in its fertility.

The fish fauna of reservoir comprises fish species like *Catla catla, Labeo rohita, Cirrhinus mrigala, Labeo calbasu, Labeo gonius* and *Gudusia chapra.* The major carps and *Labeo gonius* together contribute 51% of the total catch of which the later alone being 39 %, from the main carp fauna of the reservoir. The next important fish is the *Gudusia chapra.* It forms of about 30% of the total catch, followed by other species 11 %, Catfish 8% and Murrel 1%.

Drag net, scoop net, drift net, gill net and hooks and lines are common gears used for fishing. The dragnet fishing is most effective for catching bigger sized fishes including the carps and catfishes. Scoop nets generally used for catching the *Gudusia chapra.*

BACCHRA RESERVOIR

Bacchra reservoir (140 ha) is located in the Allahabad district. Khan *et al.* (1990) reported stocking of the Bacchra reservoir in detail.Bacchra reservoir came into being as a result of damming the Bacchra rivulet in 1980. The reservoir is located in Meja tehsil at a distance of 55 km from Allahabad. The water spread at FRL (111 m above MSL) is 140 ha with an average depth of 5.22 m. The reservoir has a storage capacity of 7.42 million m^3 at FRL and 0.03 m^3 at DSL. The earthen embankment and waste-weir are 558 m and 98 m log respectively. A single irrigation canal originates at the bed level from the reservoir. During summer season reservoir maintains a minimum waters spread area of about 4 ha.The catchment area encompassed rocky terrain, forest land and pastures. The basin of the reservoir is erratic, undulating and strewn with boulders of various dimensions. These boulders pose a lot of difficulties in operation of fishing nets.

Bacchra reservoir once harboured mostly catfishes and weed fishes up to 90% and the rest vbeing major carps as revealed by the catches of 1981. Selective fishing for catfishes and stocking with Indian major carp seed was initiated in Bacchra reservoir in 1983. The major carps firmly established in the reservoir within two years of planting as was evident from experimental fishing. The natural breeding and auto stocking of major carp also helped augmenting the population of primary consumers in the ecosystem. This was corroborated when breeding of major carps for the first time was observed on 16th July 1986 in the reservoir. An iron screen of 6 mm mesh was fixed in the irrigation canal in order to check the escapement of fingerlings.

In year 1988 fish production in reservoir was 139 kg/ha/yr. Among the fishes landed, the detritivores formed 29.54 to 31.87% followed by predatory catfishes (29.66% to 17.28%) and omnivores (22.36 to 25.40%) during 1986-88 period. The contribution of herbivores was meager being around 12% in the abovementioned period. This emphasizes the need for introduction of a suitable herbivore which can utilize this underexploited niche.

PILI RESERVOIR

Pili reservoir is situated about 3 kms north of Afzalgarh-Kashipur highway in Bijnore district. There are four flood gates to remove the excess water. It is an earthen dam with area of 1440 hecatres.The catchment area of reservoir is 60 square miles. The soil of dam is mainly sandy and clayey. The Salient features of the reservoir are depicted in Table 18.5. The exploitation of reservoir by the fisheries department started in 1966-67. The fishes commonly recorded in Pili reservoir are *Catla catla, Labeo rohita, Labeo calbasu, Labeo bata, L.dicolus, Cirrhinus mrigala, Cirrhius reba, Puntius sarana, Puntius ticto, Tor tor, Barilius bola, Ambassis* sp., *Amblypharyngodon mola, Channa* spp., *Notopterus notopterus, Wallago attu* and *Ompak pabda*. The gears used in fishery are gill net, cast net, drag net, hooks and mahajal.

Table 18.5. Salient fatures of Pili Reservoir in Uttar Pardesh

1. Rivers	Pili, Banali and Dhara
2. Type of project	Medium
3. Purpose of project	Irrigation
4. Year of completion	1966
5. Totat Area (ha)	1440
6. Area at FRL (ha)	1184
7. Area at DSL (ha)	96
8. Height of dam (m)	16.5
9. Length of dam (m)	9290
10. Maximum depth (feet)	38
11. Culturable command area (ha)	18711
12. Irrigation potential (ha)	11490
13. District Benefited	Nanital and Bijnore

Rauthan *et al* (2010) studied haematological values of fresh water fish *Mastacembelus armatus* (Lecepede) of Pilli reservoir infected with haemoflagellate. Mastacembelus armatus infected with a haemotlagellate *Trypanosoma vittai* was observed. The infection was heavy and the presence of trypanosomes in the blood caused some conspicuous changes in the Total Erythrocyte Count (TEC), Total Leucocyte Count (TLC), Differential Leucocyte Count (small lymphocytes, large lymphocytes, neutrophils, eosinophils, basophils, monocytes) and Erythrocyte Sedimentation Rate (ESR).

SARDA SAGAR RESERVOIR

Sarda Sagar, an irrigation reservoir in Pilibhit district of Uttar Pradesh, came into existence in the year 1957. Chuka nadi on which Sarda Sagar. is constructed as a storage reservoir,is a small rain-fed tributary of the Sarda river. Chuka nadi has a total lengthof about 32 km with a catchement area of about 64 sq km. It flows through alluvial soil draining the water-shed between Sarda and Chuka rivers. The annual rainfall in the area is between 60 and 70 inches of which over 70 percent is precipitated during the summer monsoon months, June to October. The annual run-oflf of Chuka nadi is not significant as compared to storage capacity of Sarda Sagar which is computed at over 11,300 million cusecs after completion of the First Stage and 20,000 million cusecs at the end of the Second Stage.Along the course of Chuka nadi there were three important *jheels* covering an area of over 6.48 sq km of which Kali Kanch *jheel* alone had an area of 3.89 sp km with Raiju *tal* about 1.3 sq km and Nautala *jheel* about 0-91 sq km. It was reported that these *jheels,* since submerged, harboured extremely rich fisheries and that stocking was natural through the Chuka nadi.The Sarda Sagar Scheme was formulated with a view to augmenting the supplies in the existing Sarda canal so as to meet the irrigation demand of eastern districts of Uttar Pradesh. The scheme envisages the storage of runoff water from the catchment of Chuka nadi and of surplus water in the Sarda river. The low weir constructed across Sarda river near Banbassa at R.L.218.24 M., ensures full supply of 10,000 cusecs of Sarda main canal during certain months, but in winter months the demand is not satisfied as the river discharge falls to about 5,000 cusecs or even less. Surplus water from Sarda river, after the rainy season when its silt content is low, is diverted through Sarda main canal into the feeder channel for filling up the reservoir.The reservoir of Sarda Sagar came into being towards the middle of 1957 as a result of the construction of an earthen bund, 22.4 km in length, from opposite km 8 of the Sarda Power Channel to opposite Sarda Canal bifurcation.The natural, rather high, cliff overlooking the *Khadars* of Chuka and Sarda rivers and along which Chuka nadi drains for most of its length, forms the western flank of the reservoir, while the 20.4 km earthen bund constructed along the water shed between Sarda and Chuka rivers, forms the eastern bank.This bund has been turned towards the west across Chuka nadi to join the high cliff so that Chuka nadi has lost its entity and the section downstream of the bund is completely cut off for all times from the portion of Chuka nadi submergedwithin the reservoir. A feeder channel to carry 4,000 cusecs from the existing Sarda main canal supplies water to the reservoir and an outlet channel of 3,500 cusecs capacity takes off to feed the existing Hardoi branch some 208 km downstream. In Sarda Sagar, species of major carps and other economic species are quite dominant in the catches.

Table 18.6. Fish fauna of Sarda sagar Reservoir

Family : NOTOPTERIDAE
1. *Notopterus notopterus* (Pallas)
Family : CYPIUNIDAE
2. *Oxygaster bacmla* (Ham)
3. *Oxygaster gora* (Ham)
4. *Barilius bola* (Ham)
5. *Barilius barila* (Ham)
6. *Barilius harna* (Ham)
7. *Barilius bendelisis* var. chedfa (Ham)
8. *Barilius vagra* (Ham)
9. *Barilius* sp.
10. *Danio devario* (Ham)
11. *Aspidoparia jaya* (Ham)
12. *Tor tor* (Ham)
13. *Tor putitora* (Ham)
14. *Puntius ambassis* (Day)
15. *Puntius chagunio* (Ham)
16. *Puntius gelius* (Ham)
17. *Puntius sarana* (Ham)
18. *Puntius sophore* (Ham)
19. *Puntius tetrarupiigus* (McClell)
20. *Puntius ticto* (Ham)
21. *Catla catla* (Ham)
22. *Cirrhinus mrigala* (Ham)
23. *Cirrhinus reba* (Ham)
24. *Garra gotyla* (Ciray)
25. *Labeo bata* (Ham)
26. *Labeo dero* (Ham)
27. *Labeo dyocheilui:* (McClell)
28. *Labeo gonius* (Ham)
29. *Labeo rohita* (Ham)
30. *Osteobrama cotic* (Ham)
31. *Crossocheilus latius latius* (Ham)
Family : COBITIDAE
32. *Noemacheilus botia* (Ham)
33. *Noemacheilus corica* (Ham)
34. *Noemacheilus savona* (Ham)
35. *Noemacheilus* sp.
36. *Botia darto* (Ham)

Contd...

Table 18.6: Contd...

37. *Lepidocephalichthys guntea* (Ham)
38. *Semileptes gongota* (Ham)
Family : SILURIDAE
39. *Ompok bimaculatus* (Bloch)
40. *Wallago attu* (Bl. & Schn.)
Family : BAGRIDAE
41. *Mystus cavasius* (Ham)
42. *Mystus vittatus* (Bloch)
43. *Mystus {Osteobagrus) aor* (Ham)
44. *Mystus (Osteobagrus) seenghala* (Sykes)
45. *Rita rita* (Ham)
Family : SISORIDAE
46. *Bagarius bagarius* (Ham)
47. *Glyptothorax cavia* (Ham)
48. *Glyptothorax horai* Shaw & Shebbeare
49. *Glyptothorax pectinopterus (McClell)*
50. *Sisor rhabdophorus* (Ham)
Family : SCILBEIDAE
51. *Clupisoma garua* (Ham)
52. *Eutropiichthys vacha* (Ham)
Family : BELONIDAE
53. *Xenentodon cartcila* (Ham)
Family : CHANNIDAE
54. *Channa gachua* (Ham)
55. *Channa marulius* (Ham)
56. *Channa punctatus* (Bloch)
Family : AMBASSIDAE
57. *Ambasis nama* (Ham)
58. *A. ranga* (Ham)
Family : NANDIDAE
59. *Nandus nandus* (Ham)
Family : ANABANTIDAE
60. *Colistt fasciatus* (Bl. & Sch.)
Family : GOBIIDAE
61. *Glossogobius giuris* (Ham)

In Sarda Sagar, species of major carps and other economic species are quite dominant in the catches.

REFERENCES

- Bisht, Deepa, Bisht, G.S. and Khulbe, R.D., 2000. Fusarium-A new threat to fish production in reservoirs of Kumaun, India. *Current Science*. 78 (10) : 1241-1245.
- Dwivedi, S.N. and Chondar, S.L.,1980. Fisheries of Keetham lake. India Today and Tomorrow. 8(4) : 182.
- Joshi, S.N. and Khanna, S.S., 1980. Relative fecundity of *Labeo gonius* (Ham.) from Nanaksagar reservoir. Proceedings: Animal Sciences. 89(5) : 493-503.
- Khan, M.A., Dwivedi, R.K., Mehrotra, S.N., Srivastava, K.P., Tyagi, R.K. and Katiha, P.K., 1990. Stocking as a management tool in optimising fish yield from a small reservoir (Bacchra) in Ganga basin. In : Jhingran,Arun G. and V.K. Unnithan (eds.). Reservoir Fisheries in India. Proceedings of the National Workshop on Reservoir Fisheries, 3-4 January, 1990. Special Publication 3, Asian Fisheries Society, Indian Bracnch, Mangalore, India, pp.71-76.
- Mishra, B.B., Chaturvedi, G.B. and Tiwari, D.D., 2008. Water quality index and suitability of water of Kohargaddi dam at District Balrampur, India. *Poll. Res.* 27 (3) : 497-500.
- Rauthan, Geeta, Rauthan, J.V.S., Singla, G.D.; Bisht, R.S.; Negi, Leela; Rawat, Sangeeta; Negi, Manjeshwari, 2010. Some haematological values of fresh water fish *Mastacembelus armatus* (Lecepede) of Pilli reservoir infected with haemoflagellate. *Aquacult* (India), 11 (1) : 119-122.
- Sharma, Neelima and Sahai, Y.N., 1990.Some observations on the plankton population of Jari reservoir near Allahabad (U.P.) and their significance to fisheries. pp.131-138. In: Jhingran, Arun G. and V.K. Unnithan (eds.). Reservoir Fisheries in India. Proceedings of the National Workshop on Reservoir Fisheries, 3-4 January, 1990. Special Publication 3, Asian Fisheries Society, Indian Branch, Mangalore, India.
- Singh, C.S., Sharma, A.P. and Deorari, B.P., 1990. Energy flow with perspective to fish production in Nanaksagar reservoir (U.P.), India. pp.155-162. In : Jhingran, Arun, G. and V.K. Unnithan (ed.). Reservoir Fisheries in India. Proceedings of the National Workshop on Reservoir Fisheries, 3-4 January, 1990. Special Publication 3, Asian Fisheries Society, Indian Branch, Mangalore, India.
- Singh, Karunesh, Singh Indu and Tripathi, R.B., 2012. Phytoplankton and zoobenthos diversity in freshwater bodies of Kohargaddi dam of district Balrampur (U.P.). *Flora and Fauna*, 18(1) : 141-144.

19

WEST BENGAL

Fig. 19.1. Map of West Bengal (Not to scale)

The important reservoirs of West Bengal are as given in Table 19.1.

Table 19.1. Important reservoirs in West Bengal

Sr.No.	Reservoir	River	Year of construction	Average Productive Area (ha)
1.	Maithon	Barakar	1957	6680
2.	Panchet	Damodar	1959	7880
3.	Kangsbati and Kumari	Kansai	1966	7400

KANGSBATI RESERVOIR

The length of Kangsbati reservoir is 11.30 km. The main aim of forming Kangsbati reservoir was storage of water and its utilization for agriculture purpose. Reservoir has 11 large lock gates and two small lock gates at the site of Mukutmanipur. During the rainy season, when the water level crosses the safety limit (beyond 440 ft), the larger gates are opened for the release of excess water and to allow its flow till the water level comes down to the safety limit. The water is utilized for agriculture purposes. The excess water flows out through two main canals which are 800 km long with their ramifications having a total length of 3000 km.The reservoir supplies water to 8.42 lakh acres of kharif and 1.50 lakh acres of rabi agricultural crops in each year. 11 blocks of Bankura district, 13 blocks under Midnapore district and 1 block of Hooghly district are benefited by this reservoir.

Table 19.2. Biodiversity of Kangsbati Reservoir

Phytoplankton: *Anabaena* sp,. *Anacystis* sp., *Oscillatoria* sp., *Chlorella* sp., *Clorococuum* sp., *Tetradron* sp., *Volvox* sp., *Synedra* sp., *Diatoma* sp., *Navicula* sp., *Cladophora* sp., *Zygnaema* sp., *Spirogyra* sp., *Microspeciesora* sp.
Zooplnkton: *Euglena* sp., *Brachionus* sp., *Filinia* sp., *Daphnia* sp., *Moina* sp., *Cyclops* sp., *Mesocyclops* sp., *Diaptomus* sp., *Bosmina* sp., *Cerodaphnia* sp., *Keratella* sp., *Testudinella* sp. *Cypris* sp., *Ascomorpha* sp.
Aquatic Vegetation: *Cyprus* sp., *Ipomea* sp., *Sesbenia* sp., *Jussiea* sp., *Nymphaea* sp., *Marsilea* sp., *Ceratophyllum* sp., *Najas* sp., *Hydrilla* sp.
Aquatic Insects: *Ranatra* sp., *Laccotrephes* sp., *Anisops* sp., *Lithocerus* sp., *Limnogonus* sp., *Corixa* sp., *Micronecta* sp., *Urothemis* sp., *Canthydrus* sp., *Cybister* sp., *Allytropus* sp., *Specieshaerodema* sp.

Seed Stocking: State Fisheries Department has taken up several steps to stock 4-6 inch Indian Major Carp fingerlings in the Kangsbati reservoir. During year 2007-08, more than 75 mt of similar sizes of fingerlings of Indian Major Carps, Pungas, Tilapia and freshwater prawn have been liberated in the reservoir through the financial assistance from National Fisheries Development Board to increase diversity of the stocks.

Fish and Fisheries: During 2006-07 and 2007-08 the yield recorded was 185 t and 118.4 t respectively. One important feature was that *Labeo calbasu, Cirrhinus mrigala, Labeo bata, Labeo fimbriatus* and *Mystus aor* are dominant species. Major and minor carps contributed 11% and 15% of the total catch respectively during 2006-07. It decreased to 7% and 12% in 2007-08. A major portion of the catch is contributed by prawn and weed fishes during both the years. The year 2004-05 witnessed the dominance of catfish followed by major and minor carps in the catch. The checklist of fish fauna of reservoir is depicted in Table 19.3 and Table 19.4 shows the catches of all fish during 2004-05 to 2007-08.

Table 19.3. Fish fauna of Kangsbati Reservoir in West Bengal

Indian Carps: *Catla catla, Cirrhinus mrigala, Labeo rohita, Labeo calbasu, Labeo bata, Labeo fimbriatus, Puntius gelius, Puntius sarana, Puntius sarana, Puntius ticto, Amblypharyngodon mola.*
Exotic carps: *Hypophthalmichthys molitirx, Ctenopharyngodon idella, Cyprinus carpio.*
Featherbacks: *Notopterus notopterus, Notopterus chitala.*
Catfishes: *Wallago attu, Ompak pabda, Mystus aor, Mystus seenghala, Mystus cavasius, Mystus vittatus, Clarias batrachus, Heteropneustes fossilis, Pangasius suchi, Etropiichthys vaha, Pangasius suchi.*
Others: *Anabus testudieus, Channa striatus, channa gachua, Channa punctatus, Glossogobius giuris, Nemacheilus corica, Nemacheilus savona, Sillanginopsis panijus, Colisa fasciata, Colisa sota, Esomus dandricus, Xenentedon cancila, Chand nama, Chanda ranga, Oreochromis mossambica, Oreochromis nylotica, Masatcembelus armatus, Mastacembelus pancalus, Mastacembelus aculeatus, Anguilla bengalensis.*

Table 19.4. Fish catch in Kangsbati reservoir durig 2004-05 to 2007-08

Sr.No.	Category	Year 2004-05 (tones)	Year 2005-06 (tones)	Year 2006-07 (tones)	Year 2007-08 (tones)
1.	Major carps	29.304	30.636	20.35	8.288
2.	Minor carps	26.862	40.848	27.75	14.208
3.	Feather backs	2.442	3.404	6.85	0.592
4.	Catfishes	34.188	37.444	14.8	8.288
5.	Eels	9.768	6.808	9.25	7.696
6.	Chingri	51.282	74.888	35.15	27.282
7.	Others	85.47	136.16	72.15	50.912

The important beneficiaries of the Kangsbati reservoir are the members of 11 primary fishermen's co-operative societies Ltd. Which are affiliated to Kangsbati Central Fishermen Co-operative Limited, Mukutmanipur, Bankura. Out of 11 societies in all, 9 PFCs are from 3 blocks of Bankura district, while the other two PFCs are from 2 blocks of Purulia district of West Bengal.

The co-operative movement in this area has been recognized as the most important force for improving the economic status of fishermen who belong to the weakest section of the society and it is revealed that fishery is the major source of income among all the various resources. Only 12.5% fishermen have their own nets larger than 100 ft. 62.5% inhabitants of this area do not have ownership rights on agricultural land through which they can earn a living. The poor earning of fishermen are due to (1) severe depletion of fish stock in the reservoir and loss of auto breeding grounds (2) abrupt changes in reservoir ecology (3) lack of awareness and indiscriminate fishing (4) absence of proper marketing facilities (5) lack of adequate economic support to the PFCs (6) major involvement of moneylenders (7) non-availability of quality fishing inputs to the PFCs (8) lack of suitable ponds for rearing fish species (9) lack of technical assistance and lack of co-ordination ad co-operation from the panchyat sector (10) fishermen needs adequate assistance from difference departments of West Bengal government (11) inactivity of the NGOs to promote fisheries.

Cage Culture: The cage culture was done with the financial assistance from a centrally sponsored scheme. Effort were made to produce table fish in a completely controlled system. This experimental culture was started from February 2006 onwards and full scale culture operation was taken up with 20 cages I 3 batteries with different species *e.g.*, Indian Major Carps, Catfishes, Air breathing fishes, pearlspot, prawns etc in different combinations and stocking densities. Major fish species shown better survivality except for a few carps. But the fry of carps like *Puntius japonicus,Catla catla* and *Labeo rohita* recorded almost 90% survivality in the cages. Among other species *Pangasius* species, *Tilapia* species, *Anabas species, Clarias* species and *Macrobrachium* species have been shown economic survival value during entire culture period. The fingerlings of Clarias and Anabas gained good weight within three months of culture period. Almost 20 times of weight gain was recorded in Pangasius, Tilapia, Lates and Anabas species. The growth of Clarias was found very much encouraging. The seedling that was 12 gm at the time of liberation, in case of Clarias fingerlings, increased to 37 gm weight after three months of farming period.

Breeding Grounds: Several auto breeding grounds have been identified in the Kangsbati reservoir. Most of these grounds are situated at extreme west of the reservoir in the Manbazar-I,Manbazar-II and Pucncha Block, Purulia district, West Bengal. The rivers Kangsbati, Kumari and Totak of the Kangsbati reservoir have their own breeding grounds for major fish species.The breeding areas of Kangsbati river are: Lapangdih, Dhanara, Madhabpur, Jituzuri, Karadhara, Akshoypur, Jorberia, Kalabani, Mahulbona, Darasole, Nunyani of Manbazar-I Block and Dhabani, Bakardi, Maisamura, Khairboni, Baksarkar, Gobarda, Denka, Joysimpur, Kulmi, Hirapur, Kalapathar, Budhpur, Kusama of Pucha block. Similarly Jamda, Sukapata, Desardih, Lagdagora, Gyidumur, Doldara ad Rangatair of Manbazar-I block and Dhanda, Cahlka and Kharidura of Manbazar-II block on river Kumari.

Fisheries department has taken various steps for building up awareness among the fishermen against the use of mosquito nets, conservation of brood fishes etc.Thorough publicity, broadcasting and awareness camps are organized by the department from time to time for proper conservation measures.

REFERENCES

- Mukherjee, Madhumita and Aloknath Praharaj, 2009. Kangsbati Reservoir Fisheries Development: New Policy Approaches Through Multidisciplinary Field Demonstrations to Rural People. *Fishing Chimes*. 29 (1) : 112-121.

INDEX

I...

J...

K...

L...

M...

N...

O...

P...

R...

S...

T...

U...

V...

W...

Y...

Z...

www.ingramcontent.com/pod-product-compliance
Ingram Content Group UK Ltd.
Pitfield, Milton Keynes, MK11 3LW, UK
UKHW021439280726
14060UKWH00001BA/147